T0348378

The Enzymes

VOLUME XXVII

STRUCTURE, FUNCTION AND REGULATION OF TOR COMPLEXES FROM YEASTS TO MAMMALS

PART A

THE ENZYMES

Edited by

Michael N. Hall

Biozentrum, University of Basel
CH-4056, Basel, Switzerland

Fuyuhiko Tamanoi

Department of Microbiology
Immunology, and Molecular Genetics
Molecular Biology Institute
University of California, Los Angeles
Los Angeles, CA 90095, USA

Volume XXVII

STRUCTURE, FUNCTION AND REGULATION OF TOR COMPLEXES FROM YEASTS TO MAMMALS

PART A

AMSTERDAM • BOSTON • HEIDELBERG • LONDON
NEW YORK • OXFORD • PARIS • SAN DIEGO
SAN FRANCISCO • SINGAPORE • SYDNEY • TOKYO
Academic Press is an imprint of Elsevier

Academic Press is an imprint of Elsevier
32 Jamestown Road, London NW1 7BY, UK
Radarweg 29, PO Box 211, 1000 AE Amsterdam, The Netherlands
Linacre House, Jordan Hill, Oxford OX2 8DP, UK
30 Corporate Drive, Suite 400, Burlington, MA 01803, USA
525 B Street, Suite 1900, San Diego, CA 92101-4495, USA

First edition 2010

Notice
No responsibility is assumed by the publisher for any injury and/or damage to persons
or property as a matter of products liability, negligence or otherwise, or from any use
or operation of any methods, products, instructions or ideas contained in the material
herein. Because of rapid advances in the medical sciences, in particular, independent
verification of diagnoses and drug dosages should be made

ISBN: 978-0-12-381539-2
ISSN: 1874-6047

For information on all Academic Press publications
visit our website at elsevierdirect.com

Printed and bound by CPI Group (UK) Ltd, Croydon, CR0 4YY

Transferred to Digital Print 2011

Working together to grow
libraries in developing countries

www.elsevier.com | www.bookaid.org | www.sabre.org

ELSEVIER   BOOK AID
            International   Sabre Foundation

Contents

4. Regulation of TOR Complex 1 by Amino Acids Through Small GTPases

JOSEPH AVRUCH, XIAOMENG LONG, YENSHOU LIN, SARA ORTIZ-VEGA,
JOSEPH RAPLEY, AND NORIKO OSHIRO

5. Rag GTPases in TORC1 Activation and Nutrient Signaling

LI LI AND KUN-LIANG GUAN

6. Amino Acid Regulation of hVps34 and mTORC1 Signaling

PAWAN GULATI AND GEORGE THOMAS

7. AGC Kinases in mTOR Signaling

ESTELA JACINTO

8. mTORC1 and Cell Cycle Control

CHRISTOPHER G. PROUD

9. TORC1 Signaling in Budding Yeast

ROBBIE LOEWITH

10. TORC2 and Sphingolipid Biosynthesis and Signaling: Lessons from Budding Yeast

TED POWERS, SOFIA ARONOVA, AND BRAD NILES

11. TORC1 Signaling in the Budding Yeast Endomembrane System and Control of Cell–Cell Adhesion in Pathogenic Fungi

ROBERT J. BASTIDAS AND MARIA E. CARDENAS

12. TOR and Sexual Development in Fission Yeast

YOKO OTSUBO AND MASAYUKI YAMAMOTO

13. Fission Yeast TOR and Rapamycin

RONIT WEISMAN

14. Structure of TOR Complexes in Fission Yeast

JUNKO KANOH AND MITSUHIRO YANAGIDA

15. The TOR Complex and Signaling Pathway in Plants

MANON MOREAU, RODNAY SORMANI, BENOIT MENAND, BRUCE VEIT,
CHRISTOPHE ROBAGLIA, AND CHRISTIAN MEYER

16. Dysregulation of TOR Signaling in Tuberous Sclerosis and Lymphangioleiomyomotosis

JANE YU AND ELIZABETH PETRI HENSKE

17. Chemistry and Pharmacology of Rapamycin and Its Derivatives

ROBERT T. ABRAHAM, JAMES J. GIBBONS, AND EDMUND I. GRAZIANI

Preface

One of the themes of *The Enzymes* series is to feature multiprotein complexes that perform central roles in cell physiology. In this and the next volume (volume 27 and volume 28), we discuss TOR (target of rapamycin), a key protein kinase conserved from yeast to human. This kinase plays critical roles in the regulation of growth in response to nutrients, growth factors and energy conditions. TOR has a structure consisting of HEAT, FAT, FRB, kinase and FATC domains. Other kinases such as DNA-dependent protein kinase, ATM, ATR have a similar structure and together they constitute the PI3K-related protein kinase (PIKK) family.

TOR was initially discovered as the target of rapamycin. Rapamycin is a natural metabolite produced by a soil bacterium, *Streptomyces hygroscopicus*, isolated from the soil of Easter Island, also known as Rapa Nui. Determination of the target of rapamycin was possible with the use of yeast genetics. Since its discovery almost twenty years ago, research on TOR has expanded to include many different organisms and topics including the role of TOR in signaling pathways, protein synthesis regulation, cell cycle regulation and autophagy. The role of TOR in development, chemotaxis, aging and a variety of cell type-specific functions such as learning and memory in neurons should also be mentioned. Another important development in the TOR field was the discovery of two structurally and functionally distinct TOR complexes, TORC1 and TORC2. The dramatic growth of the TOR field calls for a comprehensive update on the subject. Our intention with this and the next volume is to capture these dramatic developments concerning the structure and function of the two TOR complexes.

The structure and function of the TOR complexes, like TOR itself, appear to be conserved throughout eukaryotic evolution. Thus, studies in a variety of organisms synergized to deepen our knowledge of the TOR complexes. In this book, we tried to bring together studies in different model systems including budding yeast, fission yeast, *Drosophila*, *C. elegans*, *Dictyostelium*, mammals and plants.

Knowledge on TOR signaling is relevant for the understanding and treatment of human diseases. Genetic disorders such as tuberous sclerosis and Peutz-Jeghers syndrome, characterized by benign tumors, are due to the overactivation of TOR signaling. Overactivation of TOR signaling, as detected by hyperphosphorylation of S6K, S6 or Akt/PKB, has also been reported in many human malignancies including melanoma and renal cell carcinoma. Rapamycin and rapamycin analogs (rapalogs), which inhibit TORC1, are in use in the clinic for the treatment of cancer and allograft rejection, and new classes of drugs that inhibit both TORC1 and TORC1 are in development. TOR has also been implicated in metabolic disorders such as obesity and diabetes. The significance of TOR in human diseases is another topic that we tried to cover in this volume.

In volume 28, we will focus more on biological effects of TOR signaling. Since a wide range of studies are being carried out with TOR, there are many topics to cover. Unfortunately, we were unable to cover all TOR-related topics. Those will be the subject of future volumes on TOR.

We conceived the idea of putting together chapters on TOR in early 2009. Without extraordinary effort from all the contributors, we would not have been able to assemble in a relatively short period of time a volume with up-to-date knowledge. We are very grateful to the authors who sent their chapters in a timely fashion. We also thank Lisa Tickner of Elsevier for her guidance and encouragement and Gloria Lee of UCLA for her assistance in communication, preparation and editing of chapters.

Fuyuhiko Tamanoi
Michael N. Hall
January 2010

1

TOR Complexes: Composition, Structure, and Phosphorylation

VITTORIA ZINZALLA[a] • THOMAS W. STURGILL[b] •
MICHAEL N. HALL[a]

[a]*Biozentrum*
University of Basel, Basel
Switzerland

[b]*Department of Pharmacology*
University of Virginia Health Sciences Center
Charlottesville
Virginia, USA

I. Abstract

The target of rapamycin (TOR) is the central node of a highly conserved signaling network that regulates cell growth in response to nutrients, growth factors, and cellular energy. TOR is found in two functionally distinct complexes, TOR complex 1 (TORC1) and TORC2. This chapter describes the composition of yeast and mammalian TOR complexes, with special focus on the roles of structural domains and phosphorylation in TOR complex function.

II. Introduction

Target of rapamycin (TOR) is a highly conserved Ser/Thr kinase that is found in two structurally and functionally distinct complexes which regulate growth and metabolism. The two TOR complexes, like TOR itself, are also

THE ENZYMES, Vol. XXVII
© 2010 Elsevier Inc. All rights reserved.

ISSN NO: 1874-6047
DOI: 10.1016/S1874-6047(10)27001-4

highly conserved. TOR complex 1 in mammals (mTORC1) is activated in response to growth factors, such as insulin and IGF1, amino acids, and energy sufficiency. mTORC1 contains mTOR, mLST8, and raptor and is sensitive to the immunosuppressive and anticancer drug rapamycin. Rapamycin, in complex with FKBP12 (FKBP12–rapamycin), binds directly to TOR in TORC1. mTORC2 is activated by growth factors and contains mTOR, mLST8, rictor, and mSIN1. Rapamycin does not bind TOR in TORC2, and TORC2 is thus insensitive to rapamycin. The two complexes signal via different effector pathways to control distinct cellular processes.

TOR was originally identified in the budding yeast *S. cerevisiae* in a genetic selection for rapamycin resistance [1, 2]. Rapamycin is a macrolide produced by a strain of *Streptomyces hygroscopicus* isolated from a soil sample collected on Easter Island, also known as Rapa Nui and hence the name rapamycin [3]. The isolation and identification of rapamycin resistant mutants in budding yeast resulted in the identification of three genes—*FPR1*, *TOR1*, and *TOR2* [1]. *FPR1* encodes the highly conserved peptidylprolyl isomerase FKBP12. Rapamycin binds FKBP12 and this complex then binds and inhibits TOR. The TOR1 and TOR2 proteins are 67% identical and were originally thought to be lipid kinases due to their resemblance to phosphatidylinositol (PI) kinases [4]. However, TOR turned out to be the founding member of a family of atypical protein kinases known as phosphatidylinositol kinase-related kinases (PIKKs) [5], none of which is a lipid kinase. The original rapamycin resistance-conferring alleles, *TOR1-1* and *TOR2-1*, contain single missense mutations, Ser1972Arg and Ser1975Ile, respectively, which prevent binding of FKBP12–rapamycin but otherwise have little-to-no effect on TOR activity [6–8]. In subsequent studies, *TOR* homologues were identified in all eukaryotic genomes, including flies, worms, plants, and mammals. FKBP12–rapamycin binds and inhibits TOR in most eukaryotes, with plants (*A. thaliana*) and worms (*C. elegans*) being notable exceptions [9, 10].

FKBP12–rapamycin binds and inhibits both TOR1 and TOR2, but early genetic studies in yeast demonstrated that TOR2 also has a rapamycin-insensitive function. Indeed, these early studies showed that TOR has several cellular readouts, and only a subset of these readouts can be inhibited by rapamycin [11]. Rapamycin-sensitive and -insensitive TOR is also conserved in other organisms, such as mammals, in which there is only one TOR (mTOR). This originally puzzling sensitivity/insensitivity of TOR was later shown to correspond to two functionally distinct signaling branches, each mediated by a specific TOR complex that regulates its own targets and readouts [12]. Rapamycin-sensitive TORC1 mediates temporal control of cell growth by activating anabolic processes such as ribosome biogenesis, protein synthesis, transcription, and nutrient uptake, and by inhibiting catabolic processes such as autophagy and ubiquitin-dependent proteolysis. In contrast, TORC2 is rapamycin insensitive and mediates spatial control of cell growth by regulating actin cytoskeleton organization and other processes.

III.   TORC1 and TORC2 Components

A.   TORC1

In budding yeast, the core components of TORC1 are TOR (TOR1 or TOR2), KOG1, and LST8. TOR1 is not essential for cell viability. In the absence of TOR1, TOR2 can substitute for TOR1 in TORC1. In contrast, TOR2 is essential for viability, because TOR2 performs an essential function as part of TORC2, and TOR1 cannot replace TOR2 in TORC2. In mammals, mTOR is essential and binds to raptor (regulatory associated protein of mTOR) and mLST8 as part of mTORC1 (Figure 1.1).

KOG1/raptor is required for TORC1/mTORC1 activity. In mammals, raptor mediates the binding of mTOR substrates to mTORC1 [14, 15]. In response to insulin, raptor binds substrates upon release of the mTORC1 inhibitor PRAS40 (proline-rich AKT substrate of 40 kDa) which is also an mTORC1 substrate [16–19]. Raptor binds PRAS40 and other substrates (4E-BP and S6K) through a consensus sequence, the TOR signaling (TOS) motif, in these target proteins [20–22]. The TOS motif

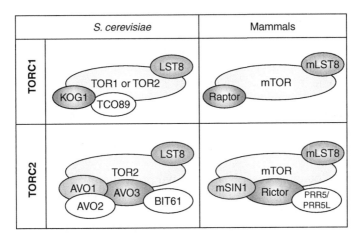

FIG. 1.1. Composition of TOR complex 1 (TORC1) and TOR complex 2 (TORC2) of *S. cerevisiae* and mammals. See Ref. [13] for list of TORC components in other species.

consists of five amino acids (FEMDI and FDIDL in the C-terminus of 4E-BP1 and in the N-terminus of S6K1, respectively) and is required for nutrient-stimulated, mTOR-dependent phosphorylation of both 4E-BP1 and S6K [21]. Mutation of residues within the TOS motif abrogates binding to raptor, indicating that the TOS motif is indeed required for this interaction [23]. Thus, the TOS motif is a functional motif via which raptor presents substrates to mTOR. Raptor also interacts with the Rag proteins which mediate mTORC1 activation by amino acids [24]. Hence, raptor integrates mTORC1 upstream and downstream signaling events.

Among the conserved TOR partners, LST8 is an essential Gβ-like propeller protein composed of seven WD40 motifs. A WD40 motif is an $\sim$40 amino acid sequence, often terminating in a Trp-Asp (W-D) dipeptide, that forms four antiparallel beta strands. LST8 binds the kinase domain in TOR, in both TORC1 and TORC2 [12, 25]. In yeast, LST8 is required for both the integrity and kinase activity of TORC2 [25], indicating that it is essential for at least TORC2 activity. In mammals, mLST8 is also found in both complexes and is required for mTOR kinase activity [26]. Interestingly, in *mLST8* knockout mouse embryonic fibroblasts (MEFs), mTORC2 signaling is strongly decreased but there is no detectable change in mTORC1 signaling [27], suggesting that mLST8 is required only for mTORC2.

B.   TORC2

In budding yeast, the core components of TORC2 are TOR2, AVO1, AVO3, and LST8. In mammals, the core components of mTORC2 are rictor, mSIN1, and mLST8 (Figure 1.1). TORC2 in budding yeast and mammals is insensitive to acute treatment with rapamycin. However, prolonged rapamycin treatment can lead to mTORC2 inhibition in a few mammalian cell lines, by preventing the binding of newly synthesized mTOR to rictor [28]. These findings indicate that, although rapamycin does not bind to preformed mTORC2, it can inhibit the assembly of mTORC2 and thereby block mTORC2 function indirectly.

In budding yeast, AVO3 (*a*dheres *vo*raciously to TOR2) is an essential protein required for TORC2 integrity and function [25]. The specific molecular role of AVO3 and of its mammalian counterpart rictor in TORC2 is unknown.

AVO1 in budding yeast is an essential protein and an integral component of TORC2. It is required for the binding of AVO3 to TOR2, but is not required for the interaction between TOR2 and LST8. It has been proposed that AVO1 binding to TOR2 within TORC2 masks the FKBP12–rapamycin binding site, leading to the rapamycin insensitivity of TORC2 [25]. AVO3 and AVO1 bind cooperatively to the N-terminal HEAT repeats of

TOR2 and, like LST8, are required for TORC2 stability. In mammals, the AVO1 ortholog mSIN1 is an integral component of mTORC2 that, like rictor, is required for complex formation and kinase activity [29–32]. Alternative splicing generates at least five isoforms of the mSIN1 protein (isoforms 1–5). mSIN1 isoforms 1, 2, and 5 assemble into mTORC2 separately and define three distinct mTORC2s [33]. Isoforms 3 and 4 are not found in mTORC2. mSIN1 isoform 3 is very weakly expressed. Isoform 4 is unique because it lacks an N-terminal 192 amino acid extension found in the other isoforms [34]. This suggests that the first 192 amino acids of mSIN1 may be responsible for the assembly of mSIN1 into mTORC2. Interestingly, insulin activates only the two mTORC2s containing mSIN1 isoforms 1 and 2, as determined by *in vitro* kinase activity. mTORC2 containing mSIN1 isoform 5 is insensitive to insulin stimulation and serum starvation. All three mTORC2s can phosphorylate the mTORC2 substrate AKT *in vitro* [33], suggesting that some mTORC2 complexes may phosphorylate AKT independently of growth factor stimulation *in vivo*. Curiously, mSIN1 isoform 5 which defines insulin-independent mTORC2 is truncated at the C-terminus and lacks part of a Raf-like Ras-binding domain (RBD) and all of a pleckstrin homology (PH) domain [33, 35] (see Section C). An interesting possibility is that the mSIN1 C-terminal domains mediate mTORC2 activation by growth factors by promoting mTORC2 membrane association.

Nonconserved proteins are also found in TORC1 and TORC2. In budding yeast, TCO89 (TOR complex one 89 kDa subunit) interacts specifically with TORC1 [36], and AVO2 and BIT61 interact specifically with TORC2 [12, 36]. All these proteins are nonessential and their function is unknown. In mammals, PRR5 (proline-rich protein 5) and PRR5-like protein bind to rictor and are weak homologs of yeast BIT61 [19, 37–39]. PRR5 and PRR5L (also known as Protor-1 and Protor-2) are not required for binding of rictor and mSIN1 to mTOR. Whether these mammalian TORC2 interactors mediate specific functions of mTORC2 remains to be determined.

IV. Domains of TOR and Its Binding Partners

A. TOR DOMAINS

TOR is a protein of approximately 300 kDa consisting of several conserved domains: the HEAT repeats and the FAT (*F*RAP, *A*TM, and *T*RRAP), FRB (*F*KBP12–*r*apamycin *b*inding), kinase, FIT (*f*ound *i*n *T*OR), and FATC (*FAT-C* terminal) domains (Figures 1.2 and 1.3). The N-terminal and central region of TOR contains 20 tandemly repeated

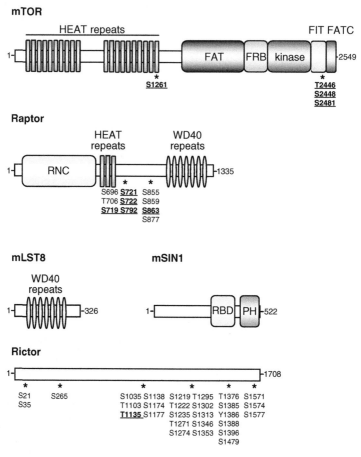

Fig. 1.2. Schematic representation of domains and phosphorylation sites in human TORC components. See text for details on the various domains. Asterisks show single or clustered phosphorylation sites. Individual phosphorylated residues are indicated. Residues shown in bold and underlined have been at least partly characterized for function (see text details).

HEAT motifs (originally identified in *H*untingtin, *e*longation factor 3 [EF3], protein phosphatase 2*A* [PP2A] and *T*OR). Each HEAT motif comprises 40–50 amino acids and forms antiparallel alpha-helices that are thought to act as a protein–protein interaction surface [46–48]. HEAT repeats, as shown in the crystal structure of PP2A, form a large extended superhelical structure [49, 50]. In mTOR, the HEAT repeats are required for association with raptor [15] and for ER and Golgi localization [51].

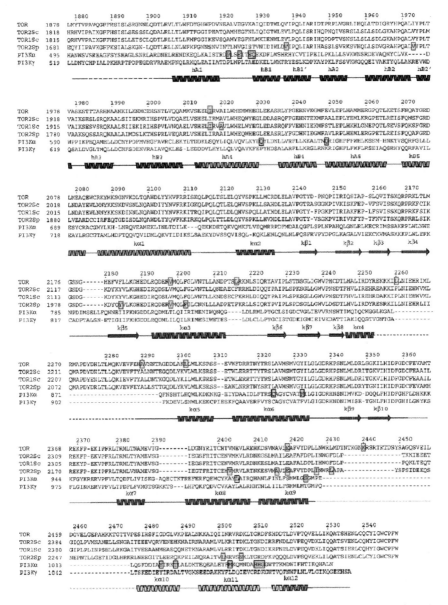

FIG. 1.3. Structural alignment of TORs and PI3Ks. The TORs are human TOR (TOR), *S. cerevisiae* TOR1 and TOR2 (TOR1Sc and TOR2Sc), and *S. pombe* TOR2 (TOR2Sp). The PI3Ks, both class I PI3Ks, are pig PI3Kγ and human PI3Kα. The structural elements, shown below the alignment, are drawn according to the crystal structure of PI3Kγ [40]. The color code of the domains is as follows: FAT (shown is residues 1906–2014 which is only part of the FAT), blue; FRB (residues 2015–2114), green; kinase (residues 2115–2426), red; FIT (residues 2427–2516); FATC (shown is residues 2517–2526 which is only part of the FATC), cyan. Sites of activating mutations in TOR and oncogenic mutations in PI3Kα [41–44] are shown in red and blue, respectively. Oncogenic mutations are particularly frequent in codons E542 and E545 (between helices hA1 and hB1) and H1047 (helix kα11), and these sites thus constitute two oncogenic hotspots. This figure is modified from Figure 1 of Ref. [45]. (See color plate section in the back of the book.)

The HEAT repeats in yeast TOR2 mediate plasma membrane localization [52, 53], likely indirectly.

A recent study described a truncated version of mTOR, mTORβ, that is generated by alternative splicing [54]. mTORβ consists of the FRB, kinase, FIT, and FATC domains fused to the 23 N-terminal amino acids of full-length mTOR. mTORβ lacks HEAT repeats but can form complexes with both raptor and rictor and appears to mediate both mTORC1 and mTORC2 downstream signaling. This suggests that the 23 residues at the N-terminus of mTOR may be sufficient for mTOR interaction with its binding partners. The function of this alternative mTOR splicing isoform in the mTOR pathway remains to be determined.

C-terminal to the HEAT repeats is the most highly conserved region of the PIKK family members. In TOR, this region comprises the FAT, FRB, kinase, FIT, and FATC domains (Figures 1.2 and 1.3). The FAT domain consists of ∼500 residues (amino acids 1513–2014 in human TOR) and is proposed to contain additional HEAT motifs [50, 55]. Overexpression of a FAT domain is toxic in yeast and this toxicity is suppressed, for unknown reasons, by overexpression of phospholipase C [56]. The FATC domain is located at the extreme C-terminus (amino acids 2517–2549 in human TOR) and is essential for TOR function. NMR structural data suggest that a disulphide bridge in the FATC domain may couple intracellular redox potential to TOR stability and function [57]. In mammals, the mTOR-raptor interaction is affected by a redox-sensitive mechanism [58].

The FRB domain consists of ∼100 residues (amino acids 2015–2114 in human TOR) and has been proposed to contain an additional pair of HEAT motifs [59]. It also contains the single missense mutations, Ser1972-Arg and Ser1975Ile, that were originally identified in the rapamycin resistant *TOR1-1* and *TOR2-1* mutants. These mutations in the FRB domain prevent the binding of FKBP12–rapamycin to TOR. In yeast, all TORC1 components can be co-precipitated with FKBP12 in the presence of rapamycin [12]. This finding suggests that the integrity of TORC1 is not affected by FKBP12–rapamycin binding. However, in mammalian cells, rapamycin decreases the binding of raptor to mTOR [26, 60]. The molecular mechanism by which rapamycin inhibits TORC1 remains to be determined.

The kinase (catalytic) domain of TOR consists of ∼300 residues (amino acids 2115–2426 in human TOR). In the N-terminal part of the mTOR catalytic domain is the binding site for the mTORC1 upstream activator Rheb, a small Ras-like GTPase. Both the active (Rheb-GTP) and inactive (Rheb-GDP) form of Rheb bind mTOR *in vitro*; however, only Rheb-GTP increases mTORC1 kinase activity [61]. According to a recent homology model of TOR (see below and Figures 1.3 and 1.4), the ATP-binding site in the kinase domain forms a novel pharmacophore space that could explain the

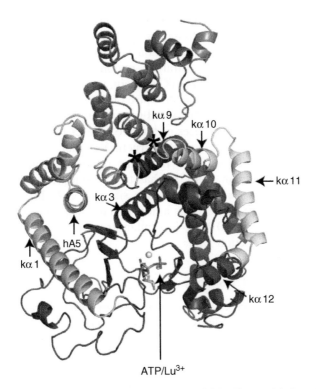

Fig. 1.4. Model of the catalytic region of human TOR. The modeled catalytic region extends from the FAT domain to near the end of the FATC domain [45]. The predicted helices are shown as ribbons colored according to domains. The color code for domains is as in Figure 1.3. ATP is shown as a backbone stick structure. Two Lu^{3+} ions are shown as two small gray spheres. The ATP and Lu^{3+} ions are positioned according to the crystal structure of PI3Kγ [40]. Asterisks indicate the insertion site of a so-far structurally undefined portion of the FIT domain (the undefined portion of the FIT domain is residues 2427–2477). The model is drawn using MacPymol (Delano Scientific). (See color plate section in the back of the book.)

specificity of TOR inhibitors currently in development [45]. The kinase domain also contains a WLK motif, in helix kα5 (Figure 1.3), that is highly conserved in TORs from different species, but is of unknown function. The conserved kinase domain ends abruptly with an invariant FxxDPL motif (Figure 1.3).

Between the kinase and FATC domains is the newly defined FIT (*found in T*OR) domain (amino acids 2427–2516 in human TOR) [45]. This domain is not highly conserved among different species. According to the recent homology model of TOR (see below and Figures 1.3 and 1.4), the FIT domain is close to the active site of TOR. This position is consistent with an earlier finding that the FIT domain is phosphorylated and may act as repressor domain (see below and Section V) [62–66].

A homology model of the catalytic region of TOR, from the FAT domain to near the end of the FATC domain, was recently described [45]. The model is based on the crystal structure of the related phosphatidylino-sitol-3-phosphate kinase γ (PI3Kγ) and was generated after manual alignment of the amino acid sequences of the two kinases (Figure 1.3). Several interesting findings emerged from this structural model. First, the FAT and FRB domains of TOR may have the same relationship in space as the equivalent helical domain of PI3Kγ has to its catalytic domain. Second, the FIT domain may lie near the active site of TOR. Third, the ATP-binding binding site in TOR forms a novel pharmacophore space. Fourth, activating mutations in TOR are in similar structures as oncogenic mutations in PI3Kα (see Section E).

B. DOMAINS OF TORC1-SPECIFIC COMPONENTS

In yeast and mammals, KOG1/raptor contain an RNC (raptor N-terminal conserved) domain (Figure 1.2). This domain binds the N-terminal region of TOR and interacts with TORC1 substrates. Furthermore, KOG1/raptor contains three internal HEAT motifs and seven C-terminal WD40 motifs that are thought to act as protein–protein interaction surfaces [12]. While the composition and domains of TORC1 are extensively described, little is known about the 3D structure of TORC1. A recent electron microscopy study reconstructed a low-resolution 3D structure of budding yeast TOR1 and the TOR1–KOG1 complex [67]. TOR1 seems to consist of a bulky C-terminal region containing the FAT, kinase, FIT, and FATC domains and an extended tubular N-terminal region that comprises the HEAT repeats. The C-terminus of KOG1 interacts with the N-terminus of TOR1, whereas the N-terminal RNC domain of KOG1 binds the C-terminus of TOR. Such an organization of the TOR–KOG1 complex would place the RNC domain of KOG1, which interacts with the TOS motif of TORC1 substrates, in close proximity to the kinase domain of TOR. This finding begins to provide a molecular view of how KOG1/raptor presents TORC1 substrates to the catalytic region of TOR.

C. DOMAINS OF TORC2-SPECIFIC COMPONENTS

The yeast TORC2 subunit AVO3 contains a RasGEFN domain in amino acids 990–1046. This domain is often a subdomain found in the N-terminal part of a larger GDP/GTP exchange domain in some guanine nucleotide exchange factors for small Ras-like GTPases. However, the function of the RasGEFN domain is unknown, and this domain is not conserved in AVO3 orthologs in higher eukaryotes [68]. Rictor in mammals does not have any defined domain.

Yeast AVO1 has a C-terminal PH-like domain that mediates TORC2 plasma membrane anchoring [69]. mSIN1, the mammalian ortholog of AVO1, contains a PH domain. mSIN1 isoforms 1 also contains a Raf-like RBD that binds the small GTPases H- and K-Ras. The PH domain and the RBD domain of mSIN1 confer membrane localization on mSIN1. The RBD domain presumably mediates membrane localization by binding to H-Ras or K-Ras [35]. These data suggest that mSIN1 isoforms containing the PH and RBD domains (isoform 1 and 2) mediate mTORC2 membrane localization.

D.   TORC1 AND TORC2 ARE MULTIMERS

Multimerization is a well-documented mechanism for the activation of protein kinases. Studies in yeast, flies, and mammals have suggested that both TORC1 and TORC2 are multimeric, likely TORC–TORC dimers [12, 25, 70–72]. TORC2 exists as two species, one of 1.5–2 MDa and a second of 0.7–0.8 MDa, which may correspond to dimeric and monomeric forms of the complex [12, 25, 70–72]. TORC multimerization is mediated by HEAT–FAT interactions between TOR proteins. In mammals, multimerization of mTOR is stimulated by amino acid sufficiency, while rapamycin treatment or growth factor stimulation has no effect [72]. Unlike mTOR, the multimerization of TOR in yeast is not affected by nutrients, raising the question of whether TOR multimerization reflects TORC activation [25]. Furthermore, mTOR multimerization appears to be inhibited by butanol treatment, which blocks phosphatidic acid (PA) production by phospholipase D (PLD). Interestingly, a recent report proposed the involvement of PLD and its metabolite PA in the regulation of mTOR activity. This study suggests that PLD and PA may regulate mTOR by facilitating the formation of mTOR complexes and that rapamycin inhibits mTOR by interfering with the PA–mTOR interaction [73]. Indeed, recent structural studies have revealed that PA can interact with the FRB domain of mTOR and cause structural changes similar to those observed when FKBP12–rapamycin binds to the FRB domain [74].

E. ACTIVATING MUTATIONS IN TOR

Mutations in TOR that increase TOR activity have been identified in budding yeast, fission yeast, and mammalian cells [75–77] (Figure 1.3). In budding yeast, TORC1 is inhibited by caffeine treatment and mutations in TOR1 were obtained that confer resistance to caffeine [75]. Sequence analysis revealed that these TOR1 mutations are within the FRB domain. In particular, a single FRB mutation (Ile1954Val in TOR1) enhances TORC1 kinase activity and modestly increases the interaction between KOG1 and TOR1. The existence of TOR1 FRB mutations that increase TORC1 activity suggests that the FRB domain plays a role in the regulation of TOR kinase activity.

In fission yeast, TORC1 is activated by the essential Rheb homolog Rhb1. Urano et al. [76] selected suppressor mutations that confer growth in the absence of Rhb1. Twenty-two single amino acid mutations in Tor2 (Tor2 is in TORC1 in S. pombe) were identified that confer Rhb1 independent growth. These Tor2 mutations are clustered mainly in the FAT and kinase domains (Figure 1.3, cluster in FAT domain not shown). In mammalian TOR, analogous mutations in the FAT and kinase domains (Leu1460-Pro and Glu2419Lys, respectively) confer constitutive mTORC1 activity in nutrient starved cells. These mutant mTORs can form multimers with wild-type mTOR and the resulting heterodimer is also constitutively active.

Ohne et al. [77] identified activating mutations in mTOR by selecting suppressors of a temperature sensitive *LST8* (*lst8^ts*) mutation, in yeast cells expressing an N-terminal TOR2 region fused to the C-terminal mTOR catalytic region. The selection was designed to identify mutations in the FRB and kinase domains of mTOR. The isolated mTOR mutants exhibited enhanced kinase activity *in vitro* and constitutive mTORC1 activity *in vivo*. Remarkably, this study identified an mTOR FRB domain mutation (Ile2017Thr) equivalent to one of the caffeine resistance mutations in budding yeast TOR1. This study also identified mTOR kinase domain mutations that correspond to fission yeast mutations that confer Rhb1 independent growth (Figure 1.3).

What is the mechanism underlying the enhanced kinase activity conferred by the activating mutations? Some of the activating mutations in the TOR kinase domain fall in the α helix kα3 (Figure 1.3). This α helix corresponds to an α helix in PI3Kγ that has been proposed to be an important mediator of conformational changes in the kinase catalytic center and is required for kinase activation [45, 78]. Thus, mutations in the mTOR kinase domain may confer increased kinase activity by allosterically altering the conformation of the mTOR kinase center. It is also interesting to note that the mTOR activator Rheb interacts with the N-terminal half of the mTOR kinase domain [61]. Hence, mutations in the kinase domain may increase kinase activity by mimicking the effect of Rheb binding. Activating mutations outside the kinase domain likely activate mTOR in a different manner. Mutations in the FIT domain, which is in close proximity to the TOR active site (Figures 1.3 and 1.4), may relieve repression of kinase activity mediated by the so-called repressor domain that is part of the FIT domain [45, 76]. Deletion of amino acids 2430–2450 in human TOR, which removes part or all of the so-far poorly delineated repressor domain within the FIT domain, results in increased phosphorylation of mTORC1 substrates [65, 66]. Finally, it is interesting to note that, according to the homology model of the catalytic region of TOR [45], some of the TOR mutations that confer Rhb1- and *lst8*-independent growth are in similar structures as oncogenic mutations in PI3Kα (Figure 1.3). In TOR, these mutations are clustered in the kinase (helices kα8 and kα9) and FIT (helix kα11) domains as well as the FRB domain. TOR mutations that confer constitutive TORC1 activity begin to shed light on how TOR domains, in particular the FAT, FRB, kinase, and FIT domains, contribute to TOR regulation.

V.   Phosphorylation of TOR and Its Binding Partners

The TORCs are highly phosphorylated and in some cases the phosphorylation sites are known. Four phosphorylation sites, Ser1261, Thr2446, Ser2448, and Ser2481, have been identified in mTOR. Ser1261 is within the N-terminal HEAT repeats (Figure 1.2) and is phosphorylated in mTORC1 and mTORC2. mTOR phosphorylation at Ser1261 seems to promote mTORC1 activity *in vivo* without affecting mTORC1 integrity [79]. The kinase responsible for Ser1261 phosphorylation and the consequence of Ser1261 phosphorylation in mTORC2 remain to be determined. The remaining three sites, Thr2446, Ser2448, and Ser2481, are clustered in the C-terminal FIT domain (Figure 1.2). Phosphorylated Ser2448 (and possibly Thr2446) is found predominantly in mTORC1 [80]. Thr2446 and Ser2448 are phosphorylated by the mTORC1 substrate S6K [62, 63] as a part of a feedback loop of unknown function. Thr2446 is also phosphorylated by AMPK, a kinase activated by energy stress [81]. Ser2481 is an mTORC2-specific autophosphorylation site whose phosphorylation is stimulated by insulin [64, 80]. According to the homology model (Figure 1.4), Ser2481 is positioned near the active site of mTOR [45]. Interestingly, Thr2446, Ser2448, and Ser2481 are all within or near the so-called repressor domain of mTOR that is part of the FIT domain, suggesting that the phosphorylation of these sites may be involved in the regulation of kinase activity. However, mutation of at least Thr2446 or Ser2448 has no effect on mTOR kinase activity [62, 63, 81]. It remains possible that several phosphorylation sites regulate TOR activity and mutation of any single site, or subset of sites, may not be sufficient to elicit a phenotype.

In addition to mTOR phosphorylation, recent studies have identified several phosphorylation sites in raptor, in a region of poor evolutionary conservation that lies between the N-terminal HEAT repeats and the WD40 repeats (Figure 1.2). These sites are Ser696, Thr706, Ser719, Ser721, Ser722, Ser792, Ser855, Ser859, Ser863, and Ser877. In response to energy stress, AMPK directly phosphorylates Ser722 and Ser792 to inhibit mTORC1 signaling [82]. AMPK-mediated phosphorylation induces 14-3-3 binding to raptor which in turn inhibits mTORC1. The MAPK-activated kinase RSK directly phosphorylates Ser719, Ser721, and Ser722 to promote mTORC1 signaling [83]. It is difficult to envisage how raptor phosphorylation by RSK or AMPK at the same site (Ser722) can have opposite effects on mTORC1 downstream signaling. mTORC1 autophosphorylates raptor at Ser859 and Ser863 to enhance mTORC1 signaling [84, 85]. The kinase(s) that phosphorylates Ser696, Thr706, Ser855, and Ser877 is unknown, as is the physiological significance of phosphorylation at these sites [85]. These sites are phosphorylated in response to insulin. Interestingly, some of the raptor phosphorylation sites are near the RNC domain that binds the TOS motif in TORC1 substrates [86]. This suggests that raptor phosphorylation may affect mTORC1 substrate recruitment. However, raptor phosphorylation does not seem to affect binding of substrates *in vitro*. The molecular mechanism that leads to increased mTORC1 signaling upon raptor phosphorylation is unknown. Taken together, the above studies reinforce the notion that raptor integrates different upstream signals to regulate mTORC1.

Rictor contains 28 identified phosphorylation sites, clustered mainly within the C-terminal half of rictor in a region that is conserved only in vertebrates (Figure 1.2). The mTORC1 effector S6K phosphorylates one of these sites, Thr1135, which in turn leads to a modest decrease in phosphorylation of the mTORC2 target site Ser473 in AKT [87–89]. Although the mechanism by which Thr1135 phosphorylation affects AKT Ser473 phosphorylation is unknown, this observation suggests that mTORC1 cross-talks to mTORC2.

VI.   Future Directions

Although research on TOR signaling, and in particular on the TOR complexes, is progressing rapidly, many questions remain unanswered. Recent structural and biochemical studies support a role for raptor in mTORC1 substrate recruitment and suggest that rictor and/or mSIN1 may have a similar role within mTORC2. It will be a major goal to determine the specific molecular functions of other TOR binding partners. Furthermore, it is of interest to determine how the TOR complexes are assembled and if their biogenesis is regulated.

mTORC1 and mTORC2 have been implicated in a number of major diseases including metabolic disorders and cancer [90–93]. The identification of activating mutations in TOR raises the interesting possibility that mTOR mutations may be found in a disease-related context. Indeed, based on structural analysis, TOR activating mutations are in similar structures as oncogenic mutations in PI3Kα [45]. However, mutationally activated mTOR does not induce cellular transformation in NIH/3T3 cells [77], suggesting that concomitant activation of an additional pathway may be required for tumorigenesis. It would also be interesting to determine why the activating mTOR mutations constitutively activate mTORC1 but not mTORC2 [94].

Despite much effort, there is relatively little structural information on TOR and the TOR complexes. A TOR structure beyond the so-far homology structure of the mTOR catalytic region will be valuable in the design and development of new drugs, in particular drugs that inhibit mTORC2.

ACKNOWLEDGMENTS

We acknowledge support from the Swiss National Science Foundation, the Swiss Cancer League, the Louis-Jeantet Foundation, and the Canton of Basel.

REFERENCES

1. Heitman, J., Movva, N.R., and Hall, M.N. (1991). Targets for cell cycle arrest by the immunosuppressant rapamycin in yeast. *Science* 253:905–909.
2. Lorberg, A., and Hall, M.N. (2004). TOR: the first 10 years. *Curr Top Microbiol Immunol* 279:1–18.
3. Sehgal, S.N. (2003). Sirolimus: its discovery, biological properties, and mechanism of action. *Transplant Proc* 35:7S–14S.
4. Kunz, J., Henriquez, R., Schneider, U., Deuter-Reinhard, M., Movva, N.R., and Hall, M.N. (1993). Target of rapamycin in yeast, TOR2, is an essential phosphatidylinositol kinase homolog required for G1 progression. *Cell* 73:585–596.
5. Keith, C.T., and Schreiber, S.L. (1995). PIK-related kinases: DNA repair, recombination, and cell cycle checkpoints. *Science* 270:50–51.
6. Cafferkey, R., *et al.* (1993). Dominant missense mutations in a novel yeast protein related to mammalian phosphatidylinositol 3-kinase and VPS34 abrogate rapamycin cytotoxicity. *Mol Cell Biol* 13:6012–6023.

7. Helliwell, S.B., Wagner, P., Kunz, J., Deuter-Reinhard, M., Henriquez, R., and Hall, M.N. (1994). TOR1 and TOR2 are structurally and functionally similar but not identical phosphatidylinositol kinase homologues in yeast. *Mol Biol Cell* 5:105–118.

8. Stan, R., McLaughlin, M.M., Cafferkey, R., Johnson, R.K., Rosenberg, M., and Livi, G.P. (1994). Interaction between FKBP12–rapamycin and TOR involves a conserved serine residue. *J Biol Chem* 269:32027–32030.

9. Long, X., Spycher, C., Han, Z.S., Rose, A.M., Muller, F., and Avruch, J. (2002). TOR deficiency in *C. elegans* causes developmental arrest and intestinal atrophy by inhibition of mRNA translation. *Curr Biol* 12:1448–1461.

10. Robaglia, C., *et al.* (2004). Plant growth: the translational connection. *Biochem Soc Trans* 32:581–584.

11. Helliwell, S.B., Howald, I., Barbet, N., and Hall, M.N. (1998). TOR2 is part of two related signaling pathways coordinating cell growth in *Saccharomyces cerevisiae*. *Genetics* 148:99–112.

12. Loewith, R., *et al.* (2002). Two TOR complexes, only one of which is rapamycin sensitive, have distinct roles in cell growth control. *Mol Cell* 10:457–468.

13. Soulard, A., Cohen, A., and Hall, M.N. (2009). TOR signaling in invertebrates. *Curr Opin Cell Biol* 21:825–836.

14. Hara, K., *et al.* (2002). Raptor, a binding partner of target of rapamycin (TOR), mediates TOR action. *Cell* 110:177–189.

15. Kim, D.H., *et al.* (2002). mTOR interacts with raptor to form a nutrient-sensitive complex that signals to the cell growth machinery. *Cell* 110:163–175.

16. Wang, L., Harris, T.E., Roth, R.A., and Lawrence, J.C., Jr. (2007). PRAS40 regulates mTORC1 kinase activity by functioning as a direct inhibitor of substrate binding. *J Biol Chem* 282:20036–20044.

17. Vander Haar, E., Lee, S.I., Bandhakavi, S., Griffin, T.J., and Kim, D.H. (2007). Insulin signalling to mTOR mediated by the Akt/PKB substrate PRAS40. *Nat Cell Biol* 9:316–323.

18. Sancak, Y., *et al.* (2007). PRAS40 is an insulin-regulated inhibitor of the mTORC1 protein kinase. *Mol Cell* 25:903–915.

19. Thedieck, K., *et al.* (2007). PRAS40 and PRR5-like protein are new mTOR interactors that regulate apoptosis. *PLoS One* 2:e1217.

20. Schalm, S.S., and Blenis, J. (2002). Identification of a conserved motif required for mTOR signaling. *Curr Biol* 12:632–639.

21. Schalm, S.S., Fingar, D.C., Sabatini, D.M., and Blenis, J. (2003). TOS motif-mediated raptor binding regulates 4E-BP1 multisite phosphorylation and function. *Curr Biol* 13:797–806.

22. Oshiro, N., *et al.* (2007). The proline-rich Akt substrate of 40 kDa (PRAS40) is a physiological substrate of mammalian target of rapamycin complex 1. *J Biol Chem* 282:20329–20339.

23. Beugnet, A., Wang, X., and Proud, C.G. (2003). Target of rapamycin (TOR)-signaling and RAIP motifs play distinct roles in the mammalian TOR-dependent phosphorylation of initiation factor 4E-binding protein 1. *J Biol Chem* 278:40717–40722.

24. Sancak, Y., *et al.* (2008). The Rag GTPases bind raptor and mediate amino acid signaling to mTORC1. *Science* 320:1496–1501.

25. Wullschleger, S., Loewith, R., Oppliger, W., and Hall, M.N. (2005). Molecular organization of target of rapamycin complex 2. *J Biol Chem* 280:30697–30704.

26. Kim, D.H., *et al.* (2003). GbetaL, a positive regulator of the rapamycin-sensitive pathway required for the nutrient-sensitive interaction between raptor and mTOR. *Mol Cell* 11:895–904.

27. Guertin, D.A., *et al.* (2006). Ablation in mice of the mTORC components raptor, rictor, or mLST8 reveals that mTORC2 is required for signaling to Akt-FOXO and PKCalpha, but not S6K1. *Dev Cell* 11:859–871.

28. Sarbassov, D.D., *et al.* (2006). Prolonged rapamycin treatment inhibits mTORC2 assembly and Akt/PKB. *Mol Cell* 22:159–168.

29. Jacinto, E., *et al.* (2006). SIN1/MIP1 maintains rictor-mTOR complex integrity and regulates Akt phosphorylation and substrate specificity. *Cell* 127:125–137.

30. Jacinto, E., *et al.* (2004). Mammalian TOR complex 2 controls the actin cytoskeleton and is rapamycin insensitive. *Nat Cell Biol* 6:1122–1128.

31. Yang, Q., Inoki, K., Ikenoue, T., and Guan, K.L. (2006). Identification of Sin1 as an essential TORC2 component required for complex formation and kinase activity. *Genes Dev* 20:2820–2832.

32. Sarbassov, D.D., *et al.* (2004). Rictor, a novel binding partner of mTOR, defines a rapamycin-insensitive and raptor-independent pathway that regulates the cytoskeleton. *Curr Biol* 14:1296–1302.

33. Frias, M.A., *et al.* (2006). mSin1 is necessary for Akt/PKB phosphorylation, and its isoforms define three distinct mTORC2s. *Curr Biol* 16:1865–1870.

34. Schroder, W., Cloonan, N., Bushell, G., and Sculley, T. (2004). Alternative polyadenylation and splicing of mRNAs transcribed from the human *Sin1* gene. *Gene* 339:17–23.

35. Schroder, W.A., *et al.* (2007). Human *Sin1* contains Ras-binding and pleckstrin homology domains and suppresses Ras signalling. *Cell Signal* 19:1279–1289.

36. Reinke, A., *et al.* (2004). TOR complex 1 includes a novel component, Tco89p (YPL180w), and cooperates with Ssd1p to maintain cellular integrity in *Saccharomyces cerevisiae*. *J Biol Chem* 279:14752–14762.

37. Hayashi, T., *et al.* (2007). Rapamycin sensitivity of the *Schizosaccharomyces pombe* tor2 mutant and organization of two highly phosphorylated TOR complexes by specific and common subunits. *Genes Cells* 12:1357–1370.

38. Woo, S.Y., *et al.* (2007). PRR5, a novel component of mTOR complex 2, regulates platelet-derived growth factor receptor beta expression and signaling. *J Biol Chem* 282:25604–25612.

39. Pearce, L.R., *et al.* (2007). Identification of Protor as a novel Rictor-binding component of mTOR complex-2. *Biochem J* 405:513–522.

40. Walker, E.H., Perisic, O., Ried, C., Stephens, L., and Williams, R.L. (1999). Structural insights into phosphoinositide 3-kinase catalysis and signalling. *Nature* 402:313–320.

41. Li, H., *et al.* (2010). PIK3CA mutations mostly begin to develop in ductal carcinoma of the breast. *Exp Mol Pathol* 88:150–155.

42. He, Y., *et al.* (2009). PIK3CA mutations predict local recurrences in rectal cancer patients. *Clin Cancer Res* 15:6956–6962.

43. Samuels, Y., *et al.* (2005). Mutant PIK3CA promotes cell growth and invasion of human cancer cells. *Cancer Cell* 7:561–573.

44. Platt, F.M., Hurst, C.D., Taylor, C.F., Gregory, W.M., Harnden, P., and Knowles, M.A. (2009). Spectrum of phosphatidylinositol 3-kinase pathway gene alterations in bladder cancer. *Clin Cancer Res* 15:6008–6017.

45. Sturgill, T., and Hall, M.N. (2009). Activating mutations in TOR in similar structures as the oncogenic mutations in PI3KCalpha. *ACS Chem Biol* 4(12):999–1015.

46. Hemmings, B.A., *et al.* (1990). Alpha- and beta-forms of the 65-kDa subunit of protein phosphatase 2A have a similar 39 amino acid repeating structure. *Biochemistry* 29:3166–3173.

47. Andrade, M.A., and Bork, P. (1995). HEAT repeats in the Huntington's disease protein. *Nat Genet* 11:115–116.

48. Andrade, M.A., Petosa, C., O'Donoghue, S.I., Muller, C.W., and Bork, P. (2001). Comparison of ARM and HEAT protein repeats. *J Mol Biol* 309:1–18.
49. Groves, M.R., and Barford, D. (1999). Topological characteristics of helical repeat proteins. *Curr Opin Struct Biol* 9:383–389.
50. Perry, J., and Kleckner, N. (2003). The ATRs, ATMs, and TORs are giant HEAT repeat proteins. *Cell* 112:151–155.
51. Liu, X., and Zheng, X.F. (2007). Endoplasmic reticulum and Golgi localization sequences for mammalian target of rapamycin. *Mol Biol Cell* 18:1073–1082.
52. Kunz, J., Schneider, U., Howald, I., Schmidt, A., and Hall, M.N. (2000). HEAT repeats mediate plasma membrane localization of Tor2p in yeast. *J Biol Chem* 275:37011–37020.
53. Sturgill, T.W., Cohen, A., Diefenbacher, M., Trautwein, M., Martin, D.E., and Hall, M.N. (2008). TOR1 and TOR2 have distinct locations in live cells. *Eukaryot Cell* 7:1819–1830.
54. Panasyuk, G., *et al.* (2009). mTORbeta splicing isoform promotes cell proliferation and tumorigenesis. *J Biol Chem* 284:30807–30814.
55. Bosotti, R., Isacchi, A., and Sonnhammer, E.L. (2000). FAT: a novel domain in PIK-related kinases. *Trends Biochem Sci* 25:225–227.
56. Alarcon, C.M., Heitman, J., and Cardenas, M.E. (1999). Protein kinase activity and identification of a toxic effector domain of the target of rapamycin TOR proteins in yeast. *Mol Biol Cell* 10:2531–2546.
57. Dames, S.A., Mulet, J.M., Rathgeb-Szabo, K., Hall, M.N., and Grzesiek, S. (2005). The solution structure of the FATC domain of the protein kinase target of rapamycin suggests a role for redox-dependent structural and cellular stability. *J Biol Chem* 280:20558–20564.
58. Sarbassov, D.D., and Sabatini, D.M. (2005). Redox regulation of the nutrient-sensitive raptor-mTOR pathway and complex. *J Biol Chem* 280:39505–39509.
59. Choi, J., Chen, J., Schreiber, S.L., and Clardy, J. (1996). Structure of the FKBP12–rapamycin complex interacting with the binding domain of human FRAP. *Science* 273:239–242.
60. Oshiro, N., *et al.* (2004). Dissociation of raptor from mTOR is a mechanism of rapamycin-induced inhibition of mTOR function. *Genes Cells* 9:359–366.
61. Long, X., Lin, Y., Ortiz-Vega, S., Yonezawa, K., and Avruch, J. (2005). Rheb binds and regulates the mTOR kinase. *Curr Biol* 15:702–713.
62. Holz, M.K., and Blenis, J. (2005). Identification of S6 kinase 1 as a novel mammalian target of rapamycin (mTOR)-phosphorylating kinase. *J Biol Chem* 280:26089–26093.
63. Chiang, G.G., and Abraham, R.T. (2005). Phosphorylation of mammalian target of rapamycin (mTOR) at Ser-2448 is mediated by p70S6 kinase. *J Biol Chem* 280:25485–25490.
64. Peterson, R.T., Beal, P.A., Comb, M.J., and Schreiber, S.L. (2000). FKBP12–rapamycin-associated protein (FRAP) autophosphorylates at serine 2481 under translationally repressive conditions. *J Biol Chem* 275:7416–7423.
65. Edinger, A.L., and Thompson, C.B. (2004). An activated mTOR mutant supports growth factor-independent, nutrient-dependent cell survival. *Oncogene* 23:5654–5663.
66. Banaszynski, L.A., Liu, C.W., and Wandless, T.J. (2005). Characterization of the FKBP.rapamycin.FRB ternary complex. *J Am Chem Soc* 127:4715–4721.
67. Adami, A., Garcia-Alvarez, B., Arias-Palomo, E., Barford, D., and Llorca, O. (2007). Structure of TOR and its complex with KOG1. *Mol Cell* 27:509–516.
68. Ho, H.L., Shiau, Y.S., and Chen, M.Y. (2005). *Saccharomyces cerevisiae* TSC11/AVO3 participates in regulating cell integrity and functionally interacts with components of the Tor2 complex. *Curr Genet* 47:273–288.
69. Berchtold, D., and Walther, T.C. (2009). TORC2 plasma membrane localization is essential for cell viability and restricted to a distinct domain. *Mol Biol Cell* 20:1565–1575.

70. Wang, L., Rhodes, C.J., and Lawrence, J.C., Jr. (2006). Activation of mammalian target of rapamycin (mTOR) by insulin is associated with stimulation of 4EBP1 binding to dimeric mTOR complex 1. *J Biol Chem* 281:24293–24303.

71. Zhang, Y., Billington, C.J., Jr., Pan, D., and Neufeld, T.P. (2006). *Drosophila* target of rapamycin kinase functions as a multimer. *Genetics* 172:355–362.

72. Takahara, T., Hara, K., Yonezawa, K., Sorimachi, H., and Maeda, T. (2006). Nutrient-dependent multimerization of the mammalian target of rapamycin through the N-terminal HEAT repeat region. *J Biol Chem* 281:28605–28614.

73. Toschi, A., Lee, E., Xu, L., Garcia, A., Gadir, N., and Foster, D.A. (2009). Regulation of mTORC1 and mTORC2 complex assembly by phosphatidic acid: competition with rapamycin. *Mol Cell Biol* 29:1411–1420.

74. Veverka, V., *et al.* (2008). Structural characterization of the interaction of mTOR with phosphatidic acid and a novel class of inhibitor: compelling evidence for a central role of the FRB domain in small molecule-mediated regulation of mTOR. *Oncogene* 27:585–595.

75. Reinke, A., Chen, J.C., Aronova, S., and Powers, T. (2006). Caffeine targets TOR complex I and provides evidence for a regulatory link between the FRB and kinase domains of Tor1p. *J Biol Chem* 281:31616–31626.

76. Urano, J., Sato, T., Matsuo, T., Otsubo, Y., Yamamoto, M., and Tamanoi, F. (2007). Point mutations in TOR confer Rheb-independent growth in fission yeast and nutrient-independent mammalian TOR signaling in mammalian cells. *Proc Natl Acad Sci USA* 104:3514–3519.

77. Ohne, Y., *et al.* (2008). Isolation of hyperactive mutants of mammalian target of rapamy-cin. *J Biol Chem* 283:31861–31870.

78. Huse, M., and Kuriyan, J. (2002). The conformational plasticity of protein kinases. *Cell* 109:275–282.

79. Acosta-Jaquez, H.A., *et al.* (2009). Site-specific mTOR phosphorylation promotes mTORC1-mediated signaling and cell growth. *Mol Cell Biol* 29:4308–4324.

80. Copp, J., Manning, G., and Hunter, T. (2009). TORC-specific phosphorylation of mam-malian target of rapamycin (mTOR): phospho-Ser2481 is a marker for intact mTOR signaling complex 2. *Cancer Res* 69:1821–1827.

81. Cheng, S.W., Fryer, L.G., Carling, D., and Shepherd, P.R. (2004). Thr2446 is a novel mammalian target of rapamycin (mTOR) phosphorylation site regulated by nutrient status. *J Biol Chem* 279:15719–15722.

82. Gwinn, D.M., *et al.* (2008). AMPK phosphorylation of raptor mediates a metabolic checkpoint. *Mol Cell* 30:214–226.

83. Carriere, A., *et al.* (2008). Oncogenic MAPK signaling stimulates mTORC1 activity by promoting RSK-mediated raptor phosphorylation. *Curr Biol* 18:1269–1277.

84. Wang, L., Lawrence, J.C., Jr., Sturgill, T.W., and Harris, T.E. (2009). Mammalian target of rapamycin complex 1 (mTORC1) activity is associated with phosphorylation of raptor by mTOR. *J Biol Chem* 284:14693–14697.

85. Foster, K.G., *et al.* (2009). Regulation of mTOR complex 1 (mTORC1) by raptor S863 and multi-site phosphorylation. *J Biol Chem* 285(1):80–94.

86. Nojima, H., *et al.* (2003). The mammalian target of rapamycin (mTOR) partner, raptor, binds the mTOR substrates p70 S6 kinase and 4E-BP1 through their TOR signaling (TOS) motif. *J Biol Chem* 278:15461–15464.

87. Dibble, C.C., Asara, J.M., and Manning, B.D. (2009). Characterization of Rictor phos-phorylation sites reveals direct regulation of mTOR complex 2 by S6K1. *Mol Cell Biol* 29:5657–5670.

88. Treins, C., Warne, P.H., Magnuson, M.A., Pende, M., and Downward, J. (2009). Rictor is a novel target of p70 S6 kinase-1. *Oncogene* (in press).

89. Julien, L.A., Carriere, A., Moreau, J., and Roux, P.P. (2010). mTORC1-activated S6K1 phosphorylates rictor on threonine 1135 and regulates mTORC2 signaling. *Mol Cell Biol* 30:908–921.
90. Polak, P., and Hall, M.N. (2009). mTOR and the control of whole body metabolism. *Curr Opin Cell Biol* 21:209–218.
91. Aoki, M., Blazek, E., and Vogt, P.K. (2001). A role of the kinase mTOR in cellular transformation induced by the oncoproteins P3k and Akt. *Proc Natl Acad Sci USA* 98:136–141.
92. Guertin, D.A., *et al.* (2009). mTOR complex 2 is required for the development of prostate cancer induced by Pten loss in mice. *Cancer Cell* 15:148–159.
93. Cybulski, N., and Hall, M.N. (2009). TOR complex 2: a signaling pathway of its own. *Trends Biochem Sci* 34:620–627.
94. Aspuria, P.J., Sato, T., and Tamanoi, F. (2007). The TSC/Rheb/TOR signaling pathway in fission yeast and mammalian cells: temperature sensitive and constitutive active mutants of TOR. *Cell Cycle* 6:1692–1695.

2

Regulation of TOR Signaling in Mammals

DUDLEY W. LAMMING • DAVID M. SABATINI

Whitehead Institute
Cambridge, Massachusetts, USA

I. Abstract

Since the initial discovery of rapamycin in the soils of Easter Island, there has been significant interest in determining the biological mechanism underlying the effects of rapamycin on mammalian tissues. These effects include changes in cell size and proliferation, as well as decreased and altered mRNA translation, and changes in autophagy. It was thus immediately clear that the mammalian target of rapamycin (mTOR) was a key regulator of cell growth and proliferation, and further work has shown that mTOR serves as a central regulator of cell processes in response to nutrients and environmental stimuli. In this chapter, we will discuss the mechanisms behind the many different functions of mTOR, with emphasis on the proteins that make up two mTOR-containing complexes and regulate mTOR signaling in response to insulin and amino acids. We will also discuss the medical relevance of these findings and the potential clinical significance of inhibiting mTOR.

II. One Enzyme, Two Complexes

Rapamycin functions by binding to the immunophilin FKBP12, and this complex then binds to and inhibits many functions of mTOR. The mTOR protein kinase was identified in 1994–1995 by three separate groups that isolated a 289-kD protein with a rapamycin-dependent interaction with FKBP12 [1–3]. mTOR was found to have approximately 40% homology to the two *Saccharomyces cerevisiae* TOR proteins, as well as high homology to the TOR proteins that would later be found in other eukaryotes. Indeed, many of the proteins that complex with mTOR to regulate and target mTOR are conserved in diverse eukaryotes, including fungi such as *S. cerevisiae,* metazoans such as *Caenorhabditis elegans*, the fruit fly *Drosophila melanogaster*, and mammals including mice and humans (see Table 2.1).

Most early work on mTOR used rapamycin extensively as a tool. However, work with kinase-dead mTOR and RNAi knockdown of mTOR made it apparent that mTOR participates in both rapamycin-dependent and independent processes. Organisms spanning complexity from humans to yeast have now been shown to have both rapamycin-sensitive and insensitive TOR-dependent processes (see Figure 2.1). Higher eukaryotes, including mammals, incorporate the same mTOR protein into both a rapamycin-sensitive mTOR complex I (mTORC1) and a complex that is relatively resistant to the effects of rapamycin, mTOR complex II (mTORC2). The core mTORC1 proteins are mTOR, Raptor, and mLST8/GβL, while the core mTORC2 proteins are mTOR, Rictor, and mLST8/GβL; the components of mTORC1 show a higher degree of evolutionary conservation than do the components of mTORC1. Yeast actually has two distinct TOR

TABLE 2.1

MTORC1 AND MTORC2 PROTEINS IN HUMANS, *DROSOPHILA MELANOGASTER*, *CAENORHABDITIS ELEGANS*, AND *SACCHAROMYCES CEREVISIAE*

	H. sapiens	*D. melanogaster*	*C. elegans*	*S. cerevisiae*
mTORC1	mTOR	dTOR	ceTOR	Tor1, Tor2
(rapamycin	mLST8	CG3004		Lst8
sensitive)	Raptor	Raptor	ceRaptor	Kog1
				Tco89
	PRAS40	Lobe	×	×
	DEPTOR	×	×	×
	RagA, RagB	dRagA	raga-1	Gtr1
	RagC, RagD	dRagC	ragc-1	Gtr2
mTORC2	mTOR	dTOR	ceTOR	Tor2
(rapamycin	mLST8	CG3004		Lst8
insensitive)	mSin1			Avo1
				Avo2
	Rictor	Rictor	CeRictor	Avo3
				Bit61
	Protor	×	×	×
	DEPTOR	×	×	×

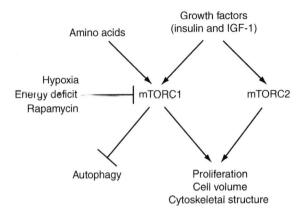

Fig. 2.1. The regulation of growth and proliferation by mTOR. mTORC1 is positively regulated by amino acids and growth factors, including insulin and IGF-1, and is inhibited by hypoxia, low levels of ATP, and rapamycin, while mTORC2 is positively regulated by growth factors. Both mTORC1 and mTORC2 signaling promote growth and proliferation, but only mTORC1 inhibits autophagy.

proteins; TOR1, which is TORC1 specific, and TOR2, which participates in both TORC1 and TORC2 (discussed in Chapter 1) [4]. Depending on cell type and nutrient conditions, numerous other proteins also associate with mTORC1, mTORC2, or both (see Table 2.1). In this chapter, we will focus on the regulation of mTOR signaling by PRAS40, DEPTOR, and the Rag family of small GTP-binding proteins.

III.   Raptor Defines mTORC1

The identification of Raptor as an mTOR associated protein came about as researchers in many labs were attempting to discover the mechanism by which nutrients and environmental stimuli regulate the phosphorylation of the mTOR targets S6K-1 and 4E-BP1. *In vivo*, these targets are phosphorylated by mTOR in response to such stimuli as amino acids, but researchers observed that the activity of the immunoprecipitated mTOR kinase against S6K-1 or 4E-BP *in vitro* was unchanged. One possible explanation for these results is that mTOR exists in a complex with additional regulatory proteins, but that these proteins were disassociating during mTOR purification. Two separate groups simultaneously investigated the properties of mTOR that was immunoprecipitated using different lysis conditions. One group approached the problem using the reversible cross-linker dithiobis (succinimidyl propionate) (DSP) [5], while another lab used an approach utilizing chromatography [6]. Both groups identified a 150-kD protein that immunoprecipitated with mTOR and showed high conservation within all eukaryotes (see Table 2.1 for a list of mTOR associated proteins and their homologues in various organisms). Both labs subsequently found that Raptor (for *r*egulatory *a*ssociated *p*rotein of m*TOR*), can also be isolated in the absence of cross-linker, indicating that the interaction is real and not an artifact of the cross-linker, so long as appropriate and gentle lysis conditions are used. Specifically, Raptor remains bound to mTOR when low concentrations of the detergent CHAPS was used instead of the more commonly used Triton X-100 or NP-40 [5].

To identify the role of Raptor in the regulation of mTOR, RNAi was used to decrease endogenous levels of Raptor in mammalian cells. Cells in which Raptor expression was thus lowered showed decreased phosphorylation of the mTOR target S6K-1, decreased proliferation, and smaller volume, all phenotypes that also occur when cells are treated with rapamycin or mTOR itself is knocked down with RNAi. RNAi against *C. elegans* homologues of Raptor and mTOR demonstrated that RNAi against either resulted in a similar set of developmental phenotypes, including delayed gonadal development, the development of large gut lysosomes, and smaller intestinal cells [6]. These experiments demonstrated that many of the phenotypes regulated by mTOR signaling are Raptor dependent.

The initial question of how nutrients and environmental stimuli regulate the phosphorylation of mTOR targets was at least partially answered by observing the properties of the mTOR–Raptor interaction. Researchers found that the stability of the mTOR–Raptor interaction was regulated by specific nutrients; specifically, the *in vivo* complex was destabilized by the addition of amino acids, leucine, and glucose, treatments that result in increased mTOR activity. Other treatments that decrease mTOR signaling to S6K-1, such as treatment with valinomycin, 2-deoxyglucose, and H_2O_2, also stabilized the mTOR–Raptor interaction. This led to the hypothesis that in the absence of nutrients, Raptor regulates mTOR by inhibiting its kinase activity, and that nutrients increase mTOR signaling by destabilizing the mTOR–Raptor interaction. This theory was supported by the finding that overexpression of Raptor suppresses the catalytic activity mTOR and inhibits phosphorylation of S6K1 and 4E-BP1 *in vivo*. However, this hypothesis does not explain everything; treatment of serum-starved cells with insulin, which results in increased mTOR signaling and increased phosphorylation of S6K1 by mTOR, does not destabilize the mTOR–Raptor interaction, suggesting that the regulation of mTOR activity by insulin and other growth factors may proceed via a different mechanism [5].

IV. Rictor Defines a Rapamycin-Insensitive mTOR Complex

Two additional proteins that complex with mTOR, Rictor and mLST8, were also discovered independently by different groups and at roughly the same time. These groups used different approaches to the same problem, some working with immunoprecipitation of mTOR followed by mass spectrometry or alternatively working up from yeast TOR and using homology to find mTOR-interacting proteins. As mentioned above, it was discovered that yeast TOR could be found in two complexes, only one of which was rapamycin sensitive (see Table 2.1 and [4]). An additional protein of unknown function, Lst8, could be found in both complexes, and shared homology with a relatively small human protein known originally as GβL and now dubbed mLST8 due to its homology to the yeast protein [7]. They found that overexpressed mLST8 and mTOR coprecipitated. A study conducted by a different group using mass spectrometry upon immunoprecipitated mTOR found that mTOR coprecipitates with a 36-kD protein known as GβL that acts to strongly increase mTOR kinase activity toward S6K1 and 4E-BP1 [8].

Yeast TOR functions in two complexes, one of which is rapamycin insensitive, and it was the subject of substantial interest to determine if mTOR similarly functioned in two complexes. Rictor was identified in 2004 as a defining component of mTORC2, a rapamycin-insensitive mTOR complex [9, 10]. Rictor was difficult to identify by homology due to its relatively low sequence conservation between mammals and lower eukaryotes such as yeast, and a lack of any domains of known function [9]. Indeed, the *C. elegans* Rictor homologue was only recently identified due to the low sequence conservation [11, 12]. The percentage of mTOR that complexes with Rictor also varies widely by cell type and is inversely correlated with the expression of Raptor. For instance, HEK293T cells contain mostly mTOR–Raptor complexes, while HeLa cells contain more Rictor than Raptor [9].

Because mTORC2 is relatively insensitive to rapamycin, the function of mTORC2 has been less clear until recently. One initial observation was that mTORC2 might regulate the cytoskeleton, although the precise mechanism behind this effect was unclear [4, 9, 10]. The deletion of mTORC2-specific subunits leads to the dephosphorylation of the hydrophobic motif site in Protein Kinase C (PKC)α. Subsequent experiments demonstrated that siRNA knockdown of either mTOR or Rictor, but not Raptor or treatment with rapamycin, also inhibited the kinase activity of PKCα, demonstrating the importance of this phosphorylation. Unfortunately, no environmental stimuli that regulate PKCα phosphorylation by mTOR have been identified, but siRNA knockdown of either Rictor or PKCα has similar effects on the actin cytoskeleton, implicating the regulation of PKCα phosphorylation as one mechanism by which mTOR signaling regulates the cytoskeleton [9].

While the effects of mTOR on the cytoskeleton remain an interesting area of study, an important step in understanding the effects of mTORC2 signaling occurred the following year with the identification of Akt as an mTORC2 substrate [13]. It had long been known that active Akt was phosphorylated on two key residues, Threonine 308 and Serine 473. The Thr308 kinase had previously been identified as PDK1, but the kinase for Serine 473 had remained elusive. While numerous kinases had been proposed as the Serine 473 kinase, including PDK1, Akt, and DNA-PK, none of these studies were fully convincing. However, in the course of experiments in *Drosophila* S2 cells and mammalian cell lines, it was observed that siRNA against mTOR or Raptor slightly increased the phosphorylation of Akt Ser473, while siRNA against Rictor resulted in the almost complete inhibition of Akt Ser473 phosphorylation. This was seen even in cell lines in which PTEN, an antagonist of PI3K signaling, had been knocked down or deleted. Subsequent experiments confirmed that mTORC2 directly phosphorylates Akt Ser473 both *in vitro and in vivo*, and that mTORC2 was the long-sought "PDK2" for Akt Ser473 [13]. Interestingly, mTORC2 appears to also regulate the activity of a related kinase, SGK1 [14]. mTORC2 regulates activation of SGK1 via direct phosphorylation of SGK1 Ser422, and may thus act as a regulator of numerous downstream targets of SGK1, including NDRG1 and FoxO proteins [14].

V.   Additional mTORC1 and mTORC2 Proteins

We can therefore define the two major mTOR complexes as the rapamycin-sensitive mTORC1, containing mTOR, Raptor, and mLST8, while the remainder of mTOR is in the relatively rapamycin-insensitive mTORC2, containing mTOR, Rictor, and mLST8 (see Figure 2.2 and Table 2.1). Subsequent work by a variety of labs has identified additional proteins that complex with either mTORC1, mTORC2, or both and isoforms of many of these proteins also exist. All or most of these isoforms can be combined together, which together likely define several different "flavors" of mTORC1 and mTORC2 that may vary between cell types and may have different sensitivity to stimuli [15].

There are at least two isoforms of mTOR; an alternative isoform of mTOR dubbed mTORβ demonstrates increased signaling to S6K, 4E-BP1, and Akt, and when overexpressed is tumorigenic in nude mice [16].

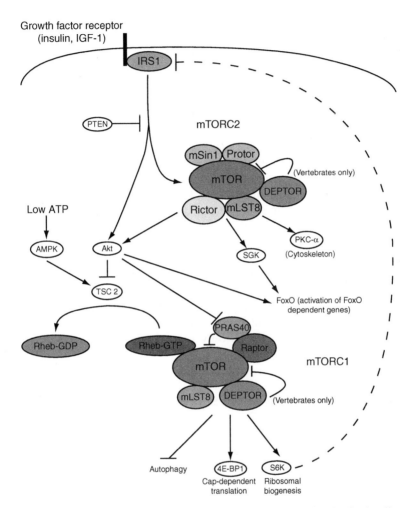

FIG. 2.2 Regulation of mTOR signaling by insulin. mTORC1 and mTORC2 signaling are activated by insulin. mTORC1 signaling via insulin is in part regulated by the tuberous sclerosis complex (TSC 1/2), which acts to inhibit mTOR by acting as the GTPase-activating protein (GAP) for Rheb-GTP, which is essential for mTORC1 activity. Akt activity is positively regulated by the activity of insulin and mTORC2, and Akt activates mTOR by (1) inhibiting the action of TSC 1/2 and (2) phosphorylating the mTORC1 inhibitor PRAS40 and thus relieving the inhibitory effects of PRAS40 on mTORC1. Finally, mTORC1 activity regulates the strength of insulin signaling at the cell surface via a feedback pathway (dashed line) by which S6K activity promotes the phosphorylation and degradation of IRS1.

There is an alternate cell-type specific isoform of Raptor; one group recently identified an isoform of Raptor which may not be able to interact with normal mTOR substrates [17]. There are several mTORC2-specific three isoforms of mSin1 [15] as well as the Protor-1 and Protor-2 proteins [18]. Finally, there are two isoforms of the mTORC1 and mTORC2 interacting protein DEPTOR, but the smaller isoform may not be significantly expressed (M. Laplante, unpublished data).

However, environmental signals which may regulate mTORC2 are largely unknown, although mTORC2 immunoprecipitated from insulin-stimulated cells has increased activity against Akt Ser473 *in vitro*, demonstrating that mTORC2 activity is regulated by insulin [15]. The proteins involved in the regulation of mTORC2 activity likewise largely remain mysterious. One exception is the recent discovery of Deptor, which interacts with and inhibits the function of both mTORC1 and mTORC2 [19]. In contrast, substantial progress has been made in understanding and defining the mechanisms by which mTORC1 is regulated. These mechanisms include the activation of mTORC1 by Rheb, the inhibition of mTOR by TSC1/2, and the inhibition of mTORC1 function by the Akt substrate PRAS40, which binds to and inhibits mTORC1 function in the absence of insulin [20–23]. Finally, while it has long been known that mTORC1 is regulated by amino acids, it was only recently discovered that the Rag proteins mediate this effect [24, 25].

VI.   The Regulation of mTOR Signaling by Insulin and PRAS40

Insulin signaling via mTORC1 is positively stimulated by the GTP-binding protein Rheb, which is itself negatively regulated by the action of the tuberous sclerosis tumor suppressor proteins TSC1/TSC2 [26]. TSC2 is a GTPase-activating protein, and the loss or mutation of TSC2 results in the constitutive loading of Rheb with GTP and the constitutive activation of mTORC1 signaling. The TSC1/2 complex serves as a signal-integration hub for a variety of nutrient-related signaling to mTORC1. Energy deprivation by AMPK, MAPK signaling, Wnt signaling, and hypoxia all regulate mTORC1 signaling via the regulation of TSC1/2 and Rheb [27–29]. AMPK also regulates mTOR signaling by directly phosphorylating Raptor and inhibiting its binding to mTOR [30].

TSC1/2 is also regulated by Akt, and as mentioned above the activity of Akt is itself regulated by mTORC2 [13, 23]. Inhibition of mTORC1 signaling by rapamycin leads to the stabilization of the interaction between IRS1 and the insulin receptor due to a feedback-loop mediated via S6K1 [31, 32] which in many cell types leads to increased signaling through mTORC2, and as mentioned above the phosphorylation of Ser473 on Akt and its activation [13, 32]. Akt then acts at three levels to regulate mTORC1 activity. First, it directly phosphorylates TSC2, disrupting the formation of a TSC1/2 complex and thus positively regulating mTORC1 activity [33]. Secondly, it again potentiates mTORC1 activity by phosphorylating and inhibiting the mTORC1-inhibitor PRAS40 [20]. Figure 2.2 provides a simplified diagram of the signaling between mTORC1, mTORC2, Akt, and TSC2. Finally, activated Akt stabilizes the surface expression of nutrient transporters, including Glut1 and amino acid transporters, which in turn promotes the uptake of nutrients and activates mTOR signaling [34].

The role of PRAS40 in the regulation of mTORC1 signaling was, much like many of the other core components of mTORC1, discovered at approximately the same time by different teams of researchers. A mass-spectrometry-based approach was used to examine mTOR immunoprecipitates [20, 21]. The team of Vander Haar *et al.* then used a direct approach to discover additional proteins bound to mTOR, while Sancak *et al.* discovered PRAS40 as a consequence of the development of an *in vitro* kinase assay for mTORC1. They found that mTORC1 immunoprecipitated from either insulin-stimulated or serum-starved cells was equally active, leading to the hypothesis that perhaps an additional factor that conferred insulin sensitivity was being lost during the purification process. They discovered that washing with low-salt buffers during the immunoprecipitation enabled them to recover complexes that had an insulin-induced activity difference, and subsequently identified the Akt-substrate PRAS40 as a salt-sensitive factor that inhibits mTORC1 during insulin deprivation [20]. While PRAS40 is an mTORC1 inhibitor, its action can be overcome *in vitro* by Rheb loaded with GTP, demonstrating that this is likely how insulin signaling to mTORC1 overcomes the effect of PRAS40 *in vivo*. While mTOR signaling is highly conserved, PRAS40 appears to be a more recent evolutionary development, as it is not found in yeast. However, a *Drosophila* homologue of PRAS40, Lobe, also functions as an mTORC1 inhibitor, demonstrating that this protein has been an important mTORC1 regulator for a substantial period of evolutionary time [20].

Subsequent work shed additional light on how PRAS40 functions and its potential clinical relevance. PRAS40 is now believed to function as a director inhibitor of substrate binding to Raptor and may itself also be an mTOR substrate [35–37]. PRAS40 has been identified as a target of Akt3 activity during malignant melanomas, and phosphorylated PRAS40 is believed to protect cancer cells from apoptosis [38]. However, this same property of PRAS40 may be beneficial in some contexts, and transfection of PRAS40 protects motor neurons from death in a mouse model of spinal cord injury [39].

VII.   DEPTOR: A Regulator of mTOR Signaling Found Only in Vertebrates

While PRAS40 homologues are found in flies, the same is not true of the recently identified DEPTOR protein, an inhibitor of mTORC1 and mTORC2 activity that is found only in vertebrates [19]. DEPTOR was discovered via mass spectrometry of mTOR immunoprecipitates in the same, low-salt conditions used to discover PRAS40. However, unlike PRAS40, DEPTOR interacts with both mTORC1 and mTORC2, via a C-terminal portion of mTOR [19]. The interplay between DEPTOR and mTOR is complex; DEPTOR inhibits the activity of both mTORC1 and mTORC2, and loss of DEPTOR results in the activation of mTORC1 and mTORC2 both *in vivo* and *in vitro*. However, mTORC1 and mTORC2 activity both negatively regulate DEPTOR at a transcriptional and post-translational level. DEPTOR thus serves as a mechanism for stabilizing a given level of mTOR activity; when mTOR activity is low, DEPTOR makes it harder to turn mTOR signaling on, but once mTOR signaling is on, DEPTOR makes it harder to turn off.

As DEPTOR functions as an inhibitor of both mTORC1 and mTORC2, it might be predicted that overexpression of DEPTOR would shut down all mTOR signaling. Interestingly, this is not the case, and overexpression of DEPTOR actually leads to increased mTORC2 signaling and the activation of Akt. DEPTOR has 13 serine and threonine phosphorylation sites, and DEPTOR is phosphorylated in an mTOR-dependent manner. Work with a nonphosphorylatable DEPTOR mutant indicates that phosphorylated DEPTOR binds preferentially to mTORC1, and that phosphorylation of DEPTOR reverses the inhibitory effects of DEPTOR on mTORC2 activity. Thus, overexpression of phosphorylated DEPTOR will inactivate mTORC1, while promoting insulin signaling via a negative feedback loop [40, 41]. Phosphorylation of IRS1, which is promoted by mTORC1 via S6K1, results in the disassociation of IRS1 from the insulin receptor and inhibition of insulin signaling. Conversely, inhibition of mTORC1 signaling by rapamycin results in increased interaction between IRS1 and the insulin receptor, and thus activates mTORC2 signaling and promotes the phosphorylation and activation of Akt.

One might predict that because overexpression of DEPTOR leads to constitutive activation of Akt, DEPTOR might be involved in certain types of cancer. Using transcriptional profiles of different cancer types, it was

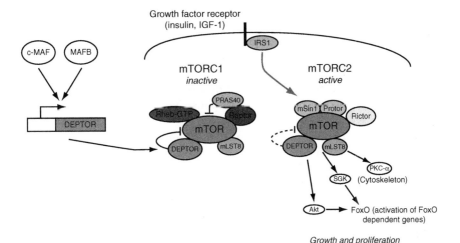

FIG. 2.3. The Roll of DEPTOR in multiple myeloma. Multiple myelomas with transloca-
tions involving the transcription factors c-MAF and MAFB have increased transcription of
DEPTOR and high levels of DEPTOR protein. This leads to the inactivation of mTORC1 and
the inactivation of the feedback loop between mTORC1 activity and signaling through IRS1.
As a result, IRS1 is stabilized, and the increased signaling through the insulin receptor drives
increased mTORC2 activity and the activation of Akt, leading to growth and proliferation.

observed that elevated DEPTOR mRNA was seen in a subset of human
multiple myeloma (MM) samples. Specifically, elevated DEPTOR is
expressed in nonhyperdiploid MM, including those which had transloca-
tions involving the c-MAF or MAFB transcription factors. Additional study
demonstrated that increased DEPTOR expression promotes survival in
MM cell lines and that the high levels of DEPTOR were driven by overexpression of c-MAF or MAFB, and resulted in hyperactivation of PI3K
signaling in general and Akt in particular. As shown in Figure 2.3, although
c-MAF or MAFB results in the upregulation of DEPTOR, only mTORC1
is inactivated, and increased signaling through mTORC2 leads to
unchecked growth and proliferation.

VIII. The Rag Proteins: Regulation of mTOR Signaling by Amino Acids

In the preceding section, we have discussed the organization of mTOR
signaling in higher eukaryotes, with particular emphasis on PRAS40 and
DEPTOR, proteins which are found only in higher eukaryotes. Raptor,

Rictor, and mLST8, which are conserved from yeast to humans, serve to target the activity of mTOR against specific endogenous targets, including S6K1, 4E-BP1, Akt1, and SGK1. However, these proteins do not link insulin signaling to mTOR activity, and this is achieved by the eukaryotic specific proteins PRAS40 and DEPTOR. However, while it has long been known that mTOR activity in yeast, flies, and humans is responsive to amino acid signaling, the mechanism behind this response has remained a mystery until recently.

As we have discussed above, mTORC1 signaling is responsive to a variety of stresses and environmental stimuli. Many of these inputs are integrated by the mTORC1 inhibitor TSC1/2; however, mTORC1 signaling remains sensitive to amino acid withdrawal even in cells lacking TSC1/2. Amino acids were therefore thought likely to regulate the mTORC1 complex directly, but previous attempts to immunoprecipitated mTORC1 in the presence or absence of amino acids did not succeed in isolating a difference. However, examination of immunoprecipitated protein lysates is complicated by the presence of a heavy-chain band that serves to obscure proteins that are similar in size. Sancak *et al.* chose to isolate mTORC1 using an approach that did not rely on antibodies; instead, Flag-tagged Raptor was isolated using a Flag affinity gel and the proteins were then eluted using Flag peptide. One of the proteins that bound only in the presence of amino acids was RagC [24]. Kim *et al.* also identified the Rag GTPase family as essential for mTORC1 activity by conducting an RNAi screen for GTPases that regulate S6 phosphorylation [25].

RagC is a member of a family of four small Ras-related GTP-binding proteins in humans (RagA through RagD). RagA and RagB are homologous to each other as well as to yeast Gtr1, while RagC and RagD are homologous to each other as well as yeast Gtr2. The mode of action of these proteins is conserved from yeast to humans, and in both Rag proteins function as heterodimers consisting of one Gtr1-like protein (Rag A or RagB) and one Gtr2-like protein (RagC or Rag D). In yeast, the Gtr1 and Gtr2 proteins have been linked to control of amino acid permeases and microautophagy, processes that are regulated by TOR signaling. Sancak *et al.* showed that Raptor copurified with a heterodimer of Rag proteins, and both Sancak *et al.* and Kim *et al.* found that the Rag proteins were essential for mTOR signaling by amino acids [24, 25]. Sancak *et al.* went on to show that in the presence of amino acids, Rag proteins function by localizing mTORC1 to a subcellular vesicular compartment marked by Rab7, indicating that mTOR is localized to the late endosomal and lysosomal compartments [24]. Therefore, in the presence of amino acids, both mTOR and its activator Rheb are present in Rab7-positive structures (see Figure 2.4) [24, 42]. Recently, it was demonstrated that TORC1 in yeast may be regulated in a similar manner by GTP-bound Gtr1, and that the

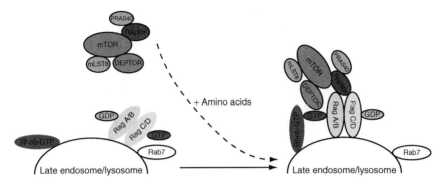

FIG. 2.4. Activation of mTORC1 in response to amino acids. In the absence of amino acids, mTORC1 is distributed throughout the cell, while its activator Rheb-GTP is localized to a late endosomal or lysosomal compartment marked by Rab7. Heterodimeric pairs of Rag proteins in Rab7-positive compartments are inactive, with Rag A/B bound to GDP and Rag C/D bound to GTP. In the presence of amino acids, the Rag proteins become active, with Rag A/B bound to GTP and Rag C/D bound to GDP. mTORC1 then relocalizes to the Rab7-positive endosomal compartments and is united with Rheb-GTP.

nucleotide binding of status of Gtr1 was regulated by the guanine nucleotide exchange factor Vam6 [43].

IX.   The Future: Remaining Mysteries of mTOR Signaling and Clinical Significance of mTOR

In this chapter, we have explored the function and structure of the two mammalian mTOR complexes, mTORC1 and mTORC2, and their control in higher eukaryotes such as human via PRAS40, DEPTOR, and the Rag proteins. We might, looking at a complex diagram of mTOR signaling, think that we now fully know everything about mTOR signaling, but in fact substantial work remains in understanding the function of the mTOR pathway. We will discuss some of these mysteries below and then discuss briefly the clinical significance of the mTOR pathway in light of recent discoveries about mTOR, cancer, and aging.

Above, we discussed how the Rag proteins activate mTORC1 in response to amino acids by localizing it to a compartment containing Rheb. However, it is still not clear how signals from various amino acids activate this relocalization, and GEFs for the mammalian Rag proteins have not yet been identified. It is also unclear exactly how localization of mTORC1 may contribute to the localization (and phosphorylation) of its substrates. In yeast, the S6K1 homologue Sch9 localizes to the vacuole and its localization is nutrient sensitive [44], so perhaps S6K1 (and other mTORC1 substrates) localize to the same compartment as mTORC1 during periods of abundant nutrients. We also do not have a good understanding of what other substrates there may be for mTORC1, or of how mTORC1 signaling regulates autophagy.

Turning our attention to mTORC2, there is still much we do not understand. At the most basic level, while we know that increased PI3K activity results in increased mTORC2 activity, we do not understand how this signal is transmitted to mTORC2. While rapamycin has allowed researchers around the world to study mTORC1 signaling, mTORC2 substrates have only been identified in the last few years. This includes such key regulatory proteins as Akt, which regulates the activity of forkhead transcription factors as well as the activity of mTORC1, and SGK [13, 14]. It is likely that as work progresses, additional substrates for mTORC2 activity will also be found. It has also recently been appreciated that rapamycin can inhibit mTORC2 function in some cell types, likely by inhibiting mTORC2 assembly [45]. As mTORC2 signaling may be important in some types of cancer, including prostate cancer and AML, the regulation of mTORC2 function by rapamycin may not only lead to the discovery of interesting biology but may also be clinically relevant [46, 47]. One factor that will help drive the discovery of more mTORC2 biology is the development of drugs that specifically inhibit mTORC2. While no mTORC2-specific drugs have yet been identified, several groups have identified compounds that will directly inhibit the activity of the mTOR kinase and thus inhibit both mTORC1 and mTORC2 [48–50].

From a clinical perspective, rapamycin has always been viewed as a promising drug candidate for the treatment of cancer—it specifically inhibits mTORC1 at nanomolar concentrations, and mTORC1 is hyperactive in many types of cancer. However, despite numerous clinical trials, rapamycin has had less of a clinical impact than hoped, although recent work suggests that rapamycin may be of use in treating some types of AML, renal and prostate cancers [46, 47, 51]. However, when applied to cancer cell lines in a tissue culture model, rapamycin often only has modest effects on proliferation. This may in part be due to the feedback loop between mTORC1 and IRS1 discussed above, resulting in the activation of PI3K signaling and subsequent phosphorylation and activation of Akt. Drugs that inhibit the mTOR kinase itself, possibly in combination with inhibition of PI3K activity, may fare better and at least one new inhibitor of mTORC1 function has been reported to induce cell cycle arrest [50].

Finally, a great deal of attention from both the scientific community and the popular press has recently been focused on the mTOR pathway as possible means of regulating the aging process. The influence of mTOR signaling on lifespan was first observed in 2003 in *C. elegans*, and was soon duplicated in

Drosophila and yeast [52–54]. Treatment of an organism with rapamycin similarly extends lifespan in yeast, and was recently shown to extend the lifespan of mice started on a diet containing rapamycin at 20 months of age [55, 56]. While the mechanisms behind this extension of lifespan in mammals have not yet been determined, it seems likely that the effect relies in part on decreased or altered translation. Several studies in *C. elegans* have demonstrated that RNAi against certain ribosomal proteins, S6K, the translation initiation factor eIF2, or other proteins required for translation such as eIF4E and eiF4G all extend lifespan and decrease protein translation [57].

However, the lifespan of worms with RNAi against ceTOR can be further extended by RNAi against ribosomal proteins, S6K, and transcription initiation factors, indicating that perhaps inhibition of TOR engages a different life extension pathway than the other interventions. Indeed, the Kenyon lab found that RNAi against many of these factors induced *daf-16/FOXO* signaling, while RNAi against ceTOR did not; also, RNAi against TOR in *C. elegans* subjected to dietary restriction (DR) reduced translation but did not further extend lifespan [57]. A recent study of *S6K1*$^{-/-}$ mice by Selman *et al.* found that female *S6K1*$^{-/-}$ mice had extended lifespan despite not demonstrating decreased protein synthesis or rates of translation, and the researchers noted that these mice had a gene expression pattern similar to that of mice subjected to CR or with an activated AMPK pathway [58]. Clearly, inhibition of mTOR signaling and DR is likely to involve the engagement of separate, but somewhat overlapping sets of pathways.

ACKNOWLEDGMENTS

We thank D.E. Cohen, M. Laplante, T. Peterson, Y. Sancak, and A. Strohecker for helpful discussions and critical reading of the manuscript. DWL was supported by a Ruth L. Kirschstein National Research Service Award (NRSA) Postdoctoral Fellowship (F32AG032833) from the National Institute of Aging, but the preceding content is solely the responsibility of the authors and does not necessarily represent the official views of the NIA or the NIH. DMS was supported by grants from the NIH (R01-AI47389 and R01-CA103866) and awards from the American Federation of Aging Research, the Keck Foundation and the LAM (Lymphangioleiomyomatosis) Foundation and is an investigator of the Howard Hughes Medical Institute.

REFERENCES

1. Brown, E.J., *et al.* (1994). A mammalian protein targeted by G1-arresting rapamycin-receptor complex. *Nature* 369:756–758.
2. Sabatini, D.M., Erdjument-Bromage, H., Lui, M., Tempst, P., and Snyder, S.H. (1994). RAFT1: a mammalian protein that binds to FKBP12 in a rapamycin-dependent fashion and is homologous to yeast TORs. *Cell* 78:35–43.

3. Sabers, C.J., *et al.* (1995). Isolation of a protein target of the FKBP12-rapamycin complex in mammalian cells. *J Biol Chem* 270:815–822.

4. Loewith, R., *et al.* (2002). Two TOR complexes, only one of which is rapamycin sensitive, have distinct roles in cell growth control. *Mol Cell* 10:457–468.

5. Kim, D.H., *et al.* (2002). mTOR interacts with raptor to form a nutrient-sensitive complex that signals to the cell growth machinery. *Cell* 110:163–175.

6. Hara, K., *et al.* (2002). Raptor, a binding partner of target of rapamycin (TOR), mediates TOR action. *Cell* 110:177–189.

7. Chen, E.J., and Kaiser, C.A. (2003). LST8 negatively regulates amino acid biosynthesis as a component of the TOR pathway. *J Cell Biol* 161:333–347.

8. Kim, D.H., *et al.* (2003). GbetaL, a positive regulator of the rapamycin-sensitive pathway required for the nutrient-sensitive interaction between raptor and mTOR. *Mol Cell* 11:895–904.

9. Sarbassov, D.D., *et al.* (2004). Rictor, a novel binding partner of mTOR, defines a rapamycin-insensitive and raptor-independent pathway that regulates the cytoskeleton. *Curr Biol* 14:1296–1302.

10. Jacinto, E., *et al.* (2004). Mammalian TOR complex 2 controls the actin cytoskeleton and is rapamycin insensitive. *Nat Cell Biol* 6:1122–1128.

11. Jones, K.T., Greer, E.R., Pearce, D., and Ashrafi, K. (2009). Rictor/TORC2 regulates *Caenorhabditis elegans* fat storage, body size, and development through sgk-1. *PLoS Biol* 7:e60.

12. Soukas, A.A., Kane, E.A., Carr, C.E., Melo, J.A., and Ruvkun, G. (2009). Rictor/TORC2 regulates fat metabolism, feeding, growth, and life span in *Caenorhabditis elegans*. *Genes Dev* 23:496–511.

13. Sarbassov, D.D., Guertin, D.A., Ali, S.M., and Sabatini, D.M. (2005). Phosphorylation and regulation of Akt/PKB by the rictor-mTOR complex. *Science* 307:1098–1101.

14. Garcia-Martinez, J.M., and Alessi, D.R. (2008). mTOR complex 2 (mTORC2) controls hydrophobic motif phosphorylation and activation of serum- and glucocorticoid-induced protein kinase 1 (SGK1). *Biochem J* 416:375–385.

15. Frias, M.A., *et al.* (2006). mSin1 is necessary for Akt/PKB phosphorylation, and its isoforms define three distinct mTORC2s. *Curr Biol* 16:1865–1870.

16. Panasyuk, G., *et al.* (2009). The mTORbeta splicing isoform promotes cell proliferation and tumorigenesis. *J Biol Chem* 284(45):30807–30814.

17. Sun, C., Southard, C., and Di Rienzo, A. (2009). Characterization of a novel splicing variant in the RAPTOR gene. *Mutat Res* 662:88–92.

18. Pearce, L.R., *et al.* (2007). Identification of Protor as a novel Rictor-binding component of mTOR complex-2. *Biochem J* 405:513–522.

19. Peterson, T.R., *et al.* (2009). DEPTOR is an mTOR inhibitor frequently overexpressed in multiple myeloma cells and required for their survival. *Cell* 137:873–886.

20. Sancak, Y., *et al.* (2007). PRAS40 is an insulin-regulated inhibitor of the mTORC1 protein kinase. *Mol Cell* 25:903–915.

21. Vander Haar, E., Lee, S.I., Bandhakavi, S., Griffin, T.J., and Kim, D.H. (2007). Insulin signalling to mTOR mediated by the Akt/PKB substrate PRAS40. *Nat Cell Biol* 9:316–323.

22. Long, X., Ortiz-Vega, S., Lin, Y., and Avruch, J. (2005). Rheb binding to mammalian target of rapamycin (mTOR) is regulated by amino acid sufficiency. *J Biol Chem* 280:23433–23436.

23. Inoki, K., Li, Y., Zhu, T., Wu, J., and Guan, K.L. (2002). TSC2 is phosphorylated and inhibited by Akt and suppresses mTOR signalling. *Nat Cell Biol* 4:648–657.

24. Sancak, Y., *et al.* (2008). The Rag GTPases bind raptor and mediate amino acid signaling to mTORC1. *Science* 320:1496–1501.
25. Kim, E., Goraksha-Hicks, P., Li, L., Neufeld, T.P., and Guan, K.L. (2008). Regulation of TORC1 by Rag GTPases in nutrient response. *Nat Cell Biol* 10:935–945.
26. Zhang, Y., Gao, X., Saucedo, L.J., Ru, B., Edgar, B.A., and Pan, D. (2003). Rheb is a direct target of the tuberous sclerosis tumour suppressor proteins. *Nat Cell Biol* 5:578–581.
27. Inoki, K., *et al.* (2006). TSC2 integrates Wnt and energy signals via a coordinated phosphorylation by AMPK and GSK3 to regulate cell growth. *Cell* 126:955–968.
28. Inoki, K., Zhu, T., and Guan, K.L. (2003). TSC2 mediates cellular energy response to control cell growth and survival. *Cell* 115:577–590.
29. Brugarolas, J., *et al.* (2004). Regulation of mTOR function in response to hypoxia by REDD1 and the TSC1/TSC2 tumor suppressor complex. *Genes Dev* 18:2893–2904.
30. Gwinn, D.M., *et al.* (2008). AMPK phosphorylation of raptor mediates a metabolic checkpoint. *Mol Cell* 30:214–226.
31. Harrington, L.S., *et al.* (2004). The TSC1-2 tumor suppressor controls insulin-PI3K signaling via regulation of IRS proteins. *J Cell Biol* 166:213–223.
32. Tremblay, F., Gagnon, A., Veilleux, A., Sorisky, A., and Marette, A. (2005). Activation of the mammalian target of rapamycin pathway acutely inhibits insulin signaling to Akt and glucose transport in 3T3-L1 and human adipocytes. *Endocrinology* 146:1328–1337.
33. Potter, C.J., Pedraza, L.G., and Xu, T. (2002). Akt regulates growth by directly phosphorylating Tsc2. *Nat Cell Biol* 4:658–665.
34. Edinger, A.L., and Thompson, C.B. (2002). Akt maintains cell size and survival by increasing mTOR-dependent nutrient uptake. *Mol Biol Cell* 13:2276–2288.
35. Oshiro, N., *et al.* (2007). The proline-rich Akt substrate of 40 kDa (PRAS40) is a physiological substrate of mammalian target of rapamycin complex 1. *J Biol Chem* 282:20329–20339.
36. Wang, L., Harris, T.E., Roth, R.A., and Lawrence, J.C., Jr (2007). PRAS40 regulates mTORC1 kinase activity by functioning as a direct inhibitor of substrate binding. *J Biol Chem* 282:20036–20044.
37. Fonseca, B.D., Smith, E.M., Lee, V.H., MacKintosh, C., and Proud, C.G. (2007). PRAS40 is a target for mammalian target of rapamycin complex 1 and is required for signaling downstream of this complex. *J Biol Chem* 282:24514–24524.
38. Madhunapantula, S.V., Sharma, A., and Robertson, G.P. (2007). PRAS40 deregulates apoptosis in malignant melanoma. *Cancer Res* 67:3626–3636.
39. Yu, F., Narasimhan, P., Saito, A., Liu, J., and Chan, P.H. (2008). Increased expression of a proline-rich Akt substrate (PRAS40) in human copper/zinc-superoxide dismutase transgenic rats protects motor neurons from death after spinal cord injury. *J Cereb Blood Flow Metab* 28:44–52.
40. Ozes, O.N., *et al.* (2001). A phosphatidylinositol 3-kinase/Akt/mTOR pathway mediates and PTEN antagonizes tumor necrosis factor inhibition of insulin signaling through insulin receptor substrate-1. *Proc Natl Acad Sci USA* 98:4640–4645.
41. Shah, O.J., Wang, Z., and Hunter, T. (2004). Inappropriate activation of the TSC/Rheb/mTOR/S6K cassette induces IRS1/2 depletion, insulin resistance, and cell survival deficiencies. *Curr Biol* 14:1650–1656.
42. Avruch, J., Long, X., Ortiz-Vega, S., Rapley, J., Papageorgiou, A., and Dai, N. (2009). Amino acid regulation of TOR complex 1. *Am J Physiol Endocrinol Metab* 296:E592–E602.
43. Binda, M., *et al.* (2009). The Vam6 GEF controls TORC1 by activating the EGO complex. *Mol Cell* 35:563–573.

44. Jorgensen, P., Rupes, I., Sharom, J.R., Schneper, L., Broach, J.R., and Tyers, M. (2004). A dynamic transcriptional network communicates growth potential to ribosome synthesis and critical cell size. *Genes Dev* 18:2491–2505.
45. Sarbassov, D.D., *et al.* (2006). Prolonged rapamycin treatment inhibits mTORC2 assembly and Akt/PKB. *Mol Cell* 22:159–168.
46. Zeng, Z., *et al.* (2007). Rapamycin derivatives reduce mTORC2 signaling and inhibit AKT activation in AML. *Blood* 109:3509–3512.
47. Guertin, D.A., *et al.* (2009). mTOR complex 2 is required for the development of prostate cancer induced by Pten loss in mice. *Cancer Cell* 15:148–159.
48. Feldman, M.E., *et al.* (2009). Active-site inhibitors of mTOR target rapamycin-resistant outputs of mTORC1 and mTORC2. *PLoS Biol* 7:e38.
49. Thoreen, C.C., *et al.* (2009). An ATP-competitive mammalian target of rapamycin inhibitor reveals rapamycin-resistant functions of mTORC1. *J Biol Chem* 284:8023–8032.
50. Garcia-Martinez, J.M., *et al.* (2009). Ku-0063794 is a specific inhibitor of the mammalian target of rapamycin (mTOR). *Biochem J* 421:29–42.
51. Oudard, S., *et al.* (2009). Everolimus (RAD001): an mTOR inhibitor for the treatment of metastatic renal cell carcinoma. *Expert Rev Anticancer Ther* 9:705–717.
52. Vellai, T., Takacs-Vellai, K., Zhang, Y., Kovacs, A.L., Orosz, L., and Muller, F. (2003). Genetics: influence of TOR kinase on lifespan in *C. elegans*. *Nature* 426:620.
53. Kaeberlein, M., *et al.* (2005). Regulation of yeast replicative life span by TOR and Sch9 in response to nutrients. *Science* 310:1193–1196.
54. Kapahi, P., Zid, B.M., Harper, T., Koslover, D., Sapin, V., and Benzer, S. (2004). Regulation of lifespan in *Drosophila* by modulation of genes in the TOR signaling pathway. *Curr Biol* 14:885–890.
55. Medvedik, O., Lamming, D.W., Kim, K.D., and Sinclair, D.A. (2007). MSN2 and MSN4 Link Calorie Restriction and TOR to Sirtuin-Mediated Lifespan Extension in *Saccharomyces cerevisiae*. *PLoS Biol* 5:e261.
56. Harrison, D.E., *et al.* (2009). Rapamycin fed late in life extends lifespan in genetically heterogeneous mice. *Nature* 460:392–395.
57. Hansen, M., Taubert, S., Crawford, D., Libina, N., Lee, S.J., and Kenyon, C. (2007). Lifespan extension by conditions that inhibit translation in *Caenorhabditis elegans*. *Aging Cell* 6:95–110.
58. Selman, C., *et al.* (2009). Ribosomal protein S6 kinase 1 signaling regulates mammalian life span. *Science* 326:140–144.

3

Rheb G-Proteins and the Activation of mTORC1

NITIKA PARMAR[a] • FUYUHIKO TAMANOI[b]

[a]*Biology Program*
California State University Channel Islands
1 University Drive, Camarillo, California, USA

[b]*Department of Microbiology, Immunology, and Molecular Genetics*
University of California, Los Angeles, California, USA

I. Abstract

Rheb belongs to a unique family within the Ras superfamily of G-proteins. Although initially identified in rat brain, this G-protein is highly conserved from yeast to human. While only one Rheb is present in lower eukaryotes, two Rheb proteins exist in mammalian cells. A number of studies establish that one of the functions of Rheb is to activate mTOR leading to growth. In particular, the ability of Rheb to activate mTORC1 *in vitro* points to direct interaction of Rheb with the mTORC1 complex. Additional functions of Rheb that are independent of mTOR have also been suggested.

II. Rheb Defines a Unique Family Within the Ras Superfamily G-Proteins

A. The Ras Superfamily G-Proteins

The Ras superfamily of G-proteins regulates a variety of signal transduction processes [1, 2]. These proteins are monomeric proteins of approximately 20–30 kDa with conserved structures and functions. These proteins

ISSN NO: 1874-6047
DOI: 10.1016/S1874-6047(10)27003-8

bind guanine nucleotides and function as a molecular switch by shuttling between GTP-bound and GDP-bound forms. The Ras superfamily consists of subfamilies of Ras, Rho, Rab, Arf, and Ran with Ras serving as the founding member. The pathways controlled by these proteins include signal transduction, cellular growth, transformation, transport, motility, traffic, and adhesion. Although these proteins have intrinsic GTPase activity, GTP hydrolysis is greatly enhanced by a set of proteins called GAPs (GTPase activator proteins) [3, 4], which maintain these proteins in the inactive GDP-bound state. On the other hand, the exchange of GDP to GTP is mediated by another set of proteins called GEFs (guanine nucleotide exchange factors) [5, 6], which maintain the proteins in the active GTP-bound state. An extra layer of regulation is provided by an additional set of proteins called GDIs (GDP dissociation inhibitors), which inhibit nucleotide exchange by binding to the GDP-bound form [7, 8].

The Ras superfamily is highly conserved across species ranging from yeast to mammals, including *Drosophila*, *C. elegans*, and plants. The proteins contain five conserved G boxes: G1, G2, G3, G4, and G5. A core effector domain called the switch I domain, which is involved in the interaction with downstream effectors, is conserved. Majority but not all of the superfamily members undergo a posttranslational modification at their C-terminus. These modifications facilitate membrane association and proper subcellular localization of the proteins which is critical for their functions. The C-terminus of many proteins end with a motif known as the CaaX box, which consists of a tetrapeptide sequence (C = Cys, a = aliphatic, X = C-terminal amino acid) [9]. This motif is modified by the addition of two types of lipid chains—farnesyl or geranylgeranyl groups—mediated by protein farnesyltransferase and protein geranylgeranyltransferase I, respectively. Other motifs such as CC, CCX, CCXX are also found to exist and they are modified by protein geranylgeranyltransferase II [10]. A few members have a modification at their N-terminus which involves the addition of a myristate fatty acid.

B. RHEB FAMILY GENES

The *Rheb* gene was first discovered in 1994 by Yamagata *et al.* [11] as a transcript which was rapidly induced in hippocampal granule cells by seizures and by NMDA-dependent synaptic activity. The product of this gene encodes a protein belonging to the Ras superfamily of small G-proteins. Rheb shares a 30–40% sequence identity with the Ras proteins. The Rheb proteins are highly conserved from yeast to mammals but are absent in prokaryotes [12, 13]. Rheb has intrinsic GTPase activity and can alternate between an active GTP-bound form and an inactive GDP-bound form [14].

In yeasts (*S. cerevisiae, S. pombe*), Rheb is encoded by a single gene [13]. Similarly, a single *Rheb* gene was identified in *Drosophila* [15]. However, the *Rheb* gene in *Drosophila* is alternatively spliced to yield two isoforms (Isoform A and Isoform B) that differ in their 5′ untranslated regions [16]. In mammals, there are two genes designated as *Rheb1* and *Rheb2* (*RhebL1*) [15]. The gene is located on chromosome 3 in *S. cerevisiae, C. elegans*, and *Drosophila* and on chromosome 2 in *S. pombe*. In humans, Rheb1 is located on chromosome 7 while Rheb2 is on chromosome 12. In addition, three *Rheb* pseudogenes exist in the human genome—two on chromosome 10 and one on chromosome 22. The genomic structure of human *Rheb1* contains seven exons while that of human *Rheb 2* consists of eight exons. At the protein level, the coding sequences of the two human *Rheb* genes share a 52% identity. *Rheb1* is ubiquitously expressed with high-level expression in skeletal muscle and cardiac muscles. On the other hand, the expression of *Rheb2* is more limited with high expression in brain [17].

C. RHEB PROTEIN STRUCTURE

The Rheb protein consists of 209 amino acids in *S. cerevisiae* (ScRheb), 185 amino acids in *S. pombe* (SpRheb), 207 amino acids in *C. elegans* (CeRheb), and 182 amino acids in *Drosophila melanogaster* (DmRheb) [13]. Sequence alignment shows that both ScRheb and CeRheb proteins have an N-terminal extension of 7–10 residues, function of which is not known. In humans, HsRheb1 consists of 184 amino acids while HsRheb2 consists of 183 amino acids and the two proteins share a 51% identity. HsRheb1 shares a 63% identity with DmRheb, 43% identity with CeRheb, 53% identity with SpRheb, and 37% identity with ScRheb. The molecular weight of the Rheb proteins ranges from 20 to 24 kD in the species mentioned above. Structure of Rheb proteins is shown in Figure 3.1.

Like the Ras proteins, the Rheb proteins contain five G boxes involved in GTP binding and hydrolysis [13]. Interestingly, most Rheb contains an arginine residue (R15) at the position equivalent to glycine (G12) in Ras [12, 13]; G12 is highly conserved in all Ras proteins and mutation to any other residue (except proline) results in the oncogenic transformation of Ras [1]. Significance of the conserved arginine is unclear at the moment. The last four residues at the C-terminus consist of the CaaX motif where the cysteine is farnesylated posttranslationally yielding a mature protein lacking the terminal three amino acids (aaX residues are removed). The CaaX motif in Rheb proteins is -CSIM (ScRheb), -CVIA (SpRheb), -CSIS (CeRheb), -CLVS (DmRheb), -CSVM (HsRheb1), and -CHLM (HsRheb2) while it is -CVLS in the Ras proteins.

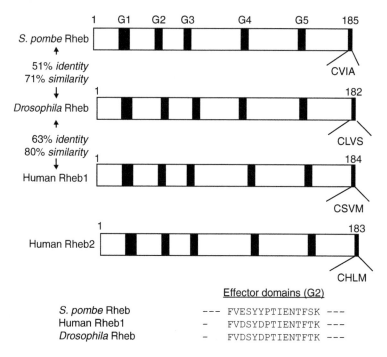

FIG. 3.1 *The Rheb family G-proteins.* Structure of Rheb proteins from human, *Drosophila,* and fission yeast are shown. G1–G5 boxes are indicated and the sequence of the effector domain (G2) is shown. Identity and similarity among different Rheb proteins are shown.

The protein structure of HsRheb1 is similar to that of Ras and Rap proteins and consists of six β-sheets and five α-helices [18] . The N-terminal 169 residues constitute the GTPase domain while the 15 residues at the C-terminus are hypervariable and show flexibility. The protein contains two switch regions (Switch I and Switch II) involved in recognition and interaction with GAPs, GEFs, and effectors. On the basis of sequence alignment, it has been observed that the switch I region in Rheb consists of residues 33–41 corresponding to residues 30–38 of Ras, while the switch II region consists of residues 63–79, corresponding to residues 60–76 of Ras. Despite the similarities, there are few interesting variations in the Rheb structure as compared to Ras [18]. The phosphate binding loop (P-loop) found in the G1 domain contains a highly conserved glycine at position 12 (G12) in the Ras family of proteins while in Rheb the equivalent position is occupied by an arginine (R15). A G12R mutation in Ras greatly decreases its intrinsic GTPase activity and renders it resistant to the action of RasGAP. Although wild-type Rheb containing R15 also has very low intrinsic GTPase activity, it is nevertheless sensitive to its GAP (Tsc2). In addition, the orientation of the side chain of R15 in Rheb proteins suggests that it is not directly involved in GTP hydrolysis and thus may not be responsible for Rheb's intrinsic low GTPase activity. Although both proteins have similar switch I regions, there are major differences in the switch II regions. In Ras, the switch II region assumes an α-helical conformation which undergoes a marked change upon GTP/GDP cycling; however, in Rheb the switch II region assumes an unraveled conformation undergoing only minor changes in response to GTP/GDP cycling. In both Ras and Rheb proteins, switch II has a highly conserved glutamine (Q61 and Q64, respectively). Q61 in Ras is directly responsible for GTP hydrolysis, but Q64 in Rheb has not been implicated to contribute to GTP hydrolysis as a result of the unique switch II region in Rheb.

D. FARNESYLATION AND LOCALIZATION OF RHEB

Farnesylation of Rheb proteins has been demonstrated with mammalian Rheb [19, 20] and the importance of farnesylation for the activation of the mTOR signaling has been reported [20–24]. On the other hand, postprenylation processes such as proteolytic cleavage and carboxylmethylation are not required for mTOR activation [22]. Effects of farneyltransferase inhibitors (FTIs) on the mTOR signaling have been reported [20, 23, 25]. Rheb can produce rapid development of aggressive and drug-resistant lymphomas in mice, and FTI treatment can block Rheb activity and induce antitumor effects in lymphoma cells [26]. Studies in yeast also suggested the significance of farnesylation on the function of Rheb [13, 27]. In *S. pombe*, Rheb is essential, as shown by the induction of cell cycle arrest at the G0/G1 phase when SpRheb is disrupted. Farnesylation of wild-type SpRheb is critical for complementing this phenotype [27]. Genetic screens to identify suppressors of the *tsc* phenotypes showed that the loss of farnesylation by the mutation in the protein farnesyltransferase β-subunit gene suppresses overactivation of the Tsc/Rheb/TOR signaling [28].

Rheb is localized to endomembranes. Buerger *et al.* [21] used EGFP-Rheb and monitored its localization in HEK293 cells. This study showed that EGFP-Rheb is localized within a defined ring around the nucleus, reminiscent of ER, and Golgi apparatus. The protein was shown to activate mTOR only when it reached the Golgi membrane. This localization was farnesylation dependent, as farnesylation deficient mutants failed to localize to the endomembrane vesicles. An intracellular localization was also observed for Rheb2 protein [29]. Sancak *et al.* [30] also used EGFP-Rheb and showed that Rheb colocalizes with *Discosoma* red fluorescent protein-labeled Rab7, suggesting that Rheb is localized to the late endosomal and lysosomal compartments. Saito *et al.* [17]

expressed EGFP tagged HsRheb1 and HsRheb2 in MDCK, HeLa, astrocytoma, and HEK293 cells and found that Rheb induced the formation of large vacuoles and that the tagged proteins were localized to the vacuolar membranes. These vacuoles were found to be late endocytic vesicles, as the late endocytic markers (Rab7 and Rab9) colocalized with Rheb. Vacuole formation did not occur when a mutant version of Rheb, lacking the C-terminal CaaX motif, was used whereby a diffuse pattern of localization was observed. Endomembrane localization of Rheb was also reported by Takahashi et al. [31]. Another study carried out by Ma et al. [32] suggested that HsRheb is localized to the mitochondria, as examined by the colocalization with mitotracker in HeLa cells. More recently, Hanker et al. [22] examined localization of GFP-Rheb1 and GFP-Rheb2 in COS-7 cells by comparing with markers for intracellular compartments. This study showed that both Rheb1 and Rheb2 localize to ER and Golgi but not to endosomes, lysosomes, or mitochondria.

E. BIOCHEMICAL ACTIVITY AND REGULATORS

Rheb is a G-protein and can bind both GTP and GDP [11, 14, 15, 33]. Biochemical studies using purified recombinant *Drosophila* Rheb (DmRheb) indicate that Rheb binds GTP preferentially over GDP and that this binding requires Mg^{2+} [15]. The binding is both time and nucleotide concentration dependent. Low but significant GTPase activity was detected with human Rheb as well as with fission yeast Rheb [14, 33]. DmRheb also exhibits low intrinsic GTPase activity [15]. The presence of arginine at the position corresponding to amino acid 12 of Ras does not completely eliminate its GTPase activity. The percentage of GTP bound to Rheb was examined as a measure of activation levels of endogenous Rheb [14]. The analysis showed that Rheb's activation state remained unusually high for a Ras superfamily member (above 20%).

A heterodimer TSC1/TSC2 consisting of the *TSC1* and *TSC2* gene products functions as a GTPase activating protein for Rheb converting GTP-bound form to a GDP-bound form. TSC2 protein contains a region (amino acids 1517–1674) of homology to the catalytic domain of Rap1-GAP at its C-terminal region and a fragment containing this domain exhibits the activity to stimulate intrinsic GTPase activity of Rheb [34]. The ability of TSC1/TSC2 to stimulate intrinsic GTPase activity of Rheb was demonstrated by expressing TSC1 and TSC2 and examining effects on Rheb [23, 24, 35–37]. In these experiments, the ratio of GTP and GDP bound to Rheb was examined after releasing guanine nucleotides from Rheb, a method described by Li et al. [38]. More recently, a novel method based on NMR was reported [39]. This method enables observation of real-time GAP activity. The study confirmed that TSC1/TSC2 accelerated GTP hydrolysis by Rheb approximately 50-fold. This is through an "asparagine-thumb" mechanism that uses a catalytic asparagine instead of arginine finger found in Ras-GAP [34, 39].

In contrast to GAP, no GEF proteins have been identified up to now. The protein TCTP (Translationally Controlled Tumor Protein) has recently been reported to function as the putative GEF for Rheb in *Drosophila* [40]. These findings were corroborated by Dong *et al.* [41], who demonstrated the interaction between human TCTP (HsTCTP) and human Rheb (HsRheb) *in vitro* and activation of the mTORC1 pathway by HsTCTP *in vivo*. A crystal structure of the E12V mutant of HsTCTP revealed that it lacked GEF activity as a result of a loss of interaction with Lys-45 of HsRheb. In addition, in *Arabidopsis thaliana* TCTP was identified as an important regulator of growth implying the plant TCTP as a mediator of TOR activity similar to that observed in nonplant systems [42]. However, contradictory reports have questioned the role of TCTP as the GEF for Rheb. Studies of HsRheb done *in vivo* and *in vitro* by Rehmann *et al.* [43] demonstrated no interaction between human Rheb and TCTP and subsequent findings by Wang *et al.* [44] also suggested that human TCTP does not interact with Rheb and has no influence on mTOR signaling.

Regulation of Rheb function by binding proteins has been reported. Bnip3, a hypoxia-inducible Bcl-2 homology 3 domain-containing protein, directly binds Rheb and inhibits the mTOR signaling [45]. Bnip3 was identified by carrying out yeast two-hybrid screen of cDNA libraries from HeLa cells and mouse embryos using Rheb as a bait. Bnip3 decreases Rheb GTP levels and inhibits the ability of Rheb to induce S6K phosphorylation. Another protein that was reported to bind Rheb is glyceraldehyde-3-phosphate dehydrogenase (GAPDH) [46]. Under low glucose conditions, GAPDH prevents Rheb from binding to mTOR and inhibits mTORC1 signaling.

F. RHEB MUTANTS

Extensive studies in *S. pombe* led to the identification of dominant-negative mutants of SpRheb, D60V, D60I, and D60K, whose expression led to growth inhibition and G1 arrest [33]. The D60V and D60I mutants preferentially bind GDP, while D60K binds neither GDP nor GTP. The D60I mutant was found to be more potent than D60V as its GTP binding capacity was severely reduced and GDP, once bound to it, could not be displaced by excess GTP. In Ras proteins, D57 is the critical residue involved in GTP/GDP binding. These mutations were introduced into mammalian Rheb to derive dominant negative forms. While they exhibit the ability to downregulate the mTOR signaling [33], expression of these mutants was low limiting their use for the analysis of the mTOR signaling.

Mutational analyses in *S. pombe* also led to the identification of hyperactive mutants of Rheb having point mutations at valine-17, serine-21, lysine-120, or asparagine-153 [47]. Cells expressing these mutants were found to exhibit resistance to canavanine and thialysine. These mutants exhibit an overall increase in the GTP/GDP ratio. Using sucrose gradient sedimentation experiments, it was determined that only the hyperactive SpRheb and not the wild-type version was found to interact with Tor2 suggesting that the interaction between these two proteins is dependent upon GTP. This idea was further supported by the observation that wild-type SpRheb could only interact with Tor2 if it was expressed in the *tsc2* mutant cells and not in wild-type cells. Hyperactive mutants of mammalian Rheb have been identified [48]. In particular, S16H mutant exhibits increased GTP loading *in vivo* and is resistant to TSC1/TSC2 GAP *in vitro*. The mutant exhibits increased ability to promote the phosphorylation of S6K1 and 4E-BP1. In addition, RhebQ64L exhibits increased GTP loading [37]. Jiang and Vogt [49] showed that HsRheb mutants Q64L and N153T were hyperactive and had oncogenic potential. The oncogenic ability of Q64L was shown to depend upon the presence of an intact farnesylation motif (CaaX) as well as mTOR activity since no cellular transformation was observed if rapamycin was used.

Mutations in the effector domain have been reported. Mutations in the switch I region have significant effect on the function of Rheb. For example, D36A, P37A, T38A, and N41A mutants of human Rheb still retain GTP binding but are defective in mTORC1 activation *in vitro* [50]. In addition, N41A mutant of human Rheb was less effective in supporting the mTOR in the absence of nutrients [51]. Significance of the switch I region of fission yeast Rheb was shown by the lack of complementation of the *rheb* mutant cells by RhebE40K [27]. Other reports suggest that regions outside of switch I also play important roles for the ability of Rheb to activate mTOR signaling. An alanine-scanning mutagenesis study identified Y54A and L56A mutants of human Rheb as mutants defective in mTOR activation [51]. In addition, a systematic change of solvent exposed residues of human Rheb identified two mutants, Y67A/I69A and I76A/D77A, as mutants defective in the ability to stimulate mTOR signaling in the absence of nutrients [52]. These mutants still retain the ability to bind guanine nucleotides. Interestingly, the mutations are located in the switch II segment.

Using fission yeast Rheb, a new type of Rheb mutants has been obtained. Murai *et al.* [53] isolated two dominant active *rhb1* mutants, *rhb1-DA4* and *rhb1-DA8* of SpRheb. The *rhb1-DA4* carries a V17A mutation within the

G1 box. This mutant is the same as the one that has been previously shown to act as a hyperactive mutant [47]. On the other hand, the *rhb1-DA8* mutant carries two mutations Q52R and I76F. While an amino acid uptake was prevented by both *rhb1-DA4* and *rhb1-DA8* mutants in a dominant fashion, only the *rhb1-DA4* mutant prevented the response to nitrogen starvation. This may suggest that these mutants act differently and can be used to genetically dissect the Rheb dependent signaling cascade.

III. Activation of mTORC1 by Rheb

A. INITIAL OBSERVATIONS

Rheb functions as a critical player in the insulin/TOR/S6K signaling and is involved with the regulation of growth and cell cycle progression [12, 54]. Interaction of Rheb with mTOR has been reported [55, 56]. Use of Rheb mutants suggested that the interaction did not depend on the binding of GTP and in fact was stronger with a mutant that exists predominantly as a GDP-bound form [55, 56]. Effects of amino acid starvation on the interaction of Rheb and mTOR were reported [57]. Expression of different mTOR fragments in mammalian cells together with Rheb mapped the Rheb interaction site to the 2148–2300 residues of mTOR [55]. The interaction of Rheb and Tor2 in fission yeast was reported [47]. In this case, the use of hyperactivated form of Rheb was critical in the detection of the interaction, suggesting that the interaction is preferential with the GTP-bound form.

B. RHEB ACTIVATES MTORC1 *IN VITRO*

More recently, Rheb was shown to activate mTOR *in vitro*, thus providing conclusive evidence that Rheb is an activator of mTOR [50, 58]. In these experiments, mTORC1 was isolated from cells starved for nutrients by using antibody against raptor. The complex is inactive, however, the addition of recombinant Rheb causes activation of mTORC1 as detected by the phosphorylation of substrate protein 4E-BP1 using antibody against phospho-4E-BP1 (Figure 3.2). The activation of mTORC1 by Rheb is dependent on the binding of GTP, as no activation was observed when GDP-bound Rheb or nucleotide free Rheb was used. The activation also was dependent on the presence of intact effector domain of Rheb. This point was shown by the use of Rheb mutants such as D36A and P37A.

The activation showed strong specificity for Rheb protein, as other members of the Ras family G-proteins such as RalA, RalB, KRas, RRas, Rad did not activate the mTORC1. Rho family G-proteins such as Rac1

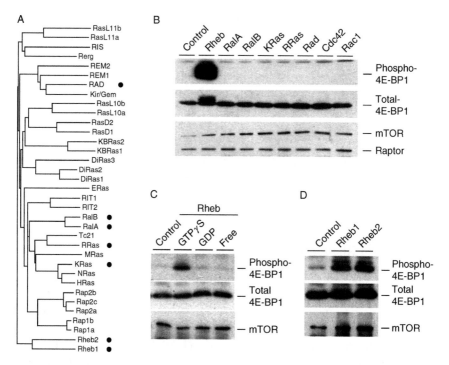

FIG. 3.2 Rheb G-protein specifically activates mTORC1 *in vitro*. (B) Recombinant Rheb but not other Ras superfamily G-proteins can activate mTORC1 isolated from HEK293T cells after nutrient starvation. 4E-BP1 was used as a substrate protein and its phosphorylation was detected on a SDS polyacrylamide gel after incubation with ATP. G-proteins tested are shown in (A). The reaction was dependent on GTP bound to Rheb (C). Both Rheb1 and Rheb2 activated mTORC1 *in vitro* (D). Reproduced from Ref. [50].

and Cdc42 also did not activate mTORC1. In addition to Rheb, mammalian cells have another Rheb called Rheb2 (also called RhebL1). We have shown that Rheb2 also activates mTORC1 in the *in vitro* system [50]. Thus these two proteins have similar function but they differ in their tissue distribution [17].

It is important to point out that Rag G-proteins do not activate mTORC1 [30]. While Rag plays critical roles in the activation of mTORC1 in response to amino acid signaling, the effect appears to occur prior to the Rheb activation, as Rag cannot activate mTORC1 in the absence of Rheb [30].

Rheb activation is specific to mTORC1, as no dramatic effects of Rheb on mTORC2 were observed. In the experiment we carried out [50], mTORC2 was isolated from cells starved for nutrients (serum and amino acids) by using antibody against Rictor. Activity of the isolated complex was examined by using Akt as a substrate. No significant stimulation was observed by the addition of Rheb, suggesting that the effect of Rheb is preferentially on mTORC1.

The ability of Rheb to activate mTORC1 *in vitro* does not depend on the farnesylation of Rheb. In the *in vitro* systems, Rheb purified after expression in *E. coli* has been used and this unmodified form was sufficient to induce mTORC1 activation. When Rheb isolated from insect cells after baculovirus infection was used, no improvement of the efficacy was observed, suggesting that farnesylation does not provide improvement in potency of Rheb to activate mTORC1.

The fact that Rheb can activate mTORC1 *in vitro* suggests that Rheb directly interacts with mTORC1. This interaction may be transient, as it has been difficult to detect Rheb associated with mTORC1 after the incubation (unpublished). Perhaps the interaction is transient and strong interaction may be counterproductive for mTORC1 activation. This may agree with the previous observations that the interaction between Rheb and mTOR is rather weak.

C. RHEB ENHANCES RECRUITMENT OF THE SUBSTRATE
    PROTEIN TO mTORC1

Activation of mTORC1 occurs in two steps. In the first step, substrate protein is recruited to mTORC1 mediated by raptor that recognizes the TOS motif present in the substrate proteins such as 4E-BP1 or S6K. In the second step, kinase activity of mTOR phosphorylates the substrate protein. We have shown that the activation of Rheb induces enhanced recruitment of substrate proteins to mTORC1. This was shown by examining the association of 4E-BP1 with mTORC1 in the presence and absence of Rheb-GTP [50]. This assay was carried out in the absence of ATP to avoid turnover of the substrate protein. The Rheb-induced enhancement of 4E-BP1 recruitment to mTORC1 was dependent on the presence of the intact effector domain of Rheb.

It is known that detergents used during the isolation of mTORC1 affect the association of raptor with mTOR [59]. While CHAPS can retain the interaction of Rheb and mTOR, NP40 disrupts the interaction. The mTORC1 isolated by using the buffer containing NP40 was not activated by Rheb, in agreement with the idea that raptor plays important roles in the Rheb-induced activation of mTORC1. PRAS40 inhibits mTORC1 activation by competing with substrate proteins, and the Rheb-induced activation of mTORC1 was inhibited by the addition of PRAS40. Intrinsic kinase

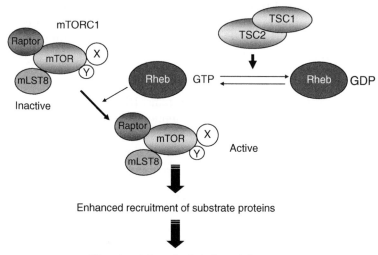

FIG. 3.3 *A model for the action of Rheb on mTORC1.* Rheb in its GTP-bound state interacts with mTORC1 and activates it. This involves enhanced recruitment of substrate proteins resulting in their phosphorylation. Rheb-GTP is converted to Rheb-GDP by the action of TSC1/TSC2 GAP.

activity of mTOR can be detected by autophosphorylation. The use of Mn^{2+} instead of Mg^{2+} in the kinase reaction promotes this reaction. We did not detect autophosphorylation of mTOR when Rheb was added. Thus, Rheb does not appear to induce intrinsic kinase activity.

These results obtained using the *in vitro* system are consistent with the idea that Rheb interacts with mTORC1 and enhances phosphorylation of substrate proteins (Figure 3.3). This involves increased recruitment of substrate proteins.

D. FKBP38 APPEARS NOT TO PLAY A MAJOR ROLE IN THE ACTIVATION OF mTORC1 BY RHEB

Bai *et al.* [60] have reported that Rheb binds FKBP38 in a GTP-dependent manner and that this relieves inhibitory effects of FKBP38 on mTORC1. The interaction between Rheb and FKBP38 was reported to be dependent on the presence of intact effector domain of Rheb [32]. However, the idea that Rheb activates mTORC1 indirectly by binding to FKBP38 is difficult to reconcile with the above observations with the *in vitro* system [50]. First, very little FKBP38 was detected in the mTORC1 preparation used for the *in vitro* reaction. Second, removal of FKBP38 from the cell extracts by the use of antibody against FKBP38 before preparing mTORC1 did not make differences in the ability of mTORC1 to be activated by Rheb. Finally, inhibition of FKBP38 expression by the use of siRNA only slightly affected the mTORC1 signaling. The mTORC1 isolated from these cells after nutrient starvation was still activated by Rheb. These results suggest that Rheb can activate mTORC1 in the absence of FKBP38. In addition, Uhlenbrock *et al.* failed to detect the interaction between Rheb and FKBP38 [61]. Conflicting results have been obtained regarding the effect of expression of FKBP38 on the mTOR signaling *in vivo* [44, 50, 61]. Taken together, it appears that FKBP38 is not a major player in the Rheb activation of mTORC1.

IV.   Functions of Rheb that Are Independent of mTOR

The Ras superfamily G-proteins generally activates multiple downstream signaling pathways. For example, downstream effectors of Ras include Raf kinase, PI3 kinase, RalGDS, and Rin1 [62, 63]. Therefore, it is expected that multiple downstream effectors of Rheb will be discovered. A number of reports suggest that this is the case.

Involvement of the TSC/Rheb signaling on perinuclear aggresome formation has been reported [64]. The aggresome represents a mechanism to dispose of misfolded proteins that exceed the degradative capacity of proteasome and autophagy systems. *TSC* mutant cells are defective in aggresome formation and undergo apoptosis upon the accumulation of misfolded proteins. Involvement of Rheb in the aggresome formation is demonstrated using Rheb mutants, Rheb-Q64L and Rheb-D60K. Use of rapamycin suggests that these effects are independent of mTORC1.

Effects of Rheb on Raf kinase have been reported. Rheb was shown to antagonize NIH3T3 transformation by H-ras^{G12V} [19]. Rheb was found to bind Raf-1 *in vitro*. Inhibition of MAPK activation by the expression of Rheb was shown in *Xenopus* oocytes. Effects of Rheb on Raf were also reported by Yee and Worley [65]. This study showed that Rheb interacts with Raf-1 kinase and regulates cAMP and growth factor signaling in conjunction with H-Ras. More recently, the effect of Rheb on Raf kinases is reported by Karbowniczek *et al.* [66, 67]. Tuberin expression increases p42/44 MAP kinase phosphorylation and B-Ras activity. Expression of Rheb inhibits wild-type B-Raf but not activated forms of B-Raf. Inhibition of B-Raf and MAPK by Rheb is resistant to rapamycin, while Rheb activation of S6K is rapamycin sensitive.

Other suggested downstream effectors of Rheb include phospholipase D1 (PLD1) [68]. Significance of PLD1 in the activation of mTOR signaling was suggested from the use of 1-butanol, an inhibitor of PLase D. Furthermore, it is shown that phosphatidic acid, the product of PLase D, is required for the formation of mTORC1 and mTORC2 complexes [69]. In this scenario, Rheb affects PLase D leading to the regulation of mTORC1 formation.

V.   Future Prospects

Studies over the years have established Rheb as a member of the Ras superfamily G-proteins. This G-protein has unique structural features and has a high GTP level. TSC1/TSC2 complex functions as a GAP for Rheb and this involves the use of the asparagine-thumb mechanism. Other regulators of Rheb need to be identified to understand the function of this G-protein in the regulation of a variety of cellular processes. A major area of study concerns identification of GEF for Rheb. While TCTP has been proposed to function as a GEF for Rheb, assignment of this protein as a Rheb GEF is premature at this point.

Involvement of Rheb in the mTORC1 signaling has been established by a number of experiments. Because Rheb can activate mTORC1 isolated from nutrient starved cells, Rheb has to interact with mTORC1. However, this interaction may be transient and may not require strong stable binding. One of the consequences of the Rheb action is to enhance recruitment of substrate proteins to mTORC1, a process that is mediated by raptor. Further investigation of this *in vitro* system is needed to understand biochemical and structural consequences of the interaction of Rheb with mTORC1. Further studies on downstream effectors of Rheb may reveal multiple downstream pathways activated by Rheb. Use of Rheb mutants that may distinguish activation of different downstream signaling pathways may be valuable for this analysis.

Acknowledgments

We thank members of the Tamanoi laboratory for discussion. This work was supported by the NIH grant CA41996 (to F.T.).

References

1. Wennerberg, K., Rossman, K.L., and Der, C.J. (2005). The Ras superfamily at a glance. *J Cell Sci* 118:843–846.
2. Bourne, H.R., Sanders, D.A., and McCormick, F. (1990). The GTPase superfamily: conserved structure and molecular mechanism. *Nature* 348:124–132.
3. Donovan, S., Shannon, K.M., and Bollag, G. (2002). GTPase activating proteins: critical regulators of intracellular signaling. *Biochim Biophys Acta* 1602:23–45.
4. Bernards, A., and Settleman, J. (2004). GAP control: regulating the regulators of small GTPases. *Trends Cell Biol* 14:377–385.
5. Cherfils, J., and Chardin, P. (1999). GEFs: structural basis for their activation of small GTP-binding proteins. *Trends Biochem Sci* 24:306–311.
6. Schmidt, A., and Hall, A. (2002). Guanine nucleotide exchange factors for Rho GTPases: turning on the switch. *Genes Dev* 16:1587–1609.
7. Olofsson, B. (1999). Rho guanine dissociation inhibitors: pivotal molecules in cellular signaling. *Cell Signal* 11:545–554.
8. Wu, S.K., Zeng, K., Wilson, I.A., and Balch, W.E. (1996). Structural insights into the function of the Rab GDI superfamily. *Trends Biochem Sci* 21:472–476.
9. Cox, A.D., and Der, C.J. (2002). Ras family signaling: therapeutic targeting. *Cancer Biol Ther* 1:599–606.
10. Maurer-Stroh, S., Washietl, S., and Eisenhaber, F. (2003). Protein prenyltransferases. *Genome Biol* 4:212.
11. Yamagata, K., Sanders, L.K., Kaufmann, W.E., Yee, W., Barnes, C.A., Nathans, D., and Worley, P.F. (1994). *Rheb*, a growth factor- and synaptic activity-regulated gene, encodes a novel Ras-related protein. *J Biol Chem* 269:16333–16339.
12. Aspuria, P.J., and Tamanoi, F. (2004). The Rheb family of GTP-binding proteins. *Cell Signal* 16:1105–1112.
13. Urano, J., Tabancay, A.P., Yang, W., and Tamanoi, F. (2000). The *Saccharomyces cerevisiae* Rheb G-protein is involved in regulating canavanine resistance and arginine uptake. *J Biol Chem* 275:11198–11206.
14. Im, E., von Lintig, F.C., Chen, J., Zhuang, S., Qui, W., Chowdhury, S., Worley, P.F., Boss, G.R., and Pilz, R.B. (2002). Rheb is in a high activation state and inhibits B-Raf kinase in mammalian cells. *Oncogene* 21:6356–6365.
15. Patel, P.H., Thapar, N., Guo, L., Martinez, M., Maris, J., Gau, C.L., Lengyel, J.A., and Tamanoi, F. (2003). *Drosophila* Rheb GTPase is required for cell cycle progression and cell growth. *J Cell Sci* 116:3601–3610.
16. Hoskins, R.A., Carlson, J.W., Kennedy, C., Acevedo, D., Evans-Holm, M., Frise, E., Wan, K.H., Park, S., Mendez-Lago, M., Rossi, F., Villasante, A., Dimitri, P., Karpen, G. H., and Celniker, S.E. (2007). Sequencing finishing and mapping of *Drosophila melanogaster* heterochromatin. *Science* 316:1625–1628.
17. Saito, K., Araki, Y., Kontani, K., Nishina, H., and Katada, T. (2005). Novel role of the small GTPase Rheb: its implication in endocytic pathway independent of the activation of mammalian target of rapamycin. *J Biochem* 137:423–430.
18. Yu, Y., Li, S., Xu, X., Li, Y., Guan, K., Arnold, E., and Ding, J. (2005). Structural basis for the unique biological function of small GTPase RHEB. *J Biol Chem* 280:17093–17100.
19. Clark, G.J., Kinch, M.S., Rogers-Graham, K., Sebti, S.M., Hamilton, A.D., and Der, C.J. (1997). The Ras-related protein Rheb is farnesylated and antagonizes Ras signaling and transformation. *J Biol Chem* 272:10608–10615.

20. Basso, A.D., Mirza, A., Liu, G., Long, B.J., Bishop, W.R., and Kirschmeier, P. (2005). The farnesyltransferase inhibitor (FTI) SCH66336 (lonafarnib) inhibits Rheb farnesylation and mTOR signaling. *J Biol Chem* 280:31101–31108.

21. Buerger, C., DeVries, B., and Stambolic, V. (2006). Localization of Rheb to the endomembrane is critical for its signaling function. *Biochem Biophys Res Commun* 344:869–880.

22. Hanker, A.B., Mitin, N., Wilder, R.S., Henske, E.P., Tamanoi, F., Cox, A.D., and Der, C.J. (2009). Requirement of CAAX-mediated posttranslational processing for Rheb localization and signaling. *Oncogene* 29:380–391.

23. Castro, A.F., Rebhun, J.F., Clark, G.J., and Quilliam, L.A. (2003). Rheb binds tuberous sclerosis complex 2 [TSC2] and promotes S6 kinase activation in a rapamycin- and farnesylation-dependent manner. *J Biol Chem* 278:32493–32496.

24. Tee, A.R., Manning, B.D., Roux, P.P., Cantley, L.C., and Blenis, J. (2003). Tuberous sclerosis complex gene products, Tuberin and Hamartin, control mTOR signaling by acting as a GTPase-activating protein complex toward Rheb. *Curr Biol* 13:1259–1268.

25. Gau, C.L., Kato-Stankiewicz, J., Jiang, C., Miyamoto, S., Guo, L., and Tamanoi, F. (2005). Farnesyltransferase inhibitors reverse altered growth and distribution of actin filaments in Tsc-deficient cells via inhibition of both rapamycin-sensitive and -insensitive pathways. *Mol Cancer Ther* 4:918–926.

26. Mavrakis, K.J., Zhu, H., Silva, R.L.A., Mills, J.R., Teruya-Feldstein, J., Lowe, S.W., Tam, W., Pelletier, J., and Wendel, H.G. (2008). Tumorigenic activity and therapeutic inhibition of Rheb GTPase. *Genes Dev* 22:2178–2188.

27. Yang, W., Tabancay, A.P., Jr., Urano, J., and Tamanoi, F. (2001). Failure to farnesylate Rheb protein contributes to the enrichment of G0/G1 phase cells in the *Schizosaccharomyces pombe* farnesyltransferase mutant. *Mol Microbiol* 6:1339–1347.

28. Nakase, Y., Fukuda, K., Chikashige, Y., Tsutsumi, C., Morita, D., Kawamoto, S., Ohnuki, M., Hiraoka, Y., and Matsumoto, T. (2006). A defect in protein farnesyltransferase suppresses a loss of *Schizosaccharomyces pombe tsc2*+, a homology of the human gene predisposing to tuberous sclerosis complex. *Genetics* 173:569–578.

29. Yuan, J., Shan, Y., Chen, X., Tang, W., Luo, K., Ni, J., Wan, B., and Yu, L. (2005). Identification and characterization of RHEBL1, a novel member of Ras family, which activates transcriptional activities of NF-Kappa B. *Mol Biol Rep* 32:205–214.

30. Sancak, Y., Peterson, T.R., Shaul, Y.D., Lindquist, R.A., Thoreen, C.C., Bar-Peled, L., and Sabatini, D.M. (2008). The Rag GTPases bind raptor and mediate amino acid signaling to mTORC1. *Science* 320:1496–1501.

31. Takahashi, K., Nakagawa, M., Young, S.G., and Yamanaka, S. (2005). Differential membrane localization of ERas and Rheb, two ras-related proteins involved in the PI3 kinase/mTOR pathway. *J Biol Chem* 280:32768–32774.

32. Ma, D., Bai, X., Guo, S., and Jiang, Y. (2008). The switch I region of Rheb is critical for its interaction with FKBP38. *J Biol Chem* 283:25963–25970.

33. Tabancay, A.P., Jr., Gau, C.L., Machado, I.M., Uhlmann, E.J., Gutmann, D.H., Guo, L., and Tamanoi, F. (2003). Identification of dominant negative mutants of Rheb GTPase and their use to implicate the involvement of human Rheb in the activation of p70S6K. *J Biol Chem* 278:39921–39930.

34. Scrima, A., Thomas, C., Diaconescu, D., and Wittinghofer, A. (2008). The Rap-RapGAP complex: GTP hydrolysis without catalytic glutamine and arginine residues. *EMBO J* 27 (7):1145–1153.

35. Zhang, Y., Gao, X., Saucedo, L.J., Ru, B., Edgar, B.A., and Pan, D. (2003). Rheb is a direct target of the tuberous sclerosis tumour suppressor proteins. *Nat Cell Biol* 5:578–581.

36. Inoki, K., Li, Y., Xu, T., and Guan, K.L. (2003). Rheb GTPase is a direct target of TSC2 GAP activity and regulates mTOR signaling. *Genes Dev* 17:1829–1834.
37. Li, Y., Inoki, K., and Guan, K.L. (2004). Biochemical and functional characterizations of small GTPase Rheb and TSC2 GAP activity. *Mol Cell Biol* 18:7965–7975.
38. Li, Y., Inoki, K., Vikis, H., and Guan, K.-L. (2006). Measurements of TSC2 GAP activity toward Rheb. *Methods Enzymol* 407:46–54.
39. Marshall, C.B., Ho, J., Buerger, C., Plevin, M., Li, G.-Y., Li, Z., Ikura, M., and Stambolic, V. (2009). Characterization of the intrinsic and TSC2-GAP regulated GTPase activity of Rheb by real-time NMR. *Sci Signal* 2:1–11.
40. Hsu, Y.C., Chern, J.J., Cai, Y., Liu, M., and Choi, K.W. (2007). *Drosophila* TCTP is essential for growth and proliferation through regulation of dRheb GTPase. *Nature* 445:785–788.
41. Dong, X., Yang, B., Li, Y., Zhong, C., and Ding, J. (2009). Molecular basis of the acceleration of the GDP-GTP exchange of human ras homolog enriched in brain by human translationally controlled tumor protein. *J Biol Chem* 284:23754–23764.
42. Berkowitz, O., Jost, R., Pollmann, S., and Masle, J. (2008). Characterization of TCTP, the translationally controlled tumor protein, from *Arabidopsis thaliana*. *Plant Cell* 20:3430–3447.
43. Rehmann, H., Brüning, M., Berghaus, C., Schwarten, M., Köhler, K., Stocker, H., Stoll, R., Zwartkruis, F.J., and Wittinghofer, A. (2008). Biochemical characterisation of TCTP questions its function as a guanine nucleotide exchange factor for Rheb. *FEBS Lett* 582:3005–3010.
44. Wang, X., Fonseca, B.D., Tang, H., Liu, R., Elia, A., Clemens, M.J., Bommer, U.A., and Proud, C.G. (2008). Re-evaluating the roles of proposed modulators of mammalian target of rapamycin complex 1 [mTORC1] signaling. *J Biol Chem* 283:30482–30492.
45. Li, Y., Wang, Y., Kim, E., Beemiller, P., Wang, C.-Y., Swanson, J., You, M., and Guan, K. L. (2007). Bnip3 mediates the hypoxia-induced inhibition on mammalian target of rapamycin by interacting with Rheb. *J Biol Chem* 282:35803–35813.
46. Lee, M.N., Ha, S.H., Kim, J., Koh, A., Lee, C.S., Kim, J.H., Jeon, H., Kim, D.-H., Suh, P.-G., and Ryu, S.H. (2009). Glycolytic flux signal to mTOR through glyceraldehyde-3-phosphate dehydrogenase-mediated regulation of Rheb. *Mol Cell Biol* 29:3991–4001.
47. Urano, J., Comiso, M.J., Guo, L., Aspuria, P.J., Deniskin, R., Tabancay, A.P., Jr, Kato-Stankiewicz, J., and Tamanoi, F. (2005). Identification of novel single amino acid changes that result in hyperactivation of the unique GTPase, Rheb, in fission yeast. *Mol Microbiol* 58:1074–1086.
48. Yan, L., Findley, G.M., Jones, R., Procter, J., Cao, Y., and Lamb, R.F. (2006). Hyperactivation of mammalian target of rapamycin (mTOR) signaling by a gain-of-function mutant of the Rheb GTPase. *J Biol Chem* 281:19793–19797.
49. Jiang, H., and Vogt, P.K. (2008). Constitutively active Rheb induces oncogenic transformation. *Oncogene* 27:5729–5740.
50. Sato, T., Nakashima, A., Guo, L., and Tamanoi, F. (2009). Specific activation of mTORC1 by Rheb G-protein *in vitro* involves enhanced recruitment of its substrate protein. *J Biol Chem* 284:12783–12791.
51. Tee, A., Blenis, J., and Proud, C.G. (2005). Analysis of mTOR signaling by the small G-proteins, Rheb and RhebL1. *FEBS Lett* 579:4763–4768.
52. Long, X., Lin, Y., Ortiz-Vega, S., Busch, S., and Avruch, J. (2007). The Rheb switch 2 segment is critical for signaling to target of rapamycin complex. *J Biol Chem* 282:18542–18551.

53. Murai, T., Nakase, Y., Fukuda, K., Chikashige, Y., Tsutsumi, C., Hiraoka, Y., and Matsumoto, T. (2009). Distinctive responses to nitrogen starvation in the dominant active mutants of the fission yeast Rheb GTPase. *Genetics* 183:517–527.

54. Aspuria, P.J., Sato, T., and Tamanoi, F. (2007). The TSC/Rheb/TOR signaling pathway in fission yeast and mammalian cells. *Cell Cycle* 6:1692–1695.

55. Long, X., Lin, Y., Ortiz-Vega, S., Yonezawa, K., and Avruch, J. (2005). Rheb binds and regulates the mTOR kinase. *Curr Biol* 15:702–713.

56. Smith, E.M., Finn, S.G., Tee, A.R., Browne, G.J., and Proud, C.G. (2005). The tuberous sclerosis protein TSC2 is not required for the regulation of the mammalian target of rapamycin by amino acids and certain cellular stresses. *J Biol Chem* 280:18717–18727.

57. Long, X., Ortiz-Vega, S., Lin, Y., and Avruch, J. (2005). Rheb binding to mammalian target of rapamycin [mTOR] is regulated by amino acid sufficiency. *J Biol Chem* 280:23433–23436.

58. Sancak, Y., Thoreen, C.C., Peterson, T.R., Lindquist, R.A., Kang, S.A., Spooner, E., Carr, S.A., and Sabatini, D.M. (2007). PRAS40 is an insulin-regulated inhibitor of the mTORC1 protein kinase. *Mol Cell* 25:903–915.

59. Kim, D.H., Sarbassov, D.D., Kim, D.H., Sarbassov, D.D., Ali, S.M., King, J.E., Latek, R. R., Erdjument-Bromage, H., Tempst, P., and Sabatini, D.M. (2002). mTOR interacts with raptor to form a nutrient-sensitive complex that signals to the cell growth machinery. *Cell* 110:163–175.

60. Bai, X., Ma, D., Liu, A., Shen, X., Wang, Q.J., Liu, Y., and Jiang, Y. (2007). Rheb activates mTOR by antagonizing its endogenous inhibitor, FKBP38. *Science* 318:977–980.

61. Uhlenbrock, K., Weiwad, M., Wetzker, R., Fischer, G., Wittinghofer, A., and Rubio, I. (2009). Reassessment of the role of FKBP38 in the Rheb/mTORC1 pathway. *FEBS Lett* 583:965–970.

62. Takai, Y., Sasaki, T., and Matozaki, T. (2001). Small GTP-binding proteins. *Physiol Rev* 81:153–208.

63. Goldfinger, L.E. (2008). Choose your own path: specificity in Ras GTPase signaling. *Mol Biosyst* 4:293–299.

64. Zhou, X., Ikenoue, T., Chen, X., Inoki, K., and Guan, K.L. (2009). Rheb controls misfolded protein metabolism by inhibiting aggresome formation and autophagy. *Proc Natl Acad Sci USA* 106:8923–8928.

65. Yee, W.M., and Worley, P.F. (1997). Rheb interacts with Raf-1 kinase and may function to integrate growth factor- and protein kinase A-dependent signals. *Mol Cell Biol* 17:921–933.

66. Karbowniczek, M., Cash, T., Cheung, M., Robertson, G.P., Astrinidis, A., and Henske, E. P. (2004). Regulation of B-Raf kinase activity by tuberin and Rheb is mammalian target of rapamycin (mTOR)-independent. *J Biol Chem* 279:29930–29937.

67. Karbowniczek, M., Robertson, G.P., and Henske, E.P. (2006). Rheb inhibits C-Raf activity and B-Raf/C-Raf heterodimerization. *J Biol Chem* 281:25447–25456.

68. Sun, Y., Fang, Y., Yoon, M.S., Zhang, C., Roccio, M., Zwartkruis, F.J., Armstrong, M., Brown, H.A., and Chen, J. (2008). Phospholipase D1 is an effector of Rheb in the mTOR pathway. *Proc Natl Acad Sci USA* 105:8286–8291.

69. Toschi, A., Lee, E., Xu, L., Garcia, A., Gadir, N., and Foster, D.A. (2009). Regulation of mTORC1 and mTORC2 complex assembly by phosphatidic acid: competition with rapamycin. *Mol Cell Biol* 29:1411–1420.

4

Regulation of TOR Complex 1 by Amino Acids Through Small GTPases

JOSEPH AVRUCH[a] • XIAOMENG LONG[b] • YENSHOU LIN[c] •
SARA ORTIZ-VEGA[a] • JOSEPH RAPLEY[a] • NORIKO OSHIRO[d]

[a]*Diabetes Research Laboratory, Department of Molecular Biology*
Diabetes Unit, Medical Services, Massachusetts General Hospital
Simches Research Building, Boston
Massachusetts, USA

[b]*SuperArray Bioscience Corp*
Executive Way, Frederick, Maryland, USA

[c]*Department of Life Sciences*
National Taiwan Normal University
Taipei, Taiwan

[d]*Biosignal Research Center*
Kobe University
1-1 Rokkodai-cho, Nada-ku
Kobe, Japan

I. Abstract

TOR complex 1 (TORC1), composed of the TOR (target of rapamycin) protein kinase, the substrate binding subunit raptor, and the 36 kDa WD propellor protein Lst8/GβL, controls cell growth in all eukaryotes in response to nutrient availability, and in metazoans, to insulin and growth factors, energy status, and stress conditions. This chapter will focus on the

DOI: 10.1016/S1874-6047(10)27004-X

mechanisms underlying mammalian TORC1 regulation by amino acids, in the context of overall TORC1 regulation in metazoan cells.

II.   Amino Acid Regulation of TORC1: Introduction

The phenotype induced by rapamycin or by TOR deficiency in *S. cerevisiae* closely resembles that caused by growth on very poor sources of N or C, an observation [1] that provided the initial indication that ScTOR participated in a nutrient-sensitive signaling pathway; the components of this putative pathway upstream of ScTOR, however, were largely unknown until the recent demonstration that the Gtr1/2 GTPase and the Gtr1 exchanger, vps39/Vam6 are required for amino acid regulation of TORC1 [2]. Early studies of rapamycin in mammalian cells focused on the dephosphorylation of S6K1 and 4E-BP [3–5], phosphoproteins originally identified as targets of insulin and mitogen activated signaling pathways. The first evidence that a nutrient-sensitive pathway of mTOR regulation operated as well in mammalian cells, independent of the insulin/mitogen signaling input, was the finding that withdrawal of amino acids from the medium of most cultured cells caused S6K1 and 4E-BP to undergo rapid dephosphorylation and to become unresponsive to insulin and mitogenic stimuli; in contrast, amino acid withdrawal had little effect on insulin/mitogen signaling to other PI-3K (phosphatidylinositol-3 kinase)/Akt targets or to Ras/MAPK [6]. Moreover, an S6K1 mutant rendered resistant to inhibition by rapamycin (Δ2-46/ΔCT104 [7]; a variant lacking the TOS motif [8]/raptor binding site that can be phosphorylated by mTORC2 but not mTORC1 [9]) proved to be resistant as well to inhibition by amino acid withdrawal although retaining insulin responsiveness. Reciprocally, increasing the medium amino acid concentration induced a selective activation of S6K1; a twofold increase over the basal amino acid levels increased S6K1 site specific phosphorylation and activity to an extent comparable to that elicited by insulin, which gave no further stimulation [6]. Many studies demonstrated subsequently that removal of extracellular amino acids inhibits the rapamycin-sensitive component of mTOR signaling, that is, mTORC1, in nearly all cultured cells; cells that are able to maintain high levels of intracellular proteolysis are resistant to the inhibitory effect of amino acid withdrawal, but in those instances inhibition of autophagy combined with amino acid withdrawal is usually sufficient to shut down signaling to S6K1 [10]. These findings pointed to the widespread operation of an amino acid-sensitive pathway that selectively regulates the mTORC1 signaling and subsequent effort has sought to identify the components of this pathway.

III.   Leucine Is the Most Potent Amino Acid Regulator of TORC1

In an effort to better define the key initiating signal, each amino acid was withdrawn individually; although each omission gave some inhibition of mTORC1, withdrawal of leucine or arginine yielded nearly as much inhibition as removal of all amino acids [6]; the potent impact of leucine withdrawal/readdition has been widely replicated [11, 12] whereas arginine has received little further study in this regard. The inhibition of mTORC1 caused by 1–2 h amino acid withdrawal from cultured cells appears to be independent of GCN2 activation, as the latter is activated comparably by deficiency of any essential amino acid [13]; moreover, mTORC1 inhibition occurs on amino acid withdrawal from GCN2 null MEFs. Nevertheless, GCN2 may be necessary for the suppression of mTORC1 at longer times, or *in vivo* in a tissue dependent manner. Normal mice fed an amino acid deficient diet inhibit both hepatic and skeletal muscle protein synthesis [14], whereas GCN2 null mice fed such diets fail to suppress hepatic protein synthesis while suppressing skeletal muscle protein synthesis, suggesting that GCN2 may be a negative regulator of mTORC1 in liver. Here we focus on the leucine regulated, GCN2-independent pathway.

Leucine appears to initiate its regulatory action on mTORC1 only after entry to the cell. TORC1 activity in *Xenopus* oocytes, which lack leucine transport capacity, is insensitive to the presence or absence of extracellular leucine, however microinjection of leucine or addition of leucine to oocytes expressing a recombinant L system transporter activates TORC1 [15]. Whether regulation is effected by leucine itself (e.g., as in leucine binding and regulation of Glutamate Dehydrogenase [16]) or by some metabolic product of leucine is unknown. Interpretation of the effects of leucinamide or alpha-ketoisocaproic acid is confounded by their rapid intracellular conversion to leucine. Leucinol, an inhibitor of leucyl tRNA synthetase, is somewhat stimulatory in *Xenopus* oocytes [15], but without effect on rat adipocytes [17]. Adipocyte mTORC1, however, is highly resistant to inhibition by amino acid withdrawal; whether this reflects a deficiency of the leucine-sensitive pathway or resistance to intracellular amino acid depletion is unknown.

IV.   Rheb Binds and Regulates TORC1

The proximate biochemical mechanisms by which leucine and amino acids regulate mTORC1 are still incompletely understood. Consequently, a consideration of proposed mechanisms needs to be preceded by a discussion of the biochemical mechanisms known with greater certainty to regulate mTORC1. The dominant proximal regulators of mTORC1 are the ras-like GTPases, Rheb, and Rhebll [18]. Although the single *Drosophila* Rheb is more similar in amino acid sequence to mouse Rheb (also called Rheb1; 64% identity) than to Rhebl1 (also called Rheb2; ~49% identity), both mammalian Rheb polypeptides, which share ~46% identity, can promote activation of mTORC1 when transiently overexpressed and important differences have not been reported. Their relative potency, however, is unexplored, and their relative contribution to mTORC1 regulation *in vivo* is as yet unknown, awaiting evaluation of tissue specific knockouts. Elimination of the single *Drosophila* Rheb abrogates mTORC1 signaling, and depletion of mammalian Rheb and/or Rhebl1 in cell culture strongly inhibits mTORC1 regardless of the intensity of activating input. Conversely, transient overexpression of Rheb or Rhebl1 will activate mTORC1 signaling despite the withdrawal or inhibition of all upstream inputs. The primacy of Rheb in the regulation of TORC1 is conserved from *S. pombe* [19] through mammals, but does not obtain in *S. cerevisae*; although *S. cerevisae* Rheb controls certain aspects of nitrogen metabolism, elimination of ScRheb does not substantially interfere with ScTORC1 function [20]. Inasmuch as nitrogen withdrawal inhibits TORC1 signaling in *S. cerevisiae*, Rheb-independent mechanisms of TORC1 regulation by N must exist; their nature and the extent to which they persist in metazoans is, however, unknown and they will not be considered further.

Rheb activation of mTORC1 requires Rheb GTP charging, and is therefore strongly determined by the regulators of the Rheb GTPase cycle. Little information is available concerning Rheb GTP binding *in vitro* or Rheb's intrinsic GTPase activity. As regards regulation of the Rheb GTPase cycle, the "Translationally Controlled Tumor Protein," a ubiquitously expressed 25 kDa polypeptide, has been identified as a Rheb guanyl nucleotide exchanger [21, 22], although this conclusion is disputed [23, 24]. Conversely, the Tuberous Sclerosis Complex (TSC), the heterodimer of Hamartin/TSC1 and Tuberin/TSC2, is securely established as the major Rheb GTPase activating protein (GAP). The regulation of the TSC is described in detail elsewhere [25] and in this volume, so it is sufficient here to state that considerable evidence indicates that control of TSC GAP activity is the primary mechanism by which mTORC1 signaling (and presumably, Rheb GTP charging) is controlled by most upstream regulators, including the mTORC1 activators, Type 1 PI-3 kinases/Akt, Ras-GTP/MAPK, TNFα via IKKβ as well as by the mTORC1 inhibitors, energy and hypoxic stress (via AMP activated protein kinase/GSK3 and REDD1/2), other stresses (p38/MAPKAPK-2), glucocorticoids, etc. Elimination of the TSC increases Rheb GTP charging from basal levels of ~5–20% to over 95%, and is accompanied by constitutive activation of mTORC1, which becomes insensitive to nearly all upstream activating/inhibitory inputs. A major exception is mTORC1 regulation by amino acids; TSC null cells remain fully inhibitable by amino acid withdrawal, despite sustained Rheb GTP charging of >95% [26, 27]. The inhibitory effect of amino acid withdrawal on mTORC1 signaling can nevertheless be reversed by overexpressed recombinant Rheb, although the levels of recombinant Rheb required to overcome the inhibitory effects of amino acid withdrawal exceed those of endogenous Rheb by 10- to 100-fold [28–30]. This behavior leads to the conclusion that amino acid withdrawal acts primarily distal to the TSC and reduces, but does not eliminate the ability of Rheb-GTP to activate mTORC1. Thus, understanding the role of amino acids in mTORC1 regulation requires an understanding of the mechanisms by which Rheb-GTP activates mTORC1, which of these is sensitive to amino acid withdrawal, and how the amino acid effect is exerted.

Considerable evidence indicates that Rheb-GTP activates mTORC1 by a direct interaction with TORC1. Long *et al.* [28] observed that recombinant overexpressed Rheb could coprecipitate endogenous mTORC1 from 293 cells, whereas Ras-GTP and Rap1, the most closely related small GTPases, show little binding to mTOR. The binding site for Rheb on mTOR was localized to AA2148-2300, which encodes the upper lobe of the mTOR catalytic domain. Surprisingly, the binding of Rheb to mTORC1 or to the mTOR catalytic domain is not enhanced by Rheb GTP charging, an almost universal feature of the interaction of Ras-like GTPases with *bona fide* effectors; GTP charging resulted in a weaker interaction between Rheb and mTOR. Recombinant Rheb also associates in a GTP-independent manner with recombinant Lst8 and with the carboxyterminal 300AA of raptor, which, like the Lst8 polypeptide, is a seven WD domain "propeller." Although the latter interactions appear specific, the proclivity of transiently expressed WD propeller domains toward incomplete folding casts doubt on the significance of the interaction of Rheb with such domains.

The initial evidence that a direct Rheb/mTOR interaction regulates mTOR kinase activity was the observation that Rheb mutants such as S20N or D60I, which lack the ability to bind any guanyl nucleotide and are unable to reactivate mTOR signaling in amino acid deprived cells, can still bind mTOR, however the bound mTOR polypeptides exhibit little or no kinase activity when assayed *in vitro*, that is, <10% that of mTOR bound to wild-type Rheb. Reciprocally, the mTOR polypeptides that coprecipitate with RhebQ64L, a mutant that is almost completely GTP charged *in vivo*, exhibit about twofold higher kinase activity *in vitro* than mTOR bound to wild-type Rheb, which achieves about 50% GTP charging when transiently overexpressed [28]. These findings indicated that although Rheb GTP charging is not required for mTOR binding, a direct interaction of mTOR with Rheb-GTP is critical for the acquisition of mTOR kinase activity. Further evidence for the functional importance of a direct Rheb-TOR interaction was provided by the properties of *S. pombe* Rheb mutants selected for their ability to confer resistance to toxic levels of certain amino acid analogs, a phenotype exhibited by TSC null *S. pombe* [31]. These Rheb point mutants showed a diminished ability to bind GDP with unimpaired binding of GTP and a reduced GTPase activity *in vitro*. Whereas an association of wild-type Rheb with SpTOR was not detectable, one of the hyperactive Rheb mutants, K120R, was shown to be largely bound to endogenous SpTOR2, the isoform found specifically in Sp TOR complex 1.

Subsequent work identified conditions that enabled activation of mTORC1 kinase activity *in vitro* by the direct addition of Rheb-GTP, whereas Rheb-GDP and other ras-like GTPases had little or no effect and mTORC2 is unresponsive [32, 33]. The activation of mTORC1 by Rheb-GTP *in vitro* appears to result, at least in part from an increase in the binding of polypeptide substrate to raptor, an effect that does not require MgATP; moreover, Rheb does not promote the autophosphorylation of an mTOR polypeptide freed of raptor, suggesting that Rheb-GTP does not enhance the catalytic function of the mTOR kinase domain [33]. The relationship of this *in vitro* activation process to the mechanism by which Rheb-GTP activates mTORC1 *in vivo* is unclear. Insulin activation of mTORC1, which requires the participation of Rheb-GTP, results in a stable activation of the mTORC1 assembly assayed *in vitro*, also characterized by an enhanced ability of mTORC1 to bind the polypeptide substrate [34]; nevertheless, endogenous Rheb is not recovered with the insulin-activated mTORC1. In mTORC1 isolated from unstimulated cells, the binding of polypeptide substrate to raptor can be greatly enhanced by removal of mTOR, whereas after insulin stimulation, removal of mTOR from raptor produces little additional increase in raptor substrate binding [34]. Thus the mTOR polypeptide in mTORC1 restricts substrate access to raptor, and insulin, acting through Rheb-GTP, diminishes this restriction. The stable nature of the enhanced substrate binding *in vitro* observed in the absence of accompanying Rheb suggests several possibilities: (1) the change in mTORC1 substrate access caused by binding Rheb-GTP *in vivo* (apparently through a noncovalent mechanism, as occurs *in vitro*) remains stable after extraction; (2) alternatively, the change in mTORC1 conformation initiated by Rheb-GTP *in vivo* is locked in by a subsequent covalent modification of raptor and/or mTOR. The occurrence of multiple insulin-stimulated mTOR and raptor phosphorylations, mediated by mTOR (and other kinases) has been observed [35, 36]; although evidence indicates that at least some of these phosphorylations may contribute to mTORC1 activation, a definitive conclusion on the importance of these phosphorylations for mTORC1 activation is not yet possible. (3) More remotely, the activation of mTORC1 by Rheb-GTP *in vitro* is unrelated in mechanism to that responsible for insulin-stimulated, Rheb-GTP dependent activation of mTORC1 *in vivo*.

V. Cross-competition Among Substrates for Raptor Can Influence TORC1 Signaling

An alternative proposal for the mechanism by which insulin enhances substrate binding to mTORC1 (which does not specify the function of Rheb in mTORC1 activation) involves the polypeptide PRAS40. First described by Harthill *et al.* [37] as p39, a polypeptide that bound to 14-3-3 in a mitogen and amino acid dependent manner, the polypeptide sequence was identified by Kovacina *et al.* [38], and shown to have a proline rich aminoterminal segment and a major Akt phosphorylation site necessary for 14-3-3 binding at Thr246 near the carboxyterminus. Subsequently, several groups isolated PRAS40 as a raptor-associated polypeptide [32, 39–43], and showed that PRAS40 competes with the mTORC1 substrates 4E-BP and S6K for binding to raptor; moreover, washing mTORC1 immunoprecipitates *in vitro* with high-salt buffers removed most PRAS40 and resulted in an increase in the assayed mTORC1 kinase activity to the level observed with mTORC1 isolated from insulin-stimulated cells [32]. It was proposed that PRAS40 functions primarily as an Akt-regulatable inhibitor of mTORC1; the insulin-stimulated, Akt catalyzed phosphorylation of PRAS40 at Thr246 was proposed to dislodge PRAS40 from raptor and thereby make available the substrate binding site, thus promoting the insulin activation of mTORC1 [39–41]. This proposal, however, overlooks the fact that PRAS40 is itself an mTORC1 substrate, and that the mTORC1-catalyzed phosphorylation of PRAS40 on Ser183 is required along with the Akt catalyzed phosphorylation at Thr246, for the binding of 14-3-3 [37], (as initially shown by the amino acid dependence of the p39/PRAS40-14-3-3 interaction) and for PRAS40 displacement from raptor [40]. The displacement of PRAS40 from raptor requires, and is therefore a result of mTORC1 (and Akt) activation, rather than its cause. Consistent with this view, phorbol esters, which activate mTORC1 primarily via a PKC/Erk&Rsk/TSC/Rheb pathway [44, 45] (although Rsk phosphorylation of raptor may also contribute [46]), fail to phosphorylate PRAS40 (Thr246) or promote PRAS40 release from mTORC1 [47]. All substrates of mTORC1 examined thus far, including PRAS40, appear to compete for the same binding site on raptor [40];

although PRAS40 is not primarily a regulator of mTORC1, access of poly-peptide substrates to mTORC1 does appear to be limiting *in vivo* in unsti-mulated cells, and is an important locus of mTORC1 regulation. Thus, in addition to the binding of Rheb-GTP and modifications of mTOR and raptor, cross-competition among mTORC1 substrates indicates that sub-strate abundance, other substrate modifications, and substrate localization are each also probably important determinants of the ability of a substrate to undergo mTORC1-catalyzed phosphorylation *in vivo*. The greater sensitiv-ity of mTORC1-catalyzed phosphorylation of S6K1 to a variety of mTOR inhibitors [48, 49], as compared with the phosphorylation of the 4E-BP and PRAS40 polypeptides, may reflect the much lower abundance of S6K1 as compared to those much more abundant polypeptides.

VI.   Phosphatidic Acid Is a Rheb-Directed Regulator of mTORC1

In addition to the role of a direct Rheb-GTP/mTOR interaction in the activation of mTORC1, two mechanisms have been proposed by which Rheb may activate mTORC1 indirectly, without binding to mTORC1. Chen and colleagues provided strong evidence that phosphatidic acid (PA), generated through Phospholipase-D (PL-D)-catalyzed cleavage of phosphatidylcholine, binds directly to the mTOR FKBP12/rapamycin-binding (FRB) domain [50]. Although the direct addition of PA to mTORC1 does not alter its kinase activity *in vitro*, depletion of PA *in vivo* substantially inhibits mTORC1 signaling. PA may therefore act to promote proper mTORC1 localization *in vivo*. Subsequently this group showed that Rheb binds directly to PL-D in a GTP-dependent fashion, and increases PL-D activity [51]. Although PL-D regulation of mTORC1 is not operative in *Drosophila*, it is likely to represent a recently evolved, secondary mechanism of Rheb-mediated mTORC1 activation.

VII.   FKBP38 as a Candidate Rheb-Controlled mTORC1 Regulator

More controversial is the role of FKBP38. This polypeptide contains an aminoterminal FKBP12-like peptidyl prolyl *cis–trans* isomerase (PPI) domain, followed by three TPR repeats, a Ca^{2+}/calmodulin (CM) binding domain, and near its carboxyterminus, a transmembrane domain that loca-lizes the protein to the ER and mitochondrial outer membrane with the PPI facing the cytoplasm [52]. The FKBP38 mRNA was found to be significantly upregulated in HeLa cells overexpressing either TSC1 or TSC2, and RNAi-induced depletion of FKBP38 reversed the (10–14%) reduction in HeLa cells size (FSC) caused by overexpression of TSC1 or TSC2 [53]. Bai *et al.* [54] retrieved FKBP38 in a two-hybrid screen with Rheb and found that Rheb, through its switch1 segment [55], binds in a GTP dependent manner to the FKBP38 PPI domain. Overexpression of FKBP38 inhibits the phosphorylation of 4E-BP caused by amino acid readdition or Rheb overexpression, whereas RNAi-induced depletion of endogenous FKBP38 upregulates basal S6K1(Thr389/412) and 4E-BP phosphorylation and opposes the inhibition of these phosphorylations that occurs with serum or amino acid withdrawal. Moreover, addition of FKBP38 *in vitro* inhibited the mTORC1 kinase activity. FKBP38 binds the mTOR fragment (1967–2191) in a rapamycin-independent manner; coexpression with Rheb displaces FKBP38 from this mTOR fragment. Based on these results, FKBP38 is proposed to be an endogenous, Rheb-GTP sensitive inhibitor of mTORC1. The site of FKBP38 binding on mTOR is quite close to the site of PA binding; whether PA affects the FKBP38/mTOR interaction has not been reported. Although the ability of FKBP38 to bind mTOR [56] and Rheb [23] has been confirmed by some (but not all, [57]), there is considerable disagreement concerning the role of FKBP38 in mTOR regulation, and we view this question as unresolved at present.

VIII. Amino Acids Control the Rheb-mTORC1 Interaction

Given the biochemical mechanisms described above that regulate mTORC1 kinase activity and signaling, how do amino acids, especially leucine, control mTORC1 function at a molecular level? As indicated above, evidence from cell-based studies indicates that the inhibitory effect of amino acid/leucine withdrawal is exerted downstream of the TSC [26, 27], and is characterized by a marked diminution in the ability of Rheb-GTP to promote mTORC1 activation; stated conversely, amino acids somehow enhance the efficacy of Rheb-GTP in mTORC1 activation. Long *et al.* [29] noted that the ability of recombinant Rheb to coprecipitate endogenous or recombinant mTOR was substantially but reversibly reduced by amino acid or leucine withdrawal prior to extraction. This inhibition of the Rheb/mTOR interaction was unaccompanied by any change in Rheb GTP charging, and was evident using Rheb mutants such as D60I which bind no guanyl nucleotide. Inhibition was also evident using the isolated mTOR catalytic domain fragment (2148–2430). Rheb binds to the upper lobe of the mTOR catalytic domain (2148–2300); surprisingly, deletion of the lower lobe of the catalytic domain did not alter Rheb binding to (2148–2300) but eliminated the ability of amino acid withdrawal to reduce Rheb binding. These results pointed to an ability of amino acids to promote the Rheb/mTOR interaction *in vivo*, however the mechanism responsible and specifically, how the lower lobe of the mTOR catalytic domain confers amino acid sensitivity to Rheb binding to the upper lobe, were not defined.

Three specific targets have been proposed to function in the pathway between amino acids and mTORC1: the rag GTPases, the type 3 PI-3 kinase/mVPS34, and the MAP4K3/GC kinase-like protein kinase (GLK); the evidence in support of each is summarized next.

IX.   Rag GTPases Mediate Amino Acid Regulation of the Rheb-TORC1 Interaction

A very concrete mechanism by which amino acids promote the interaction between Rheb and mTORC1 was provided by the elucidation of the rag GTPases as regulators on mTORC1 signaling. Kim *et al.* [58] generated shRNA against all 132 *Drosophila* small GTPases and found that depletion of the *Drosophila* rag (A/B) and (C/D) homologs, as well as of DRheb selectively reduced DS6K1[Thr398-P]. Independently, Sancak and colleagues [59] retrieved endogenous ragC in a raptor immunoprecipitate from mammalian cells. The four rag polypeptides A–D are small GTPases; ragA/B, orthologs of the *S. cerevisiae* polypeptide Gtr1, are nearly identical to each other, whereas ragC/D, orthologous to ScGtr2, are ~77% identical, and ~20% identical to A/B. The GTPase domain is situated in the aminoterminal half of these polypeptides, and each has an extended carboxyterminal segment that mediates the constitutive formation of heterodimers of A or B (Gtr1) with C or D (Gtr2). Expression of constitutively GTP-charged mutants of ragA or B promotes, whereas nucleotide deficient RagA or B suppresses S6K1 activation; opposite responses are observed with mutants of ragC/D, however when various mutant ragA/B are coexpressed with mutant ragC/D, the responses of S6K1 are driven by the state of ragA/B GTP charging. Notably, heterodimers containing constitutively active ragA/B are able to restore S6K1 activity/4E-BP phosphorylation in amino acid deprived cells. Sancak *et al.* [59] observed that amino acid readdition appeared to modestly enhance ragB GTP charging from 44% to 63%; moreover, leucine readdition promoted the association of the rag heterodimer with mTORC1, visualized after cross-linking, an interaction that appears to be mediated primarily by raptor. The highly conserved nature of rag GTPases regulation of TORC1 is illustrated by the ability of constitutively active Gtr1 to overcome the inhibition of TORC1 activity that occurs in leucine-deprived *S. cerevisiae* [2]. Importantly, rag activation of TORC1, at least in metazoans, is dependent on Rheb-GTP, as it is abolished by depletion of endogenous Rheb. The functional interaction of rag and Rheb *in vivo* is well illustrated by manipulation of the expression of these GTPases in clonal patches of the *Drosophila* fat body; expression of a constitutively active Drag in the presence of wild-type DRheb results in substantial cell enlargement, but elimination of DRheb markedly reduces cell size regardless of Drag activation; conversely, DRheb overexpression produces cell enlargement even in the Drag null background [58]. As to the basis for the Rheb requirement for rag action, Sancak *et al.* [59] observed that mTOR in amino acid replete cells exhibits a perinuclear localization whereas mTOR in amino acid deprived cells is found dispersed in the cytoplasm; the perinuclear localization of mTOR in replete cells is dependent on raptor and on the rag GTPases; and depletion of those elements results in cytoplasmic dispersion of mTOR. Conversely, perinuclear localization of mTOR in amino acid deprived cells can be restored by expression of constitutively active ragB but not by Rheb. The perinuclear membrane region appears to be a late endosomal compartment enriched in rab7 and presumably Rheb as well. The ability of amino acids to stimulate ragA/B to an active, GTP charged state, accompanied by enhanced association of the rag heterodimer with raptor, and activation of mTORC1 signaling, and the rag-dependent translocation of mTORC1 to a compartment enriched in Rheb, together provide strong evidence that the rag heterodimer, by promoting the proximity of mTORC1 to its indispensable activator Rheb-GTP, is a conserved mediator of amino acid regulation of mTORC1.

X.   Phosphatidyl 3′ Phosphate Contributes to Amino Acid Regulation of Mammalian TORC1

The type 3 PI-3 kinase, the metazoan ortholog of *S. cerevisiae*, vps34, the single phosphatidylinositol(PI)-3 kinase found in *S. cerevisiae*, has multiple functions in membrane trafficking and sorting [60, 61]. Acting as a heterodimer with the protein kinase vps15, mvps34 phosphorylates PI (but not PI4P or PI4,5P) to generate PI-3P; membrane domains enriched in PI3P serve as a docking sites for proteins containing FYVE or PX domains, which form complexes that direct the trafficking and sorting of early endosomes into later compartments and to lysosomes. Type 3 PI-3 kinase is also necessary for the initiation of macroautophagy. Two reports [62, 63] describe the ability of extracellular amino acids to regulate the activity of vps34 immunoprecipitated from mammalian cells; although the extent of regulation (approximately twofold) is modest, multiple presumably independently regulated vps34 complexes are known to exist. Overexpression of mvps34, especially with vps15, is capable of activating S6K1, an effect that requires the presence of amino acids. Conversely, depletion of mvps34 or vps15, or overexpression of a dimeric FYVE domain polypeptide inhibits the activation of S6K1 by amino acids. Together, these results indicate that mvps34 activity, through the generation of PI3P, is needed for amino acid regulation of mammalian TORC1, although strong data from *Drosophila* demonstrates that no requirement for PI3P exists for TORC1 regulation in that organism [64].

Coprecipitation of mTORC1 with mvps34 has been reported [63], however the mechanism by which PI3P acts to enable mTORC1 activation is unclear. In view of the likely importance of rag-regulated mTORC1 relocation to late endosomes in amino acid activation of mTORC1, and the numerous roles of PI3P in membrane trafficking through multiple endosomal compartments, a requirement for PI3P in mTORC1 activation is entirely plausible, and data providing a clearer picture of the specific basis for this requirement is awaited. Similarly, the mechanism by which amino acids regulates mvps34 activity requires elucidation; vps34 can bind calmodulin, however the Ca^{2+} dependence of amino acid activation [65] of mvps34 is disputed [66].

XI. MAP4K3/Glk May Participate in Amino Acid Regulation of mTORC1

Findlay *et al.* [67] examined the ability of RNAi directed at protein kinases to inhibit the slowed electrophoretic mobility of DS6K (presumably reflecting TORC1-mediated hyperphosphorylation) seen in *Drosophila* S2 cells depleted of TSC1. Depletion of three of the 196 kinases tested, *CG1776*, *CG7097*, and *CG8767*, gave faster DS6K mobility. The closest mammalian homologs correspond respectively to myosin light chain kinase, MAP4K3 and c-mos. MAP4K3, or the Germinal Center kinase-like kinase (GLK) is best known as a rab8 binding protein that acts as an upstream activator of the SAPK/Jnk MAP kinases in response to UV irradiation, TNFα, and Wnt3a [68, 69]. Findlay and coworkers [67] observed that depletion of MAP4K3 in HeLa cells reduced, whereas MAP4K3 overexpression enhanced S6K1(Thr389) phosphorylation, an indicator of mTORC1 activity. Moreover, amino acid withdrawal reduced the activity of transiently expressed recombinant MAP4K3 assayed *in vitro*. These results are consistent with the idea that MAP4K3 can regulate mTORC1 at a site distal to the TSC; the identity of the MAP4K3 substrate relevant to this regulation remains to be defined.

XII. Summary

Nitrogen sufficiency, most commonly in the form of the amino acid leucine, is a central regulator of TORC1 signaling, probably in all eukaryotes. Rheb is the proximate upstream activator of TORC1 in most eukaryotes characterized thus far, with the notable exception of *S. Cerevisiae*. In organisms where Rheb is known to be the primary governor of TORC1, the dominant regulator of Rheb is the TSC, a Rheb GTPase activator. TSC is the major target of most upstream inputs to TORC1, other than amino acids; in metazoan cells the inputs that inhibit TSC GAP/activate Rheb are initiated mostly by insulin and mitogens, whereas in *S. pombe* the inputs that regulate TSC are yet to be identified. Amino acids, primarily leucine, regulate TORC1 function through a pathway that is conserved, at least in part, in all eukaryotes. This pathway involves the rag GTPases; in *S. cerevisae*, the rag orthologs Gtr1/2 appear to activate TORC1 independently of the Rheb GTPase, whereas in metazoans, rag activation of TORC1 is entirely dependent on Rheb; the rag pathway, in response to amino acids, controls the ability of the Rheb GTPase to activate TORC1 by controlling the colocalization of mTORC1 with Rheb.

Full delineation of the amino acid/rag GTPase pathway will require the identification of the other components and the elucidation of their mode of operation. In particular, understanding how amino acids regulate rag activation, presumably at the level of rag GTP charging is especially important. The recent identification of *S. cerevisiae* vps39/Vam6p as a putative guanyl nucleotide exchanger for Gtr1 in the amino acid regulation of TORC1 may hasten progress toward this goal. The nature of the interaction of mTORC1 with the rag GTPases requires further clarification as does the mechanism of rag-mediated mTORC1 translocation in mammalian cells, which is not a feature of Gtr1/2 regulation of TORC1 in *S. cerevisiae*. Evidence supports a requirement for PI3P in amino acid regulation of TORC1, probably at one or more sites in the rag pathway that remain to be defined. A physiologic role for MAP4K3 in amino acid regulation of TORC1 is considered provisional at present.

ACKNOWLEDGMENTS

Work cited herein from the authors' laboratories was supported by NIH grants DK17776 (JA), CA73818 (JA) and by Scientific Research Funds of the Ministry of Education, Culture, Sports, Science and Technology of Japan and CREST, Japan Science and Technology Agency.

REFERENCES

1. Barbet, N.C., Schneider, U., Helliwell, S.B., Stansfield, I., Tuite, M.F., and Hall, M.N. (1996). TOR controls translation initiation and early G1 progression in yeast. *Mol Biol Cell* 7(1):25–42.
2. Binda, M., Péli-Gulli, M.P., Bonfils, G., Panchaud, N., Urban, J., Sturgill, T.W., Loewith, R., and De Virgilio, C. (2009). The Vam6 GEF controls TORC1 by activating the EGO complex. *Mol Cell* 35:563–573.
3. Price, D.J., Grove, J.R., Calvo, V., Avruch, J., and Bierer, B.E. (1992). Rapamycin-induced inhibition of the 70-kilodalton S6 protein kinase. *Science* 257:973–977.
4. Chung, J., Kuo, C.J., Crabtree, G.R., and Blenis, J. (1992). Rapamycin-FKBP specifically blocks growth-dependent activation of and signaling by the 70 kD S6 protein kinases. *Cell* 69:1227–1236.
5. von Manteuffel, S.R., Gingras, A.C., Ming, X.F., Sonenberg, N., and Thomas, G. (1996). 4E-BP1 phosphorylation is mediated by the FRAP-p70s6k pathway and is independent of mitogen-activated protein kinase. *Proc Natl Acad Sci USA* 93:4076–4080.
6. Hara, K., Yonezawa, K., Weng, Q.P., Kozlowski, M.T., Belham, C., and Avruch, J. (1998). Amino acid sufficiency and mTOR regulate p70 S6 kinase and eIF-4E BP1 through a common effector mechanism. *J Biol Chem* 273:14484–14494.
7. Weng, Q.P., Andrabi, K., Kozlowski, M.T., Grove, J.R., and Avruch, J. (1995). Multiple independent inputs are required for activation of the p70 S6 kinase. *Mol Cell Biol* 15:2333–2340.
8. Schalm, S.S., and Blenis, J. (2002). Identification of a conserved motif required for mTOR signaling. *Curr Biol* 12:632–639.
9. Ali, S.M., and Sabatini, D.M. (2005). Structure of S6 kinase 1 determines whether raptor-mTOR or rictor-mTOR phosphorylates its hydrophobic motif site. *J Biol Chem* 280:19445–19448.
10. Shigemitsu, K., Tsujishita, Y., Hara, K., Nanahoshi, M., Avruch, J., and Yonezawa, K. (1999). Regulation of translational effectors by amino acid and mammalian target of rapamycin signaling pathways. Possible involvement of autophagy in cultured hepatoma cells. *J Biol Chem* 274:1058–1065.
11. Kimball, S.R., and Jefferson, L.S. (2006). New functions for amino acids: effects on gene transcription and translation. *Am J Clin Nutr* 83:500S–507S.
12. Drummond, M.J., and Rasmussen, B.B. (2008). Leucine-enriched nutrients and the regulation of mammalian target of rapamycin signalling and human skeletal muscle protein synthesis. *Curr Opin Clin Nutr Metab Care* 11:222–226.
13. Hinnebusch, A.G. (2005). Translational regulation of GCN4 and the general amino acid control of yeast. *Annu Rev Microbiol* 59:407–450.
14. Anthony, T.G., McDaniel, B.J., Byerley, R.L., McGrath, B.C., Cavener, D.R., McNurlan, M.A., and Wek, R.C. (2004). Preservation of liver protein synthesis during dietary leucine deprivation occurs at the expense of skeletal muscle mass in mice deleted for eIF2 kinase GCN2. *J Biol Chem* 279:36553–36561.
15. Christie, G.R., Hajduch, E., Hundal, H.S., Proud, C.G., and Taylor, P.M. (2002). Intracellular sensing of amino acids in *Xenopus laevis* oocytes stimulates p70 S6 kinase in a target of rapamycin-dependent manner. *J Biol Chem* 277:9952–9957.
16. Smith, T.J., and Stanley, C.A. (2008). Untangling the glutamate dehydrogenase allosteric nightmare. *Trends Biochem Sci* 33:557–564.
17. Lynch, C.J., Fox, H.L., Vary, T.C., Jefferson, L.S., and Kimball, S.R. (2002). Regulation of amino acid-sensitive TOR signaling by leucine analogues in adipocytes. *J Cell Biochem* 77:234–251.

18. Avruch, J., Hara, K., Lin, Y., Liu, M., Long, X., Ortiz-Vega, S., and Yonezawa, K. (2006). Insulin and amino-acid regulation of mTOR signaling and kinase activity through the Rheb GTPase. *Oncogene* 25:6361–6372.

19. Aspuria, P.J., Sato, T., and Tamanoi, F. (2007). The TSC/Rheb/TOR signaling pathway in fission yeast and mammalian cells: temperature sensitive and constitutive active mutants of TOR. *Cell Cycle* 6:1692–1695.

20. Jacinto, E. (2008). What controls TOR? *IUBMB Life* 60:483–496.

21. Hsu, Y.C., Chern, J.J., Cai, Y., Liu, M., and Choi, K.W. (2007). *Drosophila* TCTP is essential for growth and proliferation through regulation of dRheb GTPase. *Nature* 445:785–788.

22. Dong, X., Yang, B., Li, Y., Zhong, C., and Ding, J. (2009). Molecular basis of the acceleration of the GDP-GTP exchange of human ras homolog enriched in brain by human translationally controlled tumor protein. *J Biol Chem* 284:23754–23764.

23. Wang, X., Fonseca, B.D., Tang, H., Liu, R., Elia, A., Clemens, M.J., Bommer, U.A., and Proud, C.G. (2008). Re-evaluating the roles of proposed modulators of mammalian target of rapamycin complex 1 (mTORC1) signaling. *J Biol Chem* 283:30482–30492.

24. Rehmann, H., Brüning, M., Berghaus, C., Schwarten, M., Köhler, K., Stocker, H., Stoll, R., Zwartkruis, F.J., and Wittinghofer, A. (2008). Biochemical characterisation of TCTP questions its function as a guanine nucleotide exchange factor for Rheb. *FEBS Lett* 582:3005–3010.

25. Huang, J., and Manning, B.D. (2008). The TSC1-TSC2 complex: a molecular switchboard controlling cell growth. *Biochem J* 412:179–190.

26. Roccio, M., Bos, J.L., and Zwartkruis, F.J. (2006). Regulation of the small GTPase Rheb by amino acids. *Oncogene* 25:657–664.

27. Smith, E.M., Finn, S.G., Tee, A.R., Browne, G.J., and Proud, C.G. (2005). The tuberous sclerosis protein TSC2 is not required for the regulation of the mammalian target of rapamycin by amino acids and certain cellular stresses. *J Biol Chem* 280:18717–18727.

28. Long, X., Lin, Y., Ortiz-Vega, S., Yonezawa, K., and Avruch, J. (2005). Rheb binds and regulates the mTOR kinase. *Curr Biol* 15:702–713.

29. Long, X., Ortiz-Vega, S., Lin, Y., and Avruch, J. (2005). Rheb binding to mammalian target of rapamycin (mTOR) is regulated by amino acid sufficiency. *J Biol Chem* 280:23433–23446.

30. Long, X., Lin, Y., Ortiz-Vega, S., Busch, S., and Avruch, J. (2007). The Rheb switch 2 segment is critical for signaling to target of rapamycin complex 1. *J Biol Chem* 282:18542–18551.

31. Urano, J., Comiso, M.J., Guo, L., Aspuria, P.J., Deniskin, R., Tabancay, A.P., Jr, Kato-Stankiewicz, J., and Tamanoi, F. (2005). Identification of novel single amino acid changes that result in hyperactivation of the unique GTPase, Rheb, in fission yeast. *Mol Microbiol* 58:1074–1086.

32. Sancak, Y., Thoreen, C.C., Peterson, T.R., Lindquist, R.A., Kang, S.A., Spooner, E., Carr, S.A., and Sabatini, D.M. (2007). PRAS40 is an insulin-regulated inhibitor of the mTORC1 protein kinase. *Mol Cell* 25:903–915.

33. Sato, T., Nakashima, A., Guo, L., and Tamanoi, F. (2009). Specific activation of mTORC1 by Rheb G-protein *in vitro* involves enhanced recruitment of its substrate protein. *J Biol Chem* 284:12783–12791.

34. Wang, L., Rhodes, C.J., and Lawrence, J.C., Jr (2006). Activation of mammalian target of rapamycin (mTOR) by insulin is associated with stimulation of 4EBP1 binding to dimeric mTOR complex 1. *J Biol Chem* 281:24293–243303.

35. Wang, L., Lawrence, J.C., Jr, Sturgill, T.W., and Harris, T.E. (2009). Mammalian target of rapamycin complex 1 (mTORC1) activity is associated with phosphorylation of raptor by mTOR. *J Biol Chem* 284:14693–14697.

36. Acosta-Jaquez, H.A., Keller, J.A., Foster, K.G., Ekim, B., Soliman, G.A., Feener, E.P., Ballif, B.A., and Fingar, D.C. (2009). Site-specific mTOR phosphorylation promotes mTORC1-mediated signaling and cell growth. *Mol Cell Biol* 29:4308–4324.
37. Harthill, J.E., Pozuelo Rubio, M., Milne, F.C., and MacKintosh, C. (2002). Regulation of the 14-3-3-binding protein p39 by growth factors and nutrients in rat PC12 pheochromocytoma cells. *Biochem J* 368:565–572.
38. Kovacina, K.S., Park, G.Y., Bae, S.S., Guzzetta, A.W., Schaefer, E., Birnbaum, M.J., and Roth, R.A. (2003). Identification of a proline-rich Akt substrate as a 14-3-3 binding partner. *J Biol Chem* 278:10189–10194.
39. Vander Haar, E., Lee, S.I., Bandhakavi, S., Griffin, T.J., and Kim, D.H. (2007). Insulin signalling to mTOR mediated by the Akt/PKB substrate PRAS40. *Nat Cell Biol* 9:316–323.
40. Oshiro, N., Takahashi, R., Yoshino, K., Tanimura, K., Nakashima, A., Eguchi, S., Miyamoto, T., Hara, K., Takehana, K., Avruch, J., Kikkawa, U., and Yonezawa, K. (2007). The proline-rich Akt substrate of 40 kDa (PRAS40) is a physiological substrate of mammalian target of rapamycin complex 1. *J Biol Chem* 282(28):20329–20339.
41. Wang, L., Harris, T.E., Roth, R.A., and Lawrence, J.C., Jr (2007). PRAS40 regulates mTORC1 kinase activity by functioning as a direct inhibitor of substrate binding. *J Biol Chem* 282:20036–20044.
42. Fonseca, B.D., Smith, E.M., Lee, V.H., MacKintosh, C., and Proud, C.G. (2007). PRAS40 is a target for mammalian target of rapamycin complex 1 and is required for signaling downstream of this complex. *J Biol Chem* 282:24514–24524.
43. Thedieck, K., Polak, P., Kim, M.L., Molle, K.D., Cohen, A., Jenö, P., Arrieumerlou, C., and Hall, M.N. (2007). PRAS40 and PRR5-like protein are new mTOR interactors that regulate apoptosis. *PLoS ONE* 2(11):e1217.
44. Ma, L., Chen, Z., Erdjument-Bromage, H., Tempst, P., and Pandolfi, P.P. (2005). Phosphorylation and functional inactivation of TSC2 by Erk implications for tuberous sclerosis and cancer pathogenesis. *Cell* 121:179–193.
45. Roux, P.P., Ballif, B.A., Anjum, R., Gygi, S.P., and Blenis, J. (2004). Tumor-promoting phorbol esters and activated Ras inactivate the tuberous sclerosis tumor suppressor complex via p90 ribosomal S6 kinase. *Proc Natl Acad Sci USA* 101:13489–13494.
46. Carrière, A., Cargnello, M., Julien, L.A., Gao, H., Bonneil, E., Thibault, P., and Roux, P.P. (2008). Oncogenic MAPK signaling stimulates mTORC1 activity by promoting RSK-mediated raptor phosphorylation. *Curr Biol* 18:1269–1277.
47. Fonseca, B.D., Lee, V.H., and Proud, C.G. (2008). The binding of PRAS40 to 14-3-3 proteins is not required for activation of mTORC1 signalling by phorbol esters/ERK. *Biochem J* 411:141–149.
48. Thoreen, C.C., Kang, S.A., Chang, J.W., Liu, Q., Zhang, J., Gao, Y., Reichling, L.J., Sim, T., Sabatini, D.M., and Gray, N.S. (2009). An ATP-competitive mammalian target of rapamycin inhibitor reveals rapamycin-resistant functions of mTORC1. *J Biol Chem* 284:8023–8032.
49. Feldman, M.E., Apsel, B., Uotila, A., Loewith, R., Knight, Z.A., Ruggero, D., and Shokat, K.M. (2009). Active-site inhibitors of mTOR target rapamycin-resistant outputs of mTORC1 and mTORC2. *PLoS Biol* 7(2):e38.
50. Fang, Y., Vilella-Bach, M., Bachmann, R., Flanigan, A., and Chen, J. (2001). Phosphatidic acid-mediated mitogenic activation of mTOR signaling. *Science* 294:1942–1945.
51. Sun, Y., Fang, Y., Yoon, M.S., Zhang, C., Roccio, M., Zwartkruis, F.J., Armstrong, M., Brown, H.A., and Chen, J. (2008). Phospholipase D1 is an effector of Rheb in the mTOR pathway. *Proc Natl Acad Sci USA* 105:8286–8291.

52. Shirane, M., and Nakayama, K.I. (2003). Inherent calcineurin inhibitor FKBP38 targets Bcl-2 to mitochondria and inhibits apoptosis. *Nat Cell Biol* 5:28–37.
53. Rosner, M., Hofer, K., Kubista, M., and Hengstschläger, M. (2003). Cell size regulation by the human TSC tumor suppressor proteins depends on PI3K and FKBP38. *Oncogene* 22:4786–4798.
54. Bai, X., Ma, D., Liu, A., Shen, X., Wang, Q.J., Liu, Y., and Jiang, Y. (2007). Rheb activates mTOR by antagonizing its endogenous inhibitor, FKBP38. *Science* 318:977–980.
55. Ma, D., Bai, X., Guo, S., and Jiang, Y. (2008). The switch I region of Rheb is critical for its interaction with FKBP38. *J Biol Chem* 283:25963–25970.
56. Dunlop, E.A., Dodd, K.M., Seymour, L.A., and Tee, A.R. (2009). Mammalian target of rapamycin complex 1-mediated phosphorylation of eukaryotic initiation factor 4E-binding protein 1 requires multiple protein–protein interactions for substrate recognition. *Cell Signal* 21:1073–1084.
57. Uhlenbrock, K., Weiwad, M., Wetzker, R., Fischer, G., Wittinghofer, A., and Rubio, I. (2009). Reassessment of the role of FKBP38 in the Rheb/mTORC1 pathway. *FEBS Lett* 583:965–970.
58. Kim, E., Goraksha-Hicks, P., Li, L., Neufeld, T.P., and Guan, K.L. (2008). Regulation of TORC1 by Rag GTPases in nutrient response. *Nat Cell Biol* 10:935–945.
59. Sancak, Y., Peterson, T.R., Shaul, Y.D., Lindquist, R.A., Thoreen, C.C., Bar-Peled, L., and Sabatini, D.M. (2008). The Rag GTPases bind raptor and mediate amino acid signaling to mTORC1. *Science* 320:1496–1501.
60. Lindmo, K., and Stenmark, H. (2006). Regulation of membrane traffic by phosphoinositide 3-kinases. *J Cell Sci* 119:605–614.
61. Backer, J.M. (2008). The regulation and function of Class III PI3Ks: novel roles for Vps34. *Biochem J* 410:1–17.
62. Byfield, M.P., Murray, J.T., and Backer, J.M. (2005). hVps34 is a nutrient-regulated lipid kinase required for activation of p70 S6 kinase. *J Biol Chem* 280:33076–33082.
63. Nobukuni, T., Joaquin, M., Roccio, M., Dann, S.G., Kim, S.Y., Gulati, P., Byfield, M.P., Backer, J.M., Natt, F., Bos, J.L., Zwartkruis, F.J., and Thomas, G. (2005). Amino acids mediate mTOR/raptor signaling through activation of class 3 phosphatidylinositol 3OH-kinase. *Proc Natl Acad Sci USA* 102:14238–14243.
64. Juhász, G., Hill, J.H., Yan, Y., Sass, M., Baehrecke, E.H., Backer, J.M., and Neufeld, T.P. (2008). The class III PI(3)K Vps34 promotes autophagy and endocytosis but not TOR signaling in *Drosophila*. *J Cell Biol* 18:655–666.
65. Gulati, P., Gaspers, L.D., Dann, S.G., Joaquin, M., Nobukuni, T., Natt, F., Kozma, S.C., Thomas, A.P., and Thomas, G. (2008). Amino acids activate mTOR complex 1 via Ca^{2+}/CaM signaling to hVps34. *Cell Metab* 7:456–465.
66. Yan, Y., Flinn, R.J., Wu, H., Schnur, R.S., and Backer, J.M. (2009). hVps15, but not Ca^{2+}/CaM, is required for the activity and regulation of hVps34 in mammalian cells. *Biochem J* 417:747–755.
67. Findlay, G.M., Yan, L., Procter, J., Mieulet, V., and Lamb, R.F. (2007). A MAP4 kinase related to Ste20 is a nutrient-sensitive regulator of mTOR signalling. *Biochem J* 403:13–20.
68. Shi, C.S., Leonardi, A., Kyriakis, J., Siebenlist, U., and Kehrl, J.H. (1999). TNF-mediated activation of the stress-activated protein kinase pathway: TNF receptor-associated factor 2 recruits and activates germinal center kinase related. *J Immunol* 163:3279–3285.
69. Shi, C.S., Huang, N.N., Harrison, K., Han, S.B., and Kehrl, J.H. (2006). The mitogen-activated protein kinase kinase kinase kinase GCKR positively regulates canonical and noncanonical Wnt signaling in B lymphocytes. *Mol Cell Biol* 26:6511–6521.

5

Rag GTPases in TORC1 Activation and Nutrient Signaling

LI LI[a,b] • KUN-LIANG GUAN[a]

[a]*Department of Pharmacology and Moores Cancer Center*
University of California at San Diego
La Jolla, California, USA

[b]*Department of Biological Chemistry*
University of Michigan
Ann Arbor, Michigan, USA

I. Abstract

Target of rapamycin (TOR) is a central regulator of cell growth conserved from yeast to mammals. TOR is a member of the phosphoinositide-3-kinase-related protein kinase family, which includes ATM, ATR, DNA-PK, and TOR [1]. In both yeast and human, TOR forms two distinct structural and functional complexes, TORCs. The mammalian TORC1 (mTORC1) consists of mTOR, raptor, mLST8, and PRAS40 [2–6] while mTORC2 consists of mTOR, rictor, sin1, and mLTS8 [7–10]. mTORC1 (at least part of the mTORC1 function) is potently inhibited by rapamycin while mTORC2 is not directly inhibited by rapamycin [11]. Characterized substrates of mTORC1 include the ribosomal S6 kinase (S6K), eukaryote initiation factor 4E binding protein (4EBP1) [12], and the autophagy regulatory kinase ATG1 [13–15]. Therefore, mTORC1 plays a critical role in translation, cell growth, and autophagy. In contrast, mTORC2 is required for phosphorylation of AKT [16], SGK [17], and conventional PKC [18, 19], suggesting that the two TOR complexes have different physiological

DOI: 10.1016/S1874-6047(10)27005-1

functions. This chapter will mainly discuss the regulation of mTORC1 by amino acids and the role of Rag GTPases in nutrient response.

II.   mTORC1 Activation by Multiple Signals, Including Amino Acids

As a key cell growth regulator, mTORC1 is tightly regulated during normal cell growth and development. Consistent with a key role in development, knockout of mTORC1 subunits, including mTOR [20, 21], raptor [22], or mLST8 [22], results in embryonic lethality. Moreover, elevated mTORC1 activation has been frequently observed in human cancer, suggesting an important role of mTORC1 in tumor development [23]. In fact, rapamycin has been approved as a drug for late stage renal cancer and currently there are multiple ongoing clinical trials of mTOR inhibitors in cancer treatment. It should be noted that mTORC1 activity is generally determined indirectly by measuring the phosphorylation of S6K and 4EBP1. This is because direct mTORC1 kinase assay is not simple and such *in vitro* kinase assay using immunoprecipitated mTORC1 is often unable to accurately reflect mTORC1 activation *in vivo*.

TSC1 and TSC2 are tumor suppressors mutated in the tuberous sclerosis, a genetic disorder characterized by benign tumor formation in a wide range of tissues [24]. Given the fact that TSC1/TSC2 is phosphorylated by multiple signaling kinases, they may serve as an integration point to receive information from multiple pathways to modulate cell growth by regulating mTORC1 activity. The TSC1/TSC2 complex has GTPase activating protein (GAP) activity toward Rheb, which is a Ras family small GTPase [25–27]. Rheb directly and potently activates mTORC1 when it is in a GTP-bound status [4, 28]. Therefore, TSC1/TSC2-Rheb-mTORC1 represents the core components of the mTORC1 signaling pathway.

mTORC1 is known to be regulated by a wide range of extracellular and intracellular signals. Activation of mTORC1 by growth factors and insulin is mainly mediated by the PI3K-AKT pathway [1]. AKT has been shown to phosphorylate PRAS40, an mTORC1 component, and TSC2, which is the major upstream negative regulator of mTORC1 [3, 4, 29, 30]. These phosphorylations are important for mTORC1 activation by AKT. ERK, activated by growth factors, may also contribute to mTORC1 activation by phosphorylating TSC2 [31]. In TSC1 or TSC2 mutant cells, mTORC1 is constitutively active and is no longer regulated by growth factors [29, 32], indicating a critical role of TSC1/TSC2 in mediating the growth factor signal to mTORC1 activation. mTORC1 activity is also tightly regulated by cellular energy status. This response is mediated by the AMP activated protein kinase, AMPK, a cellular energy sensor [1]. AMPK inhibits mTORC1 by phosphorylating TSC2 and raptor, an mTORC1 subunit, providing a link between cell growth and cellular energy status [33, 34].

One of the most potent signals that activate mTORC1 is amino acid [35]. How cell growth is coordinated with amino acids is a fundamental issue in cell biology. In the absence of amino acids, mTORC1 cannot be efficiently activated by other stimuli, such as insulin [35]. There is conflicting evidence regarding whether amino acids act through Rheb to stimulate mTORC1 [25, 36–38]. However, neither TSC1 nor TSC2 is essential for amino acid-induced mTORC1 activation because amino acid deprivation still inactivates mTORC1 in TSC1$^{-/-}$ or TSC2$^{-/-}$ cells [37]. However, the amino acid-stimulated mTORC1 activation is sensitive to wortmannin, indicating a possible role of PI3K in amino acid response. Vacuolar protein sorting 34, Vps34, is a class III PI3K, which is inhibited by wortmannin, and has been implicated in mTORC1 activation by amino acids. Knockdown of Vps34 suppressed mTOR activation by amino acids [38]. Similarly, expression of the FYVE domain, which may function dominant-negatively by binding to phosphatidylinositol 3-phosphate (the product catalyzed by Vps34), blocked mTORC1 activation by amino acids. However, the function of Vps34 in amino acid signaling is disputed by genetic studies in *Drosophila*, in which mutation of Vps34 had no effect on mTORC1 activation by amino acids [39]. Therefore, further experiments are needed to clarify whether Vps34 is indeed involved in amino acid signaling to mTORC1.

RNA interference screen using *Drosophila* cells has identified the MAP4K3 protein kinase as a potential regulator of dS6K phosphorylation in response to amino acids [40]. A similar role of MAP4K3 has also been verified in mammalian cells. Interestingly, amino acids rapidly stimulated MAP4K3 activity, further supporting its role in amino acid response. The effect of MAP4K3 on mTORC1 is independent of TSC1 and TSC2, results consistent with other reports that amino acid regulates mTORC1 in a manner independent of TSC1/TSC2. However, MAP4K3 overexpression only partially delayed dephosphorylation of S6K upon amino acid withdrawal and the effect of MAP4K3 knockdown on mTORC1 substrate phosphorylation was modest. Therefore, MAP4K3 may have a limited role in mTORC1 activation by amino acids.

III.   Rag GTPases and Amino Acid-Induced mTORC1 Activation

The Ras family GTPases function as molecular switches in many signaling pathways. As discussed earlier in this chapter, for example, Rheb directly activates mTORC1 when in GTP-bound form. These GTPases cycle between active GTP-bound state and inactive GDP-bound state.

The Ras family GTPases are regulated by GEF (Guanine nucleotide exchange factors) and GAP, which promote the activation and inhibition of target GTPases, respectively. Because the intrinsic GTPase activity of Ras family is rather low, GTP hydrolysis and inactivation of Ras family proteins are greatly enhanced by GAP. Similarly, GEF accelerates the nucleotide exchange, therefore activates GTPases. In the mTOR pathway, TSC2 inhibits mTORC1 by acting as a GAP toward Rheb [25–27].

Among the signals that activate mTORC1, amino acid is one of the most potent, but the mechanism of mTORC1 activation by amino acid is largely unknown. Rag GTPases have been brought to attention recently because of their critical role in mediating amino acid signals to mTORC1 activation [41, 42]. Rag GTPases (four in human, named as RagA, B, C, and D) are distantly related to Ras. RagA and RagB were previously identified in efforts of cloning novel members of Ras GTPases, but their physiological functions were unclear. RagA and RagB share around 97% amino acid sequence identity [43]. RagC and RagD were identified by virtue of their ability to form herterodimer with RagA or RagB in a yeast two-hybrid screen. RagC and RagD are highly homologous to each other (around 81% sequence identity) while the homology between RagA/B and RagC/D is rather low (less than 30% identity in the N-terminal GTPase domain. The overall sequence identity is even lower) [44]. Different from most of the Ras family members, the Rag GTPases are much bigger. In addition to their GTPase domain, Rag also has a C-terminal domain, which is important for dimerization. RagA/B forms heterodimer with RagC/D and the dimerization is independent of GTP/GDP loading [43]. Most Ras family GTPases are modified by isoprenylation in their C-terminal cysteine residues for membrane association. However, Rag GTPases lack the C-terminal cysteine for lipid modifications although they may associate with intracellular membrane.

Both genetic and biochemical studies support a conserved role of Rag in amino acid signaling. The function of Rag in mTOR signaling was discovered by an RNA interference-based screen for GTPases that were important for S6K phosphorylation [41]. In parallel, RagC was identified as a protein copurified with raptor [42]. Knockdown of dRagA or dRagC strongly decreased TORC1 activation in *Drosophila* S2 cells, as determined by a decreased phosphorylation of TORC1 substrate dS6K. Similarly, knockdown of RagA and RagB (both have redundant function) or RagC and RagD dramatically reduced the phosphorylation of S6K1 and 4EBP1 in mammalian cells. The Rag knockdown effect on mTORC1 was particularly dramatic in the presence of amino acids. Consistently, overexpression of the active GTP-bound form of RagA/B activated mTORC1 even in the absence of amino acids. These data suggest that active Rag GTPases can substitute the amino acid signals and are sufficient to activate mTORC1.

In contrast, expression of a dominant-negative RagA or RagB (GDP form mutant) potently inhibited mTORC1 activation even in the presence of amino acids, indicating that Rag GTPases are necessary for mTORC1 activation. Together, it is proposed that Rag GTPases mediate the signal from amino acids to mTORC1 activation [41, 42].

The function of Rag in TORC1 regulation is further supported by cell size analysis in both *Drosophila* and mammalian cells [41, 42]. Expression of active dRagA in *Drosophila* fat body significantly increased cell size, indicating activation of TORC1. The effect of active dRagA on cell size was less obvious in the presence of sufficient nutrient but was enhanced by nutrient starvation. This is likely because TORC1 is already activated in the presence of sufficient nutrients. However, under nutrient starvation condition, TORC1 activity is low; therefore, ectopic expression of active dRagA activates TORC1 and increases cell size. Genetic inactivation of dRagC also decreased fat body cell size in *Drosophila*, but the phenotype was evident only under nutrient sufficiency. Similar effect of RagB/D on mammalian cell size was observed. Autophagy is a physiological response to nutrient starvation. Further supporting a role of Rag in nutrient signaling is the observation that active dRagA suppressed autophagy even under nutrient starvation. In mammalian cells, LC3 cleavage and lipid modification is a well-characterized marker for autophagy. Ectopic expression of active RagA inhibited LC3 cleavage while expression of dominant-negative RagA promoted LC3 cleavage. These data not only indicate that the Rag GTPases play a key role in TORC1 activation but also demonstrate that the effect of Rag is specific to nutrient signaling [41].

The heterodimer formation of Rag GTPases has at least two functional consequences. Within the herterodimer, the two GTPases stabilize each other. In addition, the heterodimer is more active than each subunit to activate mTORC1. Interestingly, within a heterodimer, the two GTPases bind different guanine nucleotides, one binds GTP and the other binds GDP [45]. The active form of Rag herterodimer consists the GTP form of RagA/B and the GDP form of RagC/D. However, the contribution of RagA/B and RagC/D to mTORC1 activation is not equal [41]. RagA/B has a dominant role over RagC/D to activate mTORC1. In other words, the Rag heterodimer activates mTORC1 as long as RagA/B is in GTP form regardless of the nucleotide status of the associated RagC/D. Complementarily, the Rag herterodimer suppresses mTORC1 activity as long as RagA/B is in GDP form regardless of the nucleotide status of the associated RagC/D [41, 42].

Unlike Rheb, which can directly activate mTORC1, Rag does not activate mTORC1 *in vitro* [41]. However, Rag likely has a direct role in mTORC1 regulation because Rag interacts with raptor. Actually, Rag interaction with raptor is significantly stronger than Rheb. In Rag heterodimer, RagA/B is likely to be directly responsible for raptor binding. The interaction depends on GTP binding and the effector domain of RagA/B while the GTP binding status of RagC/D is not important for raptor interaction (unpublished observation). These results are consistent with the functional data that RagA/B acts dominantly over RagC/D in mTORC1 activation.

Rag GTPases must be regulated by amino acids if they are critical in mediating amino acid signals. Indeed, amino acids stimulate RagB GTP binding and RagD GDP binding, therefore activate the RagB/D herterodimer [42]. As predicted, the interaction between Rag and raptor was also stimulated by amino acids because RagB GTP binding is required for its interaction with raptor. The exact mechanism of Rag GTPases in mTORC1 activation is largely unclear. However, Rag GTPases are proposed to affect the subcellular localization of mTORC1. In response to amino acids, the active Rag heterodimer may tether mTORC1 and transport it to the vicinity of Rheb, where mTORC1 is activated by Rheb binding [42] (Figure 5.1). This model nicely explains the interdependent relationship of mTORC1 activation by insulin and amino acids. Insulin activates Rheb via the PI3K-AKT-TSC1/2 pathway while amino acids activate Rag by a mechanism yet to be characterized. Under physiological conditions, mTORC1 is activated only when both Rheb and Rag are active.

The function of Rag GTPases in amino acid-induced TORC1 activation is conserved in *Saccharomyces cerevisiae* [46], which has two Rag homologs, Gtr1 and Gtr2. Gtr1 exhibits approximately 52% sequence identity with RagA/B and Gtr2 exhibits 46% identity with RagC/D [44]. Similar to Rag, Gtr1 and Gtr2 also form a stable complex [47]. By genetic study, *GTR1* was initially identified as a gene that was important for phosphate transport [48]. Later, *GTR1* and *GTR2* were implicated to negatively regulate the

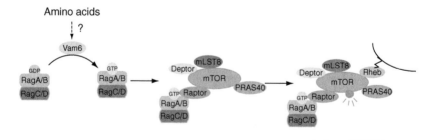

Active mTORC1

FIG. 5.1. Proposed function of Rag GTPases in TORC1 activation by amino acids. The Rag heterodimer is activated by the Vam6 GEF, which may mediate the amino acid signals by an unknown mechanism. The active Rag heterodimer binds raptor and recruits the mTORC1 to Rheb-associated vesicles, where mTORC1 is activated. This model suggests that mTORC1 activation requires both Rheb and Rag.

Ran GTPase cycle [49]. More recent studies have shown that Gtr1/2 are key components of the EGO complex, which is a protein complex associated with vacuolar membrane and is required for microautophagy during exit from rapamycin-induced growth arrest in budding yeast [50]. In another study, Gtr1/2 is shown to play important roles in proper sorting of Gap1 (general amino acid permease) from late endosome to plasma membrane [45]. Collectively, these studies imply a possible connection between *GTR1/ 2* and TOR function.

The function of Gtr1/2 in amino acid-mediated mTOR activation has recently been demonstrated in yeast [46]. Gtr1 and Gtr2 act in a manner similar to mammalian Rag GTPases. The Gtr1-GTP and Gtr2-GDP hetero-dimer was most active to stimulate phosphorylation of Sch9, which is a substrate of yeast TORC1 and homologous to the mammalian S6K. Deletion of *GTR1* or *GTR2* decreased Sch9 phosphorylation. In addition, the other two EGO complex components, *EGO1* and *EGO3*, were also required for TORC1 activation, demonstrating the function of EGO complex in TOR activation. Gtr1-GTP but not Gtr1-GDP bound to TORC1 component Tco89 and Kog1 (the yeast homolog of mammalian raptor). In addition, the interaction between wild-type Gtr1 and TORC1 was sensitive to amino acids while the interaction between the constitutively active Gtr1 mutant and TORC1 was insensitive to amino acids. Therefore, a model has been proposed that in yeast amino acids activate Gtr1/Gtr2 and promote their inter-action with TORC1, resulting in TORC1 activation [46].

Although yeast Gtr1 and Gtr2 are structurally and functionally similar to human Rag GTPases, the subcellular localization of TORC1 components, such as Tor1 and Tco89, were not affected by amino acids. In addition, Gtr1/ Gtr2 localization was not altered in response to amino acid treatments either [46]. These data are different from the observations made with mammalian Rag and mTORC1 [42], which are reported to be translocated to Rab7 positive vesicle in response to amino acid stimulation. Further studies are needed to clarify whether the mechanisms of TORC1 activation by Rag GTPases are different between yeast and mammalian cells or the amino acid-induced subcellular translocation is not critical for TORC1 activation by Rag GTPases.

IV.   Vam6 as a Rag GEF in Amino Acid-Induced TORC1 Activation

There is no evidence that Rag GTPases directly sense amino acids. Then, how is Rag activated by amino acids? A key advancement on this issue is from the yeast genetic study that identified Vam6 (also known as Vps39) as a Gtr1 GEF [46] (Figure 5.1). Vam6 is a component of HOPS complex (*ho*motypic fusion and vacuole *p*rotein *s*orting), which associates with the vacuolar membrane and is involved in vacuole fusion and vacuole protein sorting. The yeast Vam6 is known to have GEF activity toward Ypt7 (the yeast Rab7 homolog) [51]. A systematic, genome-wide synthetic dosage lethal screen showed that loss of *VAM6* yielded a strong synthetic growth defect when combined with overexpression of GTR1-GDP. Loss of *VAM6* resulted in a growth defect phenotype similar to that caused by mutation of EGO complex components. Overexpression of *Vam6* suppressed the defect caused by the dominant-negative mutant Gtr1 overexpression. Further genetic study placed *VAM6* upstream of *GTR1* in TORC1 signaling, possibly as a GEF of Gtr1. Biochemistry analysis showed that Vam6 bound to Gtr1 and facilitated GDP release from Gtr1, providing direct and convincing evidence that Vam6 has Gtr1 GEF activity. Furthermore, loss of *VAM6* decreased Gtr1-GTP level in the cell, supporting the conclusion that Vam6 acts as a Gtr1 GEF *in vivo* [46].

Vam6/Vps39 is highly conserved in other species. In fact, knockdown of Vam6 in *Drosophila* S2 cells or human HEK293 cells also decreased the phosphorylation of S6K and S6, indicating that the function of Vam6 is conserved in higher eukaryotes (unpublished observation). One may speculate that Vam6 could play a key role in mediating amino acid signals, either directly or indirectly, to Rag activation, which then in turn promotes TORC1 activation.

V.   Raptor Interacts with Both Upstream Regulators and Downstream Substrates

Raptor forms a stable complex with mTOR and is required for mTOR signaling [2, 5]. It contains a highly conserved amino-terminal region named RNC (*r*aptor *N*-terminal *c*onserved) domain, followed by three HEAT repeats and seven WD40 repeats at the C-terminal part of the protein. Raptor is ubiquitously expressed in many tissues with high levels of expression in brain, kidney, and muscle. It interacts with amino terminus HEAT repeats of mTOR, but the mTOR binding region in raptor has not been clearly defined, possibly because raptor using multiple sites to contact mTOR. Both genetic knockout and knockdown have unequivocally shown that raptor is indispensable for mTORC1 function [22].

The raptor–mTOR interaction is dynamic and very sensitive to detergent [2]. Even mild detergent, such as Triton X-100 and NP40, disrupts the mTOR complex and kinase activity toward substrate whereas mTOR autophosphorylation is enhanced by detergents. The mTORC1 complex is preserved in no detergent or CHAPS buffer and retains kinase activity toward 4EBP1 and S6K. Therefore, special care should be taken when isolating mTORC1 complex. One characterized function of raptor is to serve as a scaffold protein to recruit substrates for phosphorylation by mTOR in the mTORC1. Raptor directly binds to a short amino acid sequence called the TOS (mTOR signaling) motif found in the C-terminal region of 4EBP1 and the N-terminal region of S6 kinases [52–54]. The TOS motif in S6K and 4EBP1 is required for mTORC1 dependent phosphorylation of both proteins.

Raptor may also serve as a signal receiver for mTORC1 by directly interacting with upstream activators. For example, Rheb binds to the C-terminal WD domain of raptor and this interaction may be important for Rheb to activate mTORC1 [28]. Furthermore, RagA strongly binds to raptor [42]. This interaction depends on the RagA effector domain and GTP binding, demonstrating that raptor is a physiological downstream effector of RagA. Therefore, raptor serves a dual role, both as signal receiver and transmitter, in mTORC1. Raptor controls mTORC1 function by interacting with upstream regulators, such as Rheb and RagA, and downstream substrates, such as S6K and 4EBP1.

Recent studies have indicated that raptor function is also regulated by phosphorylation in response to upstream signals. Under energy stress condition, AMPK is activated [55]. Active AMPK can directly phosphorylate raptor on two conserved sites (Ser722, Ser792) [34]. This phosphorylation creates 14-3-3 binding sites and possibly changes mTORC1 kinase activity, as cells with phosphorylation defective raptor mutant show a defective mTORC1 inhibition by energy starvation. Raptor can also be phosphorylated by RSK on RXRXXpS/T motifs [56]. This phosphorylation likely stimulates mTORC1 activity. Thus, mitogenic growth factors stimulate mTORC1 not only through PI3K-Akt pathway to phosphorylate TSC2, but also through ERK-RSK to phosphorylate raptor. Moreover, raptor phosphorylation by mTOR has also been reported. The mTOR dependent raptor phosphorylation is stimulated by insulin and is necessary for insulin-mediated mTOR activation [57].

VI.   RalA in Nutrient-Induced mTORC1 Activation

Ral GTPases, including RalA and RalB, belong to the Ras superfamily. They have been implicated in multiple cellular processes, including cell transformation, migration, membrane dynamics, and transcription regulation [58].

For example, Ral and RalGDS, which is a Ral GEF, are known to be important for Ras-induced cellular transformation [59]. In fact, RalGDS is a direct downstream effector of Ras [60]. Maehama *et al.* reported that RalA has an essential role in amino acid-induced mTORC1 activation [61]. Knockdown of RalA or RalB decreased mTORC1 activity in Hela cells. Notably, RalA knockdown strongly blocked phosphorylation of S6K and 4EBP1 in response to amino acid or glucose but had a much weaker effect on the phosphorylation of mTORC1 substrates in response to insulin stimulation. These data indicate a specific role of RalA in nutrient signaling. Further supporting the role of Ral in nutrient response, knockdown of RalGDS also decreased mTORC1 activity. Interestingly, amino acids increased the accumulation of active GTP form of RalA, indicating a direct involvement of RalA in amino acid response.

The relationship between RalA and Rheb is different from that between Rag and Rheb. Rag and Rheb act parallel and upstream of mTORC1 [41]. Dominant-negative Rag did not block mTORC1 activation by Rheb overexpression. Consistently, mutation in dRagC did not block the growth stimulating effect caused by dRheb overexpression. In contrast, RalA knockdown suppressed Rheb S16H (a hyperactive Rheb mutant)-induced mTORC1 activation, suggesting that RalA might work downstream of Rheb [61]. The function of RalA in nutrient signal could be indirect, as no interaction between RalA and mTORC1 was detected. Furthermore, not all data are consistent with a direct role of Ral in amino acid signaling as overexpression of the active RalA did not substitute the stimulating effect of amino acids on mTORC1. Amino acid still potently activated mTORC1 in cells expressing both dominant-negative Rheb and constitutively active Ral. mTORC1 activity is likely to be affected indirectly by multiple Ras family GTPases. For example, we have observed that other Ras family GTPases, such as the Rab family, are also important for mTORC1 activation by amino acids (unpublished observations). We speculate that intracellular vesicular trafficking, which could be affected by many GTPases, such as Rag, Rab, and Ral, may play a critical role in mTORC1 activation.

The coordination between cell growth and nutrient availability is a fundamental question in cell biology. As a key cell growth regulator, mTORC1 is activated by nutrients, such as amino acids, to stimulate protein synthesis and cell growth. Recent studies have implicated possible roles of VPS34, MAP4K3, RalA, and Rag GTPases in mTORC1 activation in response to amino acids. The exact role of these proteins and the relationship among them to mediate amino acid signals to mTORC1 are largely unknown. However, the Rag GTPases emerge as the key mediator between amino acid sufficiency and mTORC1 activation. This conclusion is supported by convincing genetic and biochemical data from yeast, *Drosophila*, and mammalian cells [41, 42, 46]. However, many key questions remain to be answered. What is the amino acid sensor? How does Rag activate mTORC1 and what is the relationship between Rag and Rheb? Future research in amino acid signaling and mTORC1 activation will significantly advance our understanding not only of normal cell growth and development, but also pathological abnormality in diseases, such as diabetes and cancer.

REFERENCES

1. Wullschleger, S., Loewith, R., and Hall, M.N. (2006). TOR signaling in growth and metabolism. *Cell* 124(3):471–484.
2. Kim, D.H., *et al.* (2002). mTOR interacts with raptor to form a nutrient-sensitive complex that signals to the cell growth machinery. *Cell* 110(2):163–175.
3. Vander Haar, E., *et al.* (2007). Insulin signalling to mTOR mediated by the Akt/PKB substrate PRAS40. *Nat Cell Biol* 9(3):316–323.
4. Sancak, Y., *et al.* (2007). PRAS40 is an insulin-regulated inhibitor of the mTORC1 protein kinase. *Mol Cell* 25(6):903–915.
5. Hara, K., *et al.* (2002). Raptor, a binding partner of target of rapamycin (TOR), mediates TOR action. *Cell* 110(2):177–189.
6. Kim, D.H., *et al.* (2003). GbetaL, a positive regulator of the rapamycin-sensitive pathway required for the nutrient-sensitive interaction between raptor and mTOR. *Mol Cell* 11(4):895–904.
7. Frias, M.A., *et al.* (2006). mSin1 is necessary for Akt/PKB phosphorylation, and its isoforms define three distinct mTORC2s. *Curr Biol* 16(18):1865–1870.
8. Jacinto, E., *et al.* (2004). Mammalian TOR complex 2 controls the actin cytoskeleton and is rapamycin insensitive. *Nat Cell Biol* 6(11):1122–1128.
9. Sarbassov, D.D., *et al.* (2004). Rictor, a novel binding partner of mTOR, defines a rapamycin-insensitive and raptor-independent pathway that regulates the cytoskeleton. *Curr Biol* 14(14):1296–1302.
10. Yang, Q., *et al.* (2006). Identification of Sin1 as an essential TORC2 component required for complex formation and kinase activity. *Genes Dev* 20(20):2820–2832.
11. Loewith, R., *et al.* (2002). Two TOR complexes, only one of which is rapamycin sensitive, have distinct roles in cell growth control. *Mol Cell* 10(3):457–468.
12. Hay, N., and Sonenberg, N. (2004). Upstream and downstream of mTOR. *Genes Dev* 18(16):1926–1945.
13. Jung, C.H., *et al.* (2009). ULK-Atg13-FIP200 complexes mediate mTOR signaling to the autophagy machinery. *Mol Biol Cell* 20(7):1992–2003.
14. Hosokawa, N., *et al.* (2009). Nutrient-dependent mTORC1 association with the ULK1-Atg13-FIP200 complex required for autophagy. *Mol Biol Cell* 20(7):1981–1991.
15. Ganley, I.G., *et al.* (2009). ULK1.ATG13.FIP200 complex mediates mTOR signaling and is essential for autophagy. *J Biol Chem* 284(18):12297–12305.
16. Sarbassov, D.D., *et al.* (2005). Phosphorylation and regulation of Akt/PKB by the rictor-mTOR complex. *Science* 307(5712):1098–1101.
17. Garcia-Martinez, J.M., and Alessi, D.R. (2008). mTOR complex 2 (mTORC2) controls hydrophobic motif phosphorylation and activation of serum- and glucocorticoid-induced protein kinase 1 (SGK1). *Biochem J* 416(3):375–385.
18. Ikenoue, T., *et al.* (2008). Essential function of TORC2 in PKC and Akt turn motif phosphorylation, maturation and signalling. *EMBO J* 27(14):1919–1931.

19. Facchinetti, V., *et al.* (2008). The mammalian target of rapamycin complex 2 controls folding and stability of Akt and protein kinase C. *EMBO J* 27(14):1932–1943.
20. Murakami, M., *et al.* (2004). mTOR is essential for growth and proliferation in early mouse embryos and embryonic stem cells. *Mol Cell Biol* 24(15):6710–6718.
21. Gangloff, Y.G., *et al.* (2004). Disruption of the mouse mTOR gene leads to early postimplantation lethality and prohibits embryonic stem cell development. *Mol Cell Biol* 24(21):9508–9516.
22. Guertin, D.A., *et al.* (2006). Ablation in mice of the mTORC components raptor, rictor, or mLST8 reveals that mTORC2 is required for signaling to Akt-FOXO and PKCalpha, but not S6K1. *Dev Cell* 11(6):859–871.
23. Sarbassov, D.D., Ali, S.M., and Sabatini, D.M. (2005). Growing roles for the mTOR pathway. *Curr Opin Cell Biol* 17(6):596–603.
24. Short, M.P., *et al.* (1995). Clinical, neuropathological and genetic aspects of the tuberous sclerosis complex. *Brain Pathol* 5(2):173–179.
25. Zhang, Y., *et al.* (2003). Rheb is a direct target of the tuberous sclerosis tumour suppressor proteins. *Nat Cell Biol* 5(6):578–581.
26. Inoki, K., *et al.* (2003). Rheb GTPase is a direct target of TSC2 GAP activity and regulates mTOR signaling. *Genes Dev* 17(15):1829–1834.
27. Tee, A.R., *et al.* (2003). Tuberous sclerosis complex gene products, Tuberin and Hamartin, control mTOR signaling by acting as a GTPase-activating protein complex toward Rheb. *Curr Biol* 13(15):1259–1268.
28. Long, X., *et al.* (2005). Rheb binds and regulates the mTOR kinase. *Curr Biol* 15(8):702–713.
29. Inoki, K., *et al.* (2002). TSC2 is phosphorylated and inhibited by Akt and suppresses mTOR signalling. *Nat Cell Biol* 4(9):648–657.
30. Manning, B.D., *et al.* (2002). Identification of the tuberous sclerosis complex-2 tumor suppressor gene product tuberin as a target of the phosphoinositide 3-kinase/akt pathway. *Mol Cell* 10(1):151–162.
31. Ma, L., *et al.* (2005). Phosphorylation and functional inactivation of TSC2 by Erk implications for tuberous sclerosis and cancer pathogenesis. *Cell* 121(2):179–193.
32. Kwiatkowski, D.J., *et al.* (2002). A mouse model of TSC1 reveals sex-dependent lethality from liver hemangiomas, and up-regulation of p70S6 kinase activity in Tsc1 null cells. *Hum Mol Genet* 11(5):525–534.
33. Inoki, K., Zhu, T., and Guan, K.L. (2003). TSC2 mediates cellular energy response to control cell growth and survival. *Cell* 115(5):577–590.
34. Gwinn, D.M., *et al.* (2008). AMPK phosphorylation of raptor mediates a metabolic checkpoint. *Mol Cell* 30(2):214–226.
35. Hara, K., *et al.* (1998). Amino acid sufficiency and mTOR regulate p70 S6 kinase and eIF-4E BP1 through a common effector mechanism. *J Biol Chem* 273(23):14484–14494.
36. Long, X., *et al.* (2005). Rheb binding to mammalian target of rapamycin (mTOR) is regulated by amino acid sufficiency. *J Biol Chem* 280(25):23433–23436.
37. Roccio, M., Bos, J.L., and Zwartkruis, F.J. (2006). Regulation of the small GTPase Rheb by amino acids. *Oncogene* 25(5):657–664.
38. Nobukuni, T., *et al.* (2005). Amino acids mediate mTOR/raptor signaling through activation of class 3 phosphatidylinositol 3OH-kinase. *Proc Natl Acad Sci USA* 102(40):14238–14243.
39. Juhasz, G., *et al.* (2008). The class III PI(3)K Vps34 promotes autophagy and endocytosis but not TOR signaling in *Drosophila*. *J Cell Biol* 181(4):655–666.
40. Findlay, G.M., *et al.* (2007). A MAP4 kinase related to Ste20 is a nutrient-sensitive regulator of mTOR signalling. *Biochem J* 403(1):13–20.

41. Kim, E., *et al.* (2008). Regulation of TORC1 by Rag GTPases in nutrient response. *Nat Cell Biol* 10(8):935–945.
42. Sancak, Y., *et al.* (2008). The Rag GTPases bind raptor and mediate amino acid signaling to mTORC1. *Science* 320(5882):1496–1501.
43. Schurmann, A., *et al.* (1995). Cloning of a novel family of mammalian GTP-binding proteins (RagA, RagBs, RagB1) with remote similarity to the Ras-related GTPases. *J Biol Chem* 270(48):28982–28988.
44. Sekiguchi, T., *et al.* (2001). Novel G proteins, Rag C and Rag D, interact with GTP-binding proteins, Rag A and Rag B. *J Biol Chem* 276(10):7246–7257.
45. Gao, M., and Kaiser, C.A. (2006). A conserved GTPase-containing complex is required for intracellular sorting of the general amino-acid permease in yeast. *Nat Cell Biol* 8(7):657–667.
46. Binda, M., *et al.* (2009). The Vam6 GEF controls TORC1 by activating the EGO complex. *Mol Cell* 35(5):563–573.
47. Nakashima, N., Noguchi, E., and Nishimoto, T. (1999). *Saccharomyces cerevisiae* putative G protein, Gtr1p, which forms complexes with itself and a novel protein designated as Gtr2p, negatively regulates the Ran/Gsp1p G protein cycle through Gtr2p. *Genetics* 152(3):853–867.
48. Bun-Ya, M., Harashima, S., and Oshima, Y. (1992). Putative GTP-binding protein, Gtr1, associated with the function of the Pho84 inorganic phosphate transporter in *Saccharomyces cerevisiae. Mol Cell Biol* 12(7):2958–2966.
49. Nishimoto, T. (2000). Upstream and downstream of ran GTPase. *Biol Chem* 381 (5–6):397–405.
50. Dubouloz, F., *et al.* (2005). The TOR and EGO protein complexes orchestrate micro-autophagy in yeast. *Mol Cell* 19(1):15–26.
51. Wurmser, A.E., Sato, T.K., and Emr, S.D. (2000). New component of the vacuolar class C-Vps complex couples nucleotide exchange on the Ypt7 GTPase to SNARE-dependent docking and fusion. *J Cell Biol* 151(3):551–562.
52. Schalm, S.S., and Blenis, J. (2002). Identification of a conserved motif required for mTOR signaling. *Curr Biol* 12(8):632–639.
53. Schalm, S.S., *et al.* (2003). TOS motif-mediated raptor binding regulates 4E-BP1 multisite phosphorylation and function. *Curr Biol* 13(10):797–806.
54. Nojima, H., *et al.* (2003). The mammalian target of rapamycin (mTOR) partner, raptor, binds the mTOR substrates p70 S6 kinase and 4E-BP1 through their TOR signaling (TOS) motif. *J Biol Chem* 278(18):15461–15464.
55. Hardie, D.G. (2004). AMP-activated protein kinase: the guardian of cardiac energy status. *J Clin Invest* 114(4):465–468.
56. Carriere, A., *et al.* (2008). Oncogenic MAPK signaling stimulates mTORC1 activity by promoting RSK-mediated raptor phosphorylation. *Curr Biol* 18(17):1269–1277.
57. Wang, L., *et al.* (2009). Mammalian target of rapamycin complex 1 (mTORC1) activity is associated with phosphorylation of raptor by mTOR. *J Biol Chem* 284(22):14693–14697.
58. van Dam, E.M., and Robinson, P.J. (2006). Ral: mediator of membrane trafficking. *Int J Biochem Cell Biol* 38(11):1841–1847.
59. Yamamoto, T., Taya, S., and Kaibuchi, K. (1999). Ras-induced transformation and signaling pathway. *J Biochem* 126(5):799–803.
60. Quilliam, L.A., Rebhun, J.F., and Castro, A.F. (2002). A growing family of guanine nucleotide exchange factors is responsible for activation of Ras-family GTPases. *Prog Nucleic Acid Res Mol Biol* 71:391–444.
61. Maehama, T., *et al.* (2008). RalA functions as an indispensable signal mediator for the nutrient-sensing system. *J Biol Chem* 283(50):35053–35059.

6

Amino Acid Regulation of hVps34 and mTORC1 Signaling

PAWAN GULATI • GEORGE THOMAS

Department of Cancer and Cell Biology
Metabolic Disease Institute
University of Cincinnati
Cincinnati, Ohio, USA

I. Abstract

The mammalian target of rapamycin (mTOR) is believed to function as part of an ancient growth factor-, nutrient-, and energy-responsive signaling pathway. mTOR has been reported to reside in two distinct complexes: mTORC1 and mTORC2. Both complexes respond to hormones and growth factors, but only mTORC1 responds to nutrients and energy status. The central role of nutrients, especially branched-chain amino acids (BCAAs), in regulating mTORC1 signaling has become increasingly evident over the last decade. In addition, there is increasing evidence that excess circulating AAs play a key role in the pathogenesis of obesity, insulin resistance, and, more recently, cancer. Hence, it has become imperative to elucidate the underlying mechanism(s) by which AAs induce mTORC1 activation. In the last few years, a number of studies have identified novel proteins involved in AA-mediated regulation of mTORC1, including class 3 phosphatidyl-inositide-3OH-kinase (PI3K), or human vacuolar protein sorting 34 (hVps34); mitogen-activated protein kinase kinase kinase kinase 3 (MAP4K3); and the small GTPases termed Ras-related GTP-binding (RAG) proteins. In this chapter, we will briefly touch on the roles of MAP4K3 and the Rags, and focus mainly on a discussion of the role of

ISSN NO: 1874-6047
DOI: 10.1016/S1874-6047(10)27006-3

hVps34 in AA-mediated regulation of TORC1 signaling. Recently, it has been shown that AAs regulate hVps34 activity through an increase in intracellular calcium ($[Ca^{2+}]_i$). AA-induced increases in $[Ca^{2+}]_i$ promote calmodulin (CaM) binding to the conserved CaM-binding domain in hVps34, thus stimulating its lipid kinase activity, which leads to mTORC1 activation. Considering the fundamental importance of excessive nutrient intake in the development of obesity, insulin resistance, and cancer, these findings have the potential to contribute to the development of new intervention therapies to effectively control and treat these metabolic disorders.

II. Introduction

Mammalian target of rapamycin (mTOR) is a dominant signaling component that is responsive to both growth factor and nutrient stimulation. The mTOR protein kinase belongs to the PI3K- or PIK-related family of protein kinases [1], and is a central regulator of cell growth and metabolism. In higher vertebrates, mTOR and its downstream effectors, the initiation factor 4E binding proteins (4E-BPs) and the 40S ribosomal protein S6 kinases (S6Ks), lie at the crossroads of a nutrient/hormonal signaling network that is involved in specific pathological responses, including obesity, diabetes, and cancer (Figure 6.1) [2]. Two structurally and functionally distinct mTOR-containing multimeric protein complexes have been identified: mTORC1 and mTORC2 (Figure 6.1). mTORC1 contains regulatory-associated protein of mTOR (raptor) [3, 4] and proline-rich PKB/Akt substrate 40 kDa (PRAS40) as its signature proteins (Figure 6.1) [3, 5]. mTORC1 is rapamycin-sensitive and phosphorylates the 4E-BPs and S6Ks. In contrast, mTORC2 contains the unique proteins rapamycin-insensitive companion of mTOR (rictor) [3, 6], mammalian stress-activated protein kinase (SAPK)-interacting protein-1 (mSin1), and protein observed with rictor (protor) (Figure 6.1) [3, 7]. mTORC2 is largely insensitive to rapamycin and has been reported to phosphorylate the hydrophobic motif of protein kinase B (PKB, also known as Akt) [8]; serum- and glucocorticoid-induced protein kinase 1 (SGK1) [9]; and protein kinase C [10]. The mTORC1 pathway acts to integrate signals from growth factors, nutrients, and energy inputs to regulate many processes, including ribosome biogenesis, metabolism, and autophagy [3, 11]. Ribosome biogenesis and metabolism are both positively regulated by the presence of nutrients. Ribosome biogenesis refers to the synthesis of ribosomes, which involves the processing of preribosomal RNA (pre-rRNA) and sequential assembly of a large number of ribosomal proteins on the rRNAs [12]. This process takes place

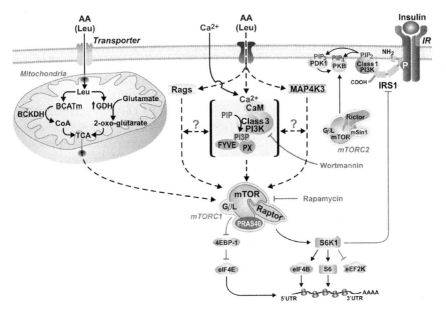

FIG. 6.1. AA-induced mTORC1 signaling and its cross talk with insulin signaling (see text for details).

largely in the nucleolus [12] and is one of the most energy-consuming biochemical processes in a proliferating cell [13]. Metabolism involves all the physical and chemical processes in the body that take up, create, and use energy, such as the digestion of food and nutrients [14]. On the other hand, autophagy is a cellular process in which portions of cytoplasm are sequestered within double-membrane vesicles known as autophagosomes before delivery to lysosomes for degradation and recycling of cellular components [15]. Autophagy is activated upon nutrient or energy depletion, and serves to degrade endogenous proteins. Studies performed by Shigemitsu *et al.* initially showed that under starvation conditions, hepatocytes generally remain insulin responsive, and are able to activate mTORC1 signaling by maintaining high levels of intracellular AAs through autophagy [16]. Moreover, in the *Drosophila* larval fat body, a nutrient-storage organ analogous to the vertebrate liver, under low TOR-signaling conditions, autophagy functions primarily to promote normal cell function and survival rather than acting as a suppressor of cell survival [17]. Autophagy takes over this role in order to maintain the integrity of the cell under prolonged starvation conditions [17]. Thus, the regulation of a bidirectional flow of nutrients between protein synthesis and degradation is achieved by cross talk between mTORC1 and these distinct cellular phenomena: ribosome biogenesis, metabolism, and autophagy.

In recent years, there has been increasing awareness of the size and complexity of the mTORC1 signaling network. This has led to a growing interest in the spatial distribution of mTOR complexes in the cell. Recently, it has been demonstrated that a small family of GTPases termed Rags regulate localization of mTORC1 from the cytoplasm to an endosomal compartment (see Chapter 5). Mammalian cells harbor four members of the Rag subfamily of Ras small GTPases (RagA, RagB, RagC, and RagD), which are present as heterodimers in the cell. These findings demonstrate that mTORC1 localization must be the crucial event in the stepwise activation of mTORC1 and that this localization might be dynamically regulated by upstream stimuli (see Chapter 5). The mechanism by which Rags translocates mTORC1 has yet to be elucidated, and it is not known if there are other protein(s) or protein complex(s) involved in the activation of mTORC1 leading to the phosphorylation of its downstream substrates. Further analysis of the temporal and spatial dynamics of endogenous mTORC1 subcellular localization will provide a better understanding of the mechanism by which AA regulates mTORC1 activation through the RAG proteins.

III. AAs as a Signaling Metabolite

Recent advances in biomedical research have revealed a key role for BCAAs as nutritional signals in the regulation of the mTORC1 pathway. Most dietary proteins consist of 20% BCAAs with BCAAs comprising some 35% of the indispensible amino acid requirements of mammals [18]. Wahren *et al.* showed that after ingestion of a protein-rich meal BCAAs accounted for >50% of the splanchnic output of AAs even though they only comprised 20% of the ingested protein [18]. It has been reported that there is a marked increased in BCAA uptake into muscle upon ingestion of a protein-rich diet [19]. Studies performed in past years have shown that cells recognize changes in BCAAs, such as leucine (leu) availability and, in response, generate alterations in mTORC1 signaling [3, 20].

Despite the reported importance of BCAAs, such as leucine, in the regulation of mTORC1 signaling, it has not yet been established whether the leucine-induced signal is generated at the membrane or in the cytoplasm. The mechanism by which this signal influences mTORC1 activity is also not known (Figure 6.1). In mammalian cells, AAs utilize different transport systems for their movement across the plasma membrane [21]. Leucine, the key BCAA regulating mTORC1 signaling, is transported by System L (L stands for "leucine-preferring") transporters [21]. The requirement for leucine transport across the cell membrane to activate mTORC1 signaling was first shown in Xenopus oocytes [22] and has been further strengthened by recent studies from Nicklin *et al.* showing the phenomenon in mammalian cells [23]. It was demonstrated that the amino acid transporter termed solute carrier family 7 member 5 (SLC7A5) is a heterodimeric bidirectional antiporter that regulates the glutamine-dependent transport of leucine into cultured mammalian cells to regulate mTORC1 signaling [23]. Currently, two mechanisms have been proposed by which leucine induces mTORC1 signaling. The first is through a putative secondary signal generated by mitochondrial metabolism of leucine and in the second mechanism leucine acting directly as a signaling metabolite. In skeletal muscle, leucine has been reported to emanate a metabolically linked secondary signal via two pathways: (a) oxidation to isovaleryl-CoA and (b) by acting as an allosteric activator of glutamate dehydrogenase (GDH), driving the production of α-ketoglutarate (Figure 6.1). Both products would be oxidized in the TCA cycle, generating a metabolically linked secondary signal, which would then drive mTORC1 signaling [24]. The mitochondrial BCAA transaminase (BCATm) isoenzyme is primarily responsible for initiating BCAA catabolism to generate α-ketoisocaproate (KIC). KIC is then acted on by the branched-chain α-keto acid dehydrogenase (BCKD), which is the rate-limiting step for BCAA catabolism. BCKD catalyzes leucine oxidation through the conversion of KIC to isovaleryl-CoA and NADH (Figure 6.1) [24]. This model of mTORC1 regulation through leucine metabolism is supported by the observations of Nicklin *et al.* showing that 2-amino-2-norborane carboxylic acid (BCH), a nonmetabolizable competitive inhibitor of leucine L-amino acid transporters, blocks mTORC1 signaling. However in the case of hepatocytes, which express little to no BCATm, leucine can still drive mTORC1 signaling [25]. Also, in a rat animal model and in adipocytes, Lynch *et al.* reported no correlation between the leucine concentrations required to activate BCKD to induce leucine metabolism and the leucine concentration that activated mTOR signaling [25]. Finally, it has been shown by Miotto *et al.* that Leu8-MAP, a nontransportable leucine analogue, can still inhibit autophagy in hepatocytes, as was observed in the case of leucine itself, supporting the notion that leucine could induce a secondary signal at the membrane [26]. Based on these studies, Miotto *et al.* proposed the presence of a leucine-recognition site(s) at or near the plasma membrane, which would act to generate an intracellular signal(s) in response to extracellular leucine in hepatocytes [26]. This set of observations lead us to a second model of mTORC1 regulation by leucine, which does not involve leucine metabolism. Elucidating the mechanism(s) by which BCAAs, particularly leucine, could act directly as a signaling metabolite in mTORC1 signaling after being transported through the cell membrane, is the objective of intense research. Earlier, it was demonstrated that growth-factor activation of S6K1, but not PKB, in the presence of AAs is a Ca^{2+}-dependent event, which indicated to us a possible role for Ca^{2+} in AA-induced mTORC1 signaling [27, 28], as AAs also fail to activate PKB [29]. In pursuing this idea, we have recently shown that AAs, as well as leucine alone, can lead to an increase in $[Ca^{2+}]_i$ levels (Figure 6.1) [30]. We showed that AA stimulation induces an increase in $[Ca^{2+}]_i$ that can be inhibited by preloading cells with the Ca^{2+} chelator BAPTA-acetoxymethyl ester (BAPTA-AM) or by adding the cell-impermeable Ca^{2+} chelator ethylene glycol tetraacetic acid (EGTA), the latter suggesting that it is an influx of extracellular Ca^{2+}, which is critical [30]. The rise in $[Ca^{2+}]_i$ is not inhibited by pretreating cells with either rapamycin or wortmannin, placing the effect of Ca^{2+} upstream of hVps34 and mTORC1 signaling. Moreover, a rise in $[Ca^{2+}]_i$, induced by thapsigargin in the absence of AAs, is sufficient to activate mTORC1 signaling. The effects of thapsigargin and AAs were partially additive, which may be attributed to the fact that together they induce a more robust and prolonged rise in $[Ca^{2+}]_i$ [30]. Critically, it was found that immunoprecipitated mTORC1 derived from cells treated with either AAs or thapsigargin alone displayed an increase in *in vitro* kinase activity, employing either S6K1 or 4E-BP1 as a substrate, a response blocked by pretreatment of cells with either BAPTA-AM or EGTA. Formally, these results indicate that an AA-induced increase in $[Ca^{2+}]_i$ acts as a second messenger, mediating the effect of AAs on the activation of mTORC1 signaling (Figure 6.1) [30].

In addition, it has been reported by Sancak *et al.* that an AA signal enhances RagB-GTP loading, leading to increased binding of Rags to raptor [31]. Following the interaction with raptor, Rags translocates mTORC1 to the subcellular compartment containing ectopically expressed Rab7, a small G-protein of the RAS-related GTP-binding (Rab) protein family that regulates vesicular transport to late endosomes [31]. Sancak *et al.* further showed that Rheb is already present in the Rab7-positive compartment, which then functions to activate mTORC1 [31]. More recently, Binda *et al.* identified Vam6 in yeast as the conserved guanine nucleotide exchange factor (GEF) for Gtr1, a yeast Rag GTPase ortholog and a component of the vacuolar-membrane-associated EGO complex (EGOC) [32]. Binda *et al.* further demonstrated that Gtr1 interacts with and activates TORC1 in an amino acid-sensitive manner, similar to mammalian cells [32].

IV.   AAs and hVps34

In the search to elucidate the mechanism by which an AA-induced increase in $[Ca^{2+}]_i$ regulates mTORC1 signaling, we analyzed the potential role of hVps34. Previously, we had shown that AA-mediated activation of mTORC1 is sensitive to wortmannin, but that this effect is independent of class 1 PI3K (Figure 6.1) [29]. The latter was demonstrated by siRNA depletion of class 1 PI3K, which blocked insulin-induced PKB and S6K1 activation, but had no effect on AA-induced S6K1 activation [29]. This led to the finding that AAs [29], as well as glucose [33], employ wortmannin-sensitive hVps34 to mediate mTOR activation. hVps34 has been reported to be involved in endosomal trafficking and autophagy [11]. Vergne *et al.* had previously shown that during *Mycobacterium tuberculosis* infection of macrophages, the bacterium produces a toxin, lipoarabinomannan (LAM), which inhibited hVps34-mediated phagosome maturation by blocking an increase in $[Ca^{2+}]_i$ [34]. They further showed that the product of hVps34, phosphatidylinositol-3-phosphate (PI(3)P), was required for this response [34]. Based on the findings of Nobukuni *et al.* [29] and Vergne *et al.* [34], it was asked whether Ca^{2+}-mediated activation of mTORC1 in the absence of AAs is sensitive to wortmannin and if this step requires hVps34. In a series of experiments, it was shown that the ability of thapsigargin to induce mTORC1 activation was blocked by wortmannin and by siRNAs directed against hVps34 [30], suggesting that hVps34 either mediates the Ca^{2+}-induced activation of mTORC1 or that it resides on a parallel pathway. To distinguish between these two possibilities, the effect of BAPTA-AM on PI(3)P production was analyzed, and it was found that preloading AA-deprived cells with BAPTA-AM was as effective as wortmannin in blocking the increase in AA-induced PI(3)P production (Figure 6.1) [30].

In their studies of *M. tuberculosis* infection of macrophages, Vergne *et al.* also demonstrated that the process of phagocytosis requires Ca^{2+}/CaM to mediate hVps34 activation [34]. Based on this finding, we explored the role of the pharmacological CaM inhibitor W7 on AA-induced activation of mTORC1 signaling. The inhibitor, as well as siRNA depletion of CaM, blocked this response [30]. This led to the finding that hVps34 contains a conserved Ca^{2+}/CaM-domain, which, when mutated, blocked binding of Ca^{2+}/CaM to hVps34 and AA-induced mTORC1 activation. Finally, it was demonstrated *in vitro*, that removal of CaM from hVps34 by EGTA abolished hVps34 activity, which could be reconstituted by addition of exogenous purified CaM, confirming the Ca^{2+}/CaM requirement of hVps34 activity [30].

V.   hVps34 and mTORC1

Previously, Nobukuni *et al.* showed that AA-induced activation of mTORC1 is dependent on hVps34 activity, indicating a crucial role for the product of hVps34 activity, PI(3)P, in regulating mTORC1 signaling [29]. It has been reported that PI(3)P acts as a docking site for Fab1/YOTB/-2K632.12/Vac1/EEA1 (FYVE) or PI(3)P-targeting phox homology (PX) domain-containing proteins leading to the formation of a large multimeric signaling platform (Figure 6.1) [11]. Consistent with this hypothesis, AA-induced PI(3)P production is sensitive to calcium chelators such as BAPTA-AM. Moreover, the *in vitro* kinase activity of endogenous hVps34 and ectopically expressed hVps34 is also regulated by AAs in a Ca^{2+}-dependent manner [30].

Despite the findings in mammalian systems, recent studies suggest fundamental differences in TOR signaling across species. Juhász *et al.* reported that, in *Drosophila*, Vps34 does not regulate TOR signaling [35], which is contrary to observations in mammalian cells. This differential effect of Vps34 on TOR signaling might simply be due to fundamental differences in the regulation of TOR signaling in lower organisms versus higher organisms. This notion is also supported by Lizcano *et al.*, who found that AA-induced TOR signaling is not sensitive to wortmannin in *Drosophila* [36], whereas it is in mammalian cell culture [29]. This observation indicates that nutrient-regulated TOR signaling in *Drosophila* does not employ a wortmannin-sensitive PI3K-dependent event, while in mammalian cells mTORC1 signaling is under the control of a wortmannin-sensitive hVps34.

VI.   Conclusions and Future Perspectives

In the last decade, there has been increasing interest in the causative role of excessive nutrient intake in the development of metabolic syndromes such as obesity and diabetes. There have been a number of studies linking high-protein diets [37–39] and elevated levels of AAs in plasma to the development of insulin resistance [40, 41]. Recently, it has been shown by Newgard *et al.* that, in human subjects, high consumption of BCAAs cooperates in the development of insulin resistance [42]. In parallel, it has been shown that an approximate twofold rise in human serum plasma AA concentrations is accompanied by an approximate 25% reduction in insulin sensitivity in human subjects [43]. Also Um *et al.* in mouse model [44] and followed by Tremblay *et al.* in the skeleton muscle [45] demonstrated that hyperactivation of S6K1 leads to the direct phosphorylation of insulin receptor substrate 1 (IRS1) S1097, leading to the inhibition of PI3K signaling and thereby impaired insulin signaling. Moreover, recent studies have shown that S6K1 can directly phosphorylate mTORC2 leading to inhibition of PKB/Akt signaling, which could then result in the development of insulin resistance [46].

The potential link between nutrient intake and tumorigenesis was first suggested by Holley, who proposed that neoplastic disease arises from membrane changes underlying the increased transport of nutrients [47]. Consistent with this model, Saier *et al.* have shown that AA transporters are one of the key targets of oncogenes and that they are involved in oncogene-mediated transformation of epithelial cells [48]. There have been a number of reports indicating cross talk between AA transporters and mTORC1 leading to cancer development. For instance, it has been shown that mTORC1 regulates the expression and localization of AA transporters, like L-type amino acid transporters 1 (LAT1) and system ASC amino acid transporter 2 (ASTC2), which have been shown to be upregulated in many cancers [49]. Furthermore, Fuchs *et al.* showed that inhibition of ASTC2 function by antisense RNA suppressed mTORC1 signaling [50]. Still, the molecular events involved in this cross talk between AA transporters and mTORC1 are not entirely clear. Elucidation of key AA transporters regulating mTORC1 signaling and further analysis of the feedback regulation involved would be helpful in gaining a better understanding of this complex web of nutrient-mediated mTORC1 signaling.

AAs are emerging (s) as dominant signaling molecules, in addition to their firmly established role as the building blocks in protein synthesis [51]. Studies performed in our laboratory, and by others, have provided increased knowledge of this new aspect of AA function. The studies on AA-mediated regulation of mTORC1 signaling described here help in providing a mechanistic explanation of how elevated levels of AAs lead to insulin resistance. Our data suggest that AA-induced Ca^{2+} and PI(3)P play important roles in the generation of a signaling complex, which could then ultimately give rise to an active signalosome [52]. The available evidence suggests a key role for Ca^{2+}/CaM and hVps34 in activating mTORC1 signaling (Figure 6.1), but there are still many questions that need to be addressed to gain a clear understanding of how nutrients regulate mTORC1 signaling. For example, it is still not known how AAs induce a Ca^{2+} signal, or which protein(s) bind to PI(3)P to serve as a signaling platform for the active mTORC1 signalosome. In addition, Findlay *et al.* reported that MAP4K3, which is under the control of AAs, mediates mTORC1 signaling [53]. It is not yet clear how AAs activate MAP4 kinase or whether MAP4K3 and RAGs lie upstream or downstream of hVps34 (Figure 6.1). More extensive studies are required to answer these questions, and this will serve in providing a better understanding of how nutrient-regulated mTORC1 signaling contributes to pathophysiological conditions like obesity, diabetes, and cancer.

ACKNOWLEDGMENTS

We are indebted to the members of the Kozma/Thomas laboratory for discussions, M. Daston for editing of the manuscript and G. Doerman for assisting us in the computer graphics. G. T. is supported by the NIH Mouse Models for Human Cancer Consortium, U01 CA84292-06, NIH NIDDK grants DK73802 and DK078019 and the Strauss Chair in Cancer Research.

REFERENCES

1. Herman, P.K., Stack, J.H., DeModena, J.A., and Emr, S.D. (1991). A novel protein kinase homolog essential for protein sorting to the yeast lysosome-like vacuole. *Cell* 64:425–437.
2. Um, S.H., D'Alessio, D., and Thomas, G. (2006). Nutrient overload, insulin resistance, and ribosomal protein S6 kinase 1, S6K1. *Cell Metab* 3:393–402.
3. Dann, S.G., Selvaraj, A., and Thomas, G. (2007). mTOR complex1–S6K1 signaling: at the crossroads of obesity, diabetes and cancer. *Trends Mol Med*.
4. Kim, D.H., *et al.* (2002). mTOR interacts with raptor to form a nutrient-sensitive complex that signals to the cell growth machinery. *Cell* 110:163–175.
5. Sancak, Y., *et al.* (2007). PRAS40 is an insulin-regulated inhibitor of the mTORC1 protein kinase. *Mol Cell* 25:903–915.
6. Sarbassov, D.D., *et al.* (2004). Rictor, a novel binding partner of mTOR, defines a rapamycin-insensitive and raptor-independent pathway that regulates the cytoskeleton. *Curr Biol* 14:1296–1302.
7. Frias, M.A., *et al.* (2006). mSin1 is necessary for Akt/PKB phosphorylation, and its isoforms define three distinct mTORC2s. *Curr Biol* 16:1865–1870.
8. Sarbassov, D.D., Guertin, D.A., Ali, S.M., and Sabatini, D.M. (2005). Phosphorylation and regulation of Akt/PKB by the rictor-mTOR complex. *Science* 307:1098–1101.
9. Yan, L., Mieulet, V., and Lamb, R.F. (2008). mTORC2 is the hydrophobic motif kinase for SGK1. *Biochem J* 416:e19–e21.
10. Ikenoue, T., Inoki, K., Yang, Q., Zhou, X., and Guan, K.L. (2008). Essential function of TORC2 in PKC and Akt turn motif phosphorylation, maturation and signalling. *EMBO J* 27:1919–1931.
11. Nobukuni, T., Kozma, S.C., and Thomas, G. (2007). hvps34, an ancient player, enters a growing game: mTOR Complex1/S6K1 signaling. *Curr Opin Cell Biol* 19:135–141.
12. Boisvert, F.M., van Koningsbruggen, S., Navascues, J., and Lamond, A.I. (2007). The multifunctional nucleolus. *Nat Rev Mol Cell Biol* 8:574–585.
13. Rudra, D., and Warner, J.R. (2004). What better measure than ribosome synthesis? *Genes Dev* 18:2431–2436.
14. Jones, R.G., and Thompson, C.B. (2009). Tumor suppressors and cell metabolism: a recipe for cancer growth. *Genes Dev* 23:537–548.
15. Klionsky, D.J., and Emr, S.D. (2000). Autophagy as a regulated pathway of cellular degradation. *Science* 290:1717–1721.

16. Shigemitsu, K., Tsujishita, Y., Hara, K., Nanahoshi, M., Avruch, J., and Yonezawa, K. (1999). Regulation of translational effectors by amino acid and mammalian target of rapamycin signaling pathways. Possible involvement of autophagy in cultured hepatoma cells. *J Biol Chem* 274:1058–1065.

17. Scott, R.C., Schuldiner, O., and Neufeld, T.P. (2004). Role and regulation of starvation-induced autophagy in the *Drosophila* fat body. *Dev Cell* 7:167–178.

18. Wahren, J., Felig, P., and Hagenfeldt, L. (1976). Effect of protein ingestion on splanchnic and leg metabolism in normal man and in patients with diabetes mellitus. *J Clin Invest* 57:987–999.

19. Hutson, S.M., Sweatt, A.J., and Lanoue, K.F. (2005). Branched-chain [corrected] amino acid metabolism: implications for establishing safe intakes. *J Nutr* 135:1557S–1564S.

20. Gulati, P., and Thomas, G. (2007). Nutrient sensing in the mTOR/S6K1 signalling pathway. *Biochem Soc Trans* 35:236–238.

21. Christensen, H.N. (1990). Role of amino acid transport and countertransport in nutrition and metabolism. *Physiol Rev* 70:43–77.

22. Christie, G.R., Hajduch, E., Hundal, H.S., Proud, C.G., and Taylor, P.M. (2002). Intracellular sensing of amino acids in Xenopus laevis oocytes stimulates p70 S6 kinase in a target of rapamycin-dependent manner. *J Biol Chem* 277:9952–9957.

23. Nicklin, P., *et al.* (2009). Bidirectional transport of amino acids regulates mTOR and autophagy. *Cell* 136:521–534.

24. Harris, R.A., Joshi, M., and Jeoung, N.H. (2004). Mechanisms responsible for regulation of branched-chain amino acid catabolism. *Biochem Biophys Res Commun* 313:391–396.

25. Lynch, C.J., *et al.* (2003). Potential role of leucine metabolism in the leucine-signaling pathway involving mTOR. *Am J Physiol Endocrinol Metab* 285:E854–E863.

26. Miotto, G., Venerando, R., Marin, O., Siliprandi, N., and Mortimore, G.E. (1994). Inhibition of macroautophagy and proteolysis in the isolated rat hepatocyte by a non-transportable derivative of the multiple antigen peptide Leu8-Lys4-Lys2-Lys-beta Ala. *J Biol Chem* 269:25348–25353.

27. Conus, N.M., Hemmings, B.A., and Pearson, R.B. (1998). Differential regulation by calcium reveals distinct signaling requirements for the activation of Akt and p70S6k. *J Biol Chem* 273:4776–4782.

28. Hannan, K.M., Thomas, G., and Pearson, R.B. (2003). Activation of S6K1 (p70 ribosomal protein S6 kinase 1) requires an initial calcium-dependent priming event involving formation of a high-molecular-mass signalling complex. *Biochem J* 370:469–477.

29. Nobukuni, T., *et al.* (2005). Amino acids mediate mTOR/raptor signaling through activation of class 3 phosphatidylinositol 3OH-kinase. *Proc Natl Acad Sci USA* 102:14238–14243.

30. Gulati, P., *et al.* (2008). Amino acids activate mTOR complex 1 via Ca2+/CaM signaling to hVps34. *Cell Metab* 7:456–465.

31. Sancak, Y., *et al.* (2008). The Rag GTPases bind raptor and mediate amino acid signaling to mTORC1. *Science* 320:1496–1501.

32. Binda, M., *et al.* (2009). The Vam6 GEF controls TORC1 by activating the EGO complex. *Mol Cell* 35:563–573.

33. Byfield, M.P., Murray, J.T., and Backer, J.M. (2005). hVps34 Is a nutrient-regulated lipid kinase required for activation of p70 S6 kinase. *J Biol Chem* 280:33076–33082.

34. Vergne, I., Chua, J., and Deretic, V. (2003). Tuberculosis toxin blocking phagosome maturation inhibits a novel Ca2+/calmodulin-PI3K hVPS34 cascade. *J Exp Med* 198:653–659.

35. Juhasz, G., *et al.* (2008). The class III PI(3)K Vps34 promotes autophagy and endocytosis but not TOR signaling in *Drosophila*. *J Cell Biol* 181:655–666.

36. Lizcano, J.M., Alrubaie, S., Kieloch, A., Deak, M., Leevers, S.J., and Alessi, D.R. (2003). Insulin-induced *Drosophila* S6 kinase activation requires phosphoinositide 3-kinase and protein kinase B. *Biochem J* 374:297–306.

37. Linn, T., Geyer, R., Prassek, S., and Laube, H. (1996). Effect of dietary protein intake on insulin secretion and glucose metabolism in insulin-dependent diabetes mellitus. *J Clin Endocrinol Metab* 81:3938–3943.

38. Linn, T., *et al.* (2000). Effect of long-term dietary protein intake on glucose metabolism in humans. *Diabetologia* 43:1257–1265.

39. Schulze, M.B., Manson, J.E., Willett, W.C., and Hu, F.B. (2003). Processed meat intake and incidence of Type 2 diabetes in younger and middle-aged women. *Diabetologia* 46:1465–1473.

40. Felig, P., Marliss, E., and Cahill, G.F., Jr. (1969). Plasma amino acid levels and insulin secretion in obesity. *N Engl J Med* 281:811–816.

41. Felig, P., Marliss, E., and Cahill, G.F., Jr. (1970). Are plasma amino acid levels elevated in obesity? *N Engl J Med* 282:166.

42. Newgard, C.B., *et al.* (2009). A branched-chain amino acid-related metabolic signature that differentiates obese and lean humans and contributes to insulin resistance, *in press*.

43. Krebs, M., *et al.* (2002). Mechanism of amino acid-induced skeletal muscle insulin resistance in humans. *Diabetes* 51:599–605.

44. Um, S.H., *et al.* (2004). Absence of S6K1 protects against age- and diet-induced obesity while enhancing insulin sensitivity. *Nature* 431:200–205.

45. Tremblay, F., *et al.* (2007). Identification of IRS-1 Ser-1101 as a target of S6K1 in nutrient- and obesity-induced insulin resistance. *Proc Natl Acad Sci USA* 104:14056–14061.

46. Dibble, C.C., Asara, J.M., and Manning, B.D. (2009). Characterization of Rictor phosphorylation sites reveals direct regulation of mTOR complex 2 by S6K1. *Mol Cell Biol.*

47. Holley, R.W. (1972). A unifying hypothesis concerning the nature of malignant growth. *Proc Natl Acad Sci USA* 69:2840–2841.

48. Saier, M.H., Jr., Daniels, G.A., Boerner, P., and Lin, J. (1988). Neutral amino acid transport systems in animal cells: potential targets of oncogene action and regulators of cellular growth. *J Membr Biol* 104:1–20.

49. Fuchs, B.C., and Bode, B.P. (2005). Amino acid transporters ASCT2 and LAT1 in cancer: partners in crime? *Semin Cancer Biol* 15:254–266.

50. Fuchs, B.C., Finger, R.E., Onan, M.C., and Bode, B.P. (2007). ASCT2 silencing regulates mammalian target-of-rapamycin growth and survival signaling in human hepatoma cells. *Am J Physiol Cell Physiol* 293:C55–C63.

51. Patti, M.E. (1999). Nutrient modulation of cellular insulin action. *Ann NY Acad Sci* 892:187–203.

52. Bai, X., *et al.* (2007). Rheb activates mTOR by antagonizing its endogenous inhibitor, FKBP38. *Science* 318:977–980.

53. Findlay, G.M., Yan, L., Procter, J., Mieulet, V., and Lamb, R.F. (2007). A MAP4 kinase related to Ste20 is a nutrient-sensitive regulator of mTOR signalling. *Biochem J* 403:13–20.

7

AGC Kinases in mTOR Signaling

ESTELA JACINTO

Department of Physiology and Biophysics
UMDNJ-Robert Wood Johnson Medical School, Piscataway
New Jersey, USA

I. Abstract

The mammalian target of rapamycin (mTOR), a protein kinase with homology to lipid kinases, orchestrates cellular responses to growth and stress signals. Various extracellular and intracellular inputs to mTOR are known. mTOR processes these inputs as part of two mTOR protein complexes, mTORC1 or mTORC2. Surprisingly, despite the many cellular functions that are linked to mTOR, there are very few direct mTOR substrates identified to date. With the recent discovery of mTORC2, mounting evidence point to mTOR as a central regulator of members of a family of protein kinases, the AGC (protein kinases A/PKG/PKC) family. The AGC kinases are one of the most well-characterized among the eukaryotic protein kinase family. A multitude of cellular functions and substrates has been ascribed to these kinases and their deregulation underlies numerous pathological conditions. mTOR phosphorylates common motifs in a number of these AGC kinases that could lead to their allosteric activation. AGC kinase activation triggers the phosphorylation of diverse targets that ultimately control cellular response to stimuli. This review will focus on the recent findings on how mTOR regulates AGC kinases. I will discuss how these kinases are wired to the mTOR signaling circuit, including examples of mTOR-dependent outputs consequent to AGC kinase activation.

ISSN NO: 1874-6047
DOI: 10.1016/S1874-6047(10)27007-5

II.   Introduction

Mammalian target of rapamycin (mTOR) is linked to diverse cellular and physiological functions that ultimately control cell and body growth [1, 2]. In the whole organism, mTOR plays a role in development, metabolism, memory, and aging. At the cellular level, mTOR responds to the presence of nutrients and other growth cues. It functions to regulate translation initiation, ribosome biogenesis, autophagy, actin cytoskeleton reorganization, and transcription. Deregulation of the mTOR signaling pathway leads to pathological conditions including cancer, immune-related diseases, diabetes, cardiovascular, and neurological disorders. mTOR is part of a cellular signaling network and it integrates multiple intracellular signals. But what are its direct targets and what exactly does it do to its targets?

Understanding TOR/mTOR signaling emerged with the use of rapamycin, a macrolide that binds and allosterically inhibits mTOR in the presence of the immunophilin FKBP12 [3]. Although more recent studies revealed that not all functions of mTOR are inhibited by rapamycin, this drug has been instrumental in defining the downstream targets and functions of mTOR. The new generation of mTOR active site inhibitors are also unraveling important clues on both the rapamycin sensitive and insensitive functions of mTOR [4]. But perhaps genetic studies that have disrupted TOR/mTOR functions are providing key pieces in assembling the mTOR signaling puzzle [2, 5]. Recently, the availability of tissue-specific knockouts and cell lines that genetically disrupt mTOR complex components have confirmed the involvement of mTORCs in regulating AGC kinases and have also led to the identification of new mTOR downstream targets. Whether most of the cellular functions of mTOR are due to its direct regulation of AGC kinases remains to be elucidated. This review will discuss the accumulating evidence that reveal how mTOR and AGC kinases coregulate each other. Although there are other kinase families, such as the CMGC and CAMK families [6], that feed signals to the mTOR pathway, so far only the AGC kinases have been shown to be direct mTOR targets as well. mTOR promotes the optimal activation of several AGC kinases and in turn the AGC kinases not only mediate mTOR signals via phosphorylation of numerous intracellular targets that lead to a cellular response but also fine tune the response via feedback regulation of upstream mTOR signals.

III.   mTOR, an Atypical Protein Kinase

mTOR is a member of a family of protein kinases termed the PIKKs (phosphatidylinositol-3-kinase-related kinases), a subgroup of the atypical protein kinases [6]. Members of this family share homology with lipid kinases but possess protein kinase activity. The kinase domain is located near the C-terminus and is flanked by two regions that are common to PIKKs called the FAT (FRAP, ATM, TRRAP) and FATC (FAT C-terminus) domains. Since these domains flank the kinase region, it is believed that they participate in kinase regulation [7]. The N-terminal region consists of HEAT (Huntingtin, Elongation factor 3, A subunit of protein phosphatase 2A, TOR1) repeats, most likely a region that mediates protein protein interactions. PIKKs form protein complexes and seem to recognize their targets via their binding partners [8]. From yeast to man, TOR forms two distinct protein complexes. Mammalian TOR complex 1 (mTORC1) is composed of mTOR and its conserved partners, raptor, and mLST8. mTORC2 consists of mTOR, rictor, SIN1, and mLST8 [1]. While there is likely extensive signal cross talk between the two mTOR complexes, they perform distinct functions.

Despite the myriad cellular roles of mTOR, little is known on its direct substrates. S6K (S6 kinase), a protein kinase belonging to the AGC kinase family was first to be identified as a rapamycin-sensitive target of mTOR [9, 10]. Another *in vitro* and *in vivo* substrate of mTOR is 4EBP, which is not an enzyme but a small regulatory protein that is highly phosphorylated. S6K and 4EBP phosphorylation is inhibited by rapamycin and both proteins function to regulate translation initiation. Three other mTOR substrates were recently identified, namely Akt/PKB, PKC, and SGK1. These substrates are also members of the AGC kinase family.

Most PIKKs show preference for the phosphorylation of Ser/Thr-Gln residues [11] but mTOR does not appear to phosphorylate such motif. Instead, it phosphorylates Ser/Thr- residues followed by -Tyr/Phe (bulky hydrophobic residue) or -Pro (Table 7.1). The 4EBP sites phosphorylated by mTOR *in vitro* are Ser/Thr-Pro sites. Because these recognition motifs are common for a number of kinases, particularly *in vitro*, other factors such

TABLE 7.1

Substrates and Sequence Motifs Phosphorylated by mTOR Complexes

mTORC1 substrates		mTORC2 substrates	
S6K1	S^{371}PDD	Akt/PKB	T^{450}PPD
	FLGFT^{389}YVAPS		FPQFS^{473}YSASG
	EKFS^{401}FEP		
	IRS^{411}PRRFIGS^{418}PRT^{421}PVS^{424}PVK	cPKCα	T^{638}PPD
			FEGFS^{657}YVNPQ
4E-BP1	DYSTT^{37}PGG	SGK1	S^{397}IGKS^{401}PD (nd)
			(not determined)
	FSTT^{46}PGG		FLGFS^{422}YAPP
	CRNS^{65}PVTKT^{70}PPR		

(nd) = not determined

as docking interactions, cellular localization, and/or scaffolding proteins could contribute to the substrate specificity of mTOR.

IV.   AGC Kinase, the "Prototype" of Protein Kinases

The AGC kinase family belongs to the conventional ePKs (eukaryotic protein kinases). The 63 members of this family in the human genome share high homology, roughly 40% sequence identity, in their kinase domain [12]. They phosphorylate their targets on Serine or Threonine residues and include some of the most familiar regulatory enzymes that have key roles on a number of important intracellular signaling pathways. The prototypical kinase structure was first resolved for an AGC kinase, the cAMP-dependent protein kinase (PKA) [13]. The X-ray structure of PKA revealed a bilobal architecture that turned out to be common to all protein kinases. Activation of AGC kinases, like other protein kinases, involves conformational changes in the key regulatory C-helix in the N-lobe [14]. Phosphorylation of the activation loop (T-loop) either by itself or by another AGC kinase, PDK1, stabilizes the kinase in a conformation that is optimal for substrate binding. In AGC kinases, repositioning of the C-helix is mediated by its carboxyl-terminal tail (C-tail) [15]. The C-tail plays a critical regulatory role and distinguishes the AGC kinases from other ePKs. It is defined by three segments: N-lobe tether (NLT) that includes the hydrophobic motif (HM), C-lobe tether (CLT) that interacts with the C-lobe and interlobe linker spanning the N and C lobes, and the active-site tether (AST) that interacts with the ATP binding pocket [15]. The C-tail serves as a docking site for *trans*-acting cellular components, in addition to its *cis*-acting function, and is thereby essential for activity and allostery.

The HM plays a pivotal role in regulating AGC kinase activity. Based on the PKA structure, the HM folds back into a hydrophobic pocket in the N-lobe. The HM is present in most AGC kinases (with the notable exception of PDK1) and contains a Ser/Thr residue that may or may not be phosphorylated. The phosphorylation of this conserved residue could also be regulated in some of the family members. Alternatively, in some AGC kinases, a phosphomimetic residue is in place. In the absence of a phosphorylatable residue or if the HM is truncated, the AGC kinase adopts other mechanisms for additional regulation [12]. In the inactive state, the HM serves as the docking site for PDK1, which promotes its phosphorylation of the AGC kinase activation loop and consequently activates the AGC kinase.

Another conserved motif in the C-tail that precedes the HM is the turn motif (TM). One or more phosphorylated or phosphomimetic residues are nestled in this motif in most of the AGC kinases [16]. The phosphorylated residue sits at the apex of a tight turn in PKA, hence the name. The structures of PKA and PKC*i* hint that the TM contributes to stabilizing the active conformation by anchoring the C-terminus to the top of the kinase domain [17]. Molecular modeling, genetic, and biochemical studies of other AGC kinases confirm that the TM phosphate promotes association between the tail and kinase domain and that this regulation is shared by a number of the AGC kinase members [18, 19].

A conserved sequence motif (PXXP; where P is Pro and X is any amino acid) that is not phosphorylated lies within the CLT [15]. The exact function of this motif remains to be determined but initial studies show that it acts as an SH3 binding site in Akt [20]. In PKC, sequences adjacent to this motif are important for binding Hsp90, which plays a role in the maturation of the protein kinase [21]. Given the high conservation of this motif and Hsp90 binding to several AGC kinases, this motif along with Hsp90 could play a role in kinase maturation and/or prolongation of kinase activity.

Because the kinase domain displays high homology between members of this family, there is significant overlapping substrate specificity. In Akt, the minimal substrate recognition sequence was determined to be RXRXXS/TB where X is any amino acid and B is a bulky hydrophobic residue [22] (Table 7.2). A similar motif is recognized by SGK, S6K, and Rsk. In PKC, the consensus motif was determined earlier to be RXXS/TXRX but a more detailed analysis has revealed more specific sequence preferences among different PKC isozymes [23].

The AGC kinases display divergent sequences outside of the catalytic domain and C-tail, hence adding another layer of specificity in their regulation and function. They could be regulated at different levels such as by transcription, posttranslational modification, subcellular localization, protein stability, and protein–protein interaction. This review will focus mainly on regulation of AGC kinases by phosphorylation and how such phosphorylation affects their function.

V.   Phosphorylation of AGC Kinases by mTOR

mTOR can phosphorylate Ser/Thr residues in the AGC kinase conserved motifs, the TM and HM. mTOR is also implicated in the phosphorylation of critical sites at the autoinhibitory domain of S6K. By phosphorylating AGC kinases at these C-tail sites, mTOR functions in the allosteric activation of these kinases. A more comprehensive review on how mTOR regulates AGC kinases was recently published [24], hence this current review will only summarize some of the critical studies and additionally discuss more recent findings.

TABLE 7.2

mTOR Signaling Molecules and Their Phosphorylation Sites Targeted by AGC Kinases

mTORC components			
mTOR	RSRTRT^{2446}DS^{2448}YS	Raptor	RLRS^{719}VS^{721}S^{722}YG
	PESIHS^{2481}FIGDG (*)	Rictor	RIRTLT^{1135}EPSVD
	KKLHVS^{1261}TINLQ (**)		
mTORC regulators			
TSC2	RARSTS^{939}LNE	IRS1 (mouse)	RKPKS^{24}MHK
	RCRSIS^{981}VSE		RPRSKS^{265}QS^{267}SS
	RDRVRS^{1130}MS^{1132}GGH		RSRTES^{302}ITATS^{307}PAS
	RPRGYT^{1462}ISD		GGKPGS^{318}FRVR
	RKRLIS^{1798}SVE		MSRPAS^{336}VDGS
	RVVS^{1364}SEGGRP		RHRGS^{357}SRL
PRAS40	RPRLNT^{246}SDF		GYMPMS^{612}PGVA
			GDYMPMS^{632}PKSV
		(H.s)	RRRHSS^{1101}ETF
AGC kinase substrates			
eEF2k	RVRTLS^{366}GSR	FOXO1	RPRSCT^{24}WPL
PDCD4	RLRKNS^{67}SRDS^{71}GRGDS^{76}VS	GSK3α	RARTSS^{21}FAE
S6	RRRLS^{235}S^{236}LR	p27 Kip1	RKRPAT^{157}D
	RRRLS^{235}S^{236}LRAS^{240}TSKS^{244}E		KPGLRRRQT^{198}
eIF4B	RSRTGS^{422}ESS		
eIF6	DELSS^{174}LLQP		

* autophosphorylation
**not determined

A. S6K

Ribosomal S6 kinase (S6K) was the first AGC kinase to be identified as a substrate of mTOR [24]. There are two homologues of S6K in mammals, S6K1 and S6K2, distinguished by divergent sequences in the N- and C-terminal regulatory regions. S6K contains a TOR signaling (TOS) motif in the N-terminus, a conserved five amino acid sequence found in mTORC1 substrates and is crucial for mTOR regulation [25]. Truncation of the N-terminus including the TOS motif prevents kinase activation of S6K1. The kinase domain is followed by the linker domain that includes the TM and HM. At the end of the C-tail is an autoinhibitory pseudosubstrate domain that is unique to S6K among AGC kinases. S6K is phosphorylated at Thr389 of the HM and Ser404 of the linker domain [26]. Both sites are highly sensitive to rapamycin treatment and HM site phosphorylation was defective in cells where mTORC1 components are knocked down [27]. *In vivo*, the adipose tissue-specific knockout of raptor also displayed defective Thr389 phosphorylation and shared similar phenotype with whole-body knockout of

S6K1 [28]. In a muscle-specific knockout of raptor, the phosphorylation of S6K was not reported although the activation of S6K is presumably defective due to decreased S6 and 4E-BP1 phosphorylation [29]. Furthermore, there was increased IRS-1 protein levels, consistent with a defective S6K-mediated feedback loop [30]. These *in vivo* findings highlight the importance of mTORC1 in S6K regulation.

S6K is also phosphorylated at seven other residues with a Ser/Thr-Pro motif. Ser371 at the TM is mitogen-inducible and is required for kinase activity along with HM and T-loop phosphorylation [31]. The TM is phosphorylated by mTOR *in vitro* [31]. There are conflicting reports as to whether this phosphorylation is rapamycin-sensitive [31–34]. Whereas the loss of the mTOR upstream regulators TSC1 or TSC2 led to hyperphosphorylation of Thr389 and the T-loop site Thr229, the phosphorylation of Ser371 was not significantly altered [33]. These findings suggest that the HM and TM sites in S6K may be differentially regulated. The increased HM and T-loop phosphorylation in the TSC-null cells are also consistent with the notion that increased phosphorylation at the HM site (presumably due to increased mTORC1 activity) increases the efficiency with which PDK1 can phosphorylate the T-loop [61].

Four phosphorylation sites (Ser411, Ser418, Thr421, and Ser424) situated in the autoinhibitory domain contain a Pro residue at the $+1$ position and a hydrophobic residue at the -2 position [26]. These sites are hypophosphorylated in quiescent cells but undergo hyperphosphorylation upon serum stimulation. Loss of either TSC1 or TSC2 increased basal phosphorylation of Ser411 but not the adjacent sites Ser421/Ser424 [33]. Because these phosphorylation sites are rapamycin sensitive, they are presumably phosphorylated by mTORC1. Both PDK1 and NEK6/7 can also promote phosphorylation of Ser411 (Ser412) [35, 36]. Mutation of the autoinhibitory sites to Ala suppresses S6K activation. When the autoinhibitory region is deleted or when all four residues are mutated to acidic residues, the effect of N-terminal truncation of S6K1 is reversed and leads to kinase activation [37–39]. This active mutant allele of S6K1 becomes rapamycin-insensitive because HM phosphorylation at Thr389 is not abolished in the presence of the drug. Instead, this mutant allele has been shown to be dependent on mTORC2 for phosphorylation [40]. It is interesting to note that this autoinhibitory domain is unique to S6K. When devoid of this domain as in other AGC kinases, mTORC2 can fulfill the function of mediating HM site phosphorylation.

Although it is quite established that the HM site of S6K is regulated by mTORC1, other kinases can phosphorylate this motif including PDK1 and S6K itself [41, 42] (Table 7.3). As discussed below for Akt and PKC, the HM site can be autophosphorylated or *trans*-phosphorylated by other

TABLE 7.3

AGC Kinases C-Tail Phosphorylation Sites and Implicated Kinases

	Phosphosite	mTOR complex/condition or evidence	Other kinases/condition or evidence
S6K1	Ser371 (TM)	mTORC1 (?)/ Mitogens;insulin	
	Thr389 (HM)	mTORC1/ Amino acids/growth factors	S6K1/coexpression of PDK1 and PKCς PDK1/PIF-bound PDK1
		mTORC2/ Truncated S6K1	
	Ser404 (linker)	mTORC1/ Amino acids/growth factors	
	Ser411 (autoinhib)	mTORC1/ Serum	NEK6/7/in vitro and overexpression PDK1/in vitro and overexpression
	Ser418 (autoinhib)	mTORC1/ Serum	
	Thr421, Ser424 (autoinhib)	mTORC1/ Serum/TSC1 or TSC2 loss	
Akt1	Thr450 (TM)	mTORC2/ Constitutive; mTORC2 disruption	
	Ser473 (HM)	mTORC2/ Growth factors; mTORC2 disruption	DNA-PK/DNA damage Akt/kinase-inactive Akt
			ILK/ILK conditional knockout PKCβ/PKCβ$^{-/-}$ mast cells ATM/gamma irradiation, siRNA
cPKCα	Ser638 (TM)	mTORC2/ Constitutive; mTORC2 disruption	cPKC/mutagenesis
	Ser657 (HM)	mTORC2/ Constitutive; mTORC2 disruption	cPKC/mutagenesis
SGK1	Ser397 (TM)	?	Akt/overexpression
	Ser401 (TM)	?	
	Ser422 (HM)	mTORC2/ Serum; mTORC2 disruption mTORC1/ Rapamycin sensitivity	

kinases. The exact context for this phosphorylation versus mTOR-mediated HM phosphorylation is still unresolved.

B. Akt/PKB

Akt/PKB (protein kinase B) is phosphorylated and activated upon its translocation to the membrane where its pleckstrin homology (PH) domain interacts with the lipid products of the phosphatidylinositol 3 (PI3)-kinase (PI3K), PtdIns $(3,4,5)P_3$, and $PI(3,4)P_2$. Akt is primarily activated by phosphorylation of the T-loop at Thr308 by PDK1. Using live cell imaging, it was shown that a conformational change in Akt is critically dependent on phosphorylated Thr308. Furthermore, a phosphomimetic at this site showed diminished membrane association, suggesting a role for promoting dissociation of activated Akt from the membrane [43]. However, phosphorylation at this site is insufficient to allow the reorganization of the catalytic site [44]. HM site (Ser473) phosphorylation, which increases the binding affinity of the HM to the hydrophobic groove, is important for optimal Akt kinase activity. Akt is phosphorylated by mTORC2 at two conserved motifs, the HM and TM [19, 45, 46]. In mTORC2-disrupted cells such as $SIN1^{-/-}$, $rictor^{-/-}$, or $mLST8^{-/-}$ murine embryonic fibroblasts (MEFs), Akt HM phosphorylation was abrogated [47–49]. In mice with adipose tissue-specific knockout of rictor, Akt HM phosphorylation was also ablated [50]. Akt remained phosphorylated at the T-loop by PDK1 further supporting that HM phosphorylation of Akt is not a prerequisite for T-loop phosphorylation in this particular AGC kinase [51]. Both *in vitro* and *in vivo*, Akt from mTORC2-disrupted cells possessed suboptimal kinase activity and only a subset of its substrates such as FOXO1/3a was defectively phosphorylated [47, 49].

Studies using mTORC2-disrupted cells also led to the discovery that mTORC2 controls TM site phosphorylation of Akt and cPKC. The TM was undetected in the early crystal structure of Akt2 [44] and its phosphorylation was demonstrated to be constitutive [52], hence its significance was unclear. The cooperative action of the HM and the TM phosphates was recently reported for several AGC kinases including Akt [18]. These studies unveiled that the TM phosphorylation promotes a zipper-like association of the tail with the kinase domain that stabilizes the HM in its kinase-activating binding site. The stabilizing function of the TM phosphorylation at Thr450 of Akt is further demonstrated *in vivo* using mTORC2-disrupted cells. In $SIN1^{-/-}$ or $rictor^{-/-}$ MEFs, Akt TM phosphorylation is completely abolished [19, 46]. A slight diminution of Akt expression was observed in these cells but Akt levels were rapidly down-regulated upon inhibition of the folding chaperone Hsp90. Specific phosphorylation of the TM but not the HM site facilitates carboxyl-terminal folding and stabilizes newly synthesized Akt. Phosphorylation of the TM is also independent of HM phosphorylation and occurs during or shortly after Akt translation. When a myristylated form of Akt in mTORC2-disrupted cells was expressed, TM phosphorylation remained absent, whereas HM phosphorylation is partially restored [19]. These findings indicate that the TM and HM sites of Akt are controlled by mTORC2 under different conditions. Indeed, whereas HM site phosphorylation can be abolished using mTOR active site inhibitors, under similar conditions the TM site phosphorylation was not diminished [53]. Where and how TM site phosphorylation occurs may provide clues as to how mTORC2 can have distinct functions in Akt regulation.

Other kinases have also been proposed to directly phosphorylate the HM and TM sites of Akt. DNA-PK mediated the HM site phosphorylation of Akt in stress situations such as DNA damage but is not required for growth factor- or insulin-induced phosphorylation at this site [54, 55]. In the muscle specific knockout of raptor, Akt Ser473 phosphorylation was upregulated and remained elevated despite co-deletion of rictor [29]. These results support the presence of compensatory mechanisms, such as other Ser473 kinases, that could phosphorylate Akt at this site when mTORC2 is disrupted. Because the Thr450 of the TM is followed by a Proline, it was speculated that a Pro-directed kinase may target this site [5]. It is conceivable that both sites, HM and TM, are phosphorylated by other protein kinases under different cellular contexts or conditions. Nevertheless, based on studies on MEFs with mTORC2 disruption, mTORC2 plays a predominant role in controlling phosphorylation of these sites in Akt under normal growing conditions [19, 46–49].

C. PKC

There are three classes of PKC (protein kinase C) in mammals: the conventional type (cPKC) that includes α, $\beta1$, βII, and γ; the novel PKC (nPKC) consisting of δ, ε, θ, and η/L; and the atypical PKC (aPKC) comprising of ζ and ι/λ [16, 56]. Like other AGC kinases, the T-loop is phosphorylated by PDK1 but this phosphorylation appears to be constitutive and does not require phosphoinositides [57]. The HM and TM site phosphorylation of cPKC and nPKC isoforms are considered to have a priming role but whether they occur via autophosphorylation or by trans-phosphorylation by another kinase is controversial [16, 56]. Among the PKC isotypes, the involvement of mTORC for phosphorylation at the conserved HM and/or TM has been demonstrated for the cPKCs and may also include nPKCε [19, 46, 49, 58]. Phosphorylation of the HM and

TM sites of cPKCs are also constitutive and insensitive to rapamycin. Genetic studies using rictor-, SIN1-, or mLST8-knockout cells revealed that the HM site phosphorylation (Ser657 in cPKCα) was abolished and this was accompanied by a dramatic reduction in cPKCα protein expression [19, 46, 49]. Intriguingly, co-deletion of raptor and rictor in the muscle appeared to rescue the phosphorylation and protein levels of cPKCα that were defective in the muscle-specific rictor knockout [29]. Whether this could be due to the increased IRS-1 signaling that may promote an mTORC2-independent phosphorylation of cPKC remains to be determined.

The TM site phosphorylation was also abrogated in the mTORC2-deficient cells [19, 46]. Phosphorylation at the TM is necessary for subsequent HM site phosphorylation. *In vitro* studies have shown that TM phosphorylation could play a role in the stabilization of the protein [59]. Consistent with this, mutation of the TM site specifically increased the binding of cPKC to the folding chaperone Hsp70. This binding of Hsp70 to the TM-site PKC mutant stabilizes the protein and prolongs the signaling capacity of the kinase [60]. Loss of SIN1 or rictor led to severe attenuation of cPKC but not Akt protein levels [19, 46]. Furthermore, inhibition of Hsp90 led to rapid degradation of remaining cPKC. Why would the absence of TM site phosphorylation have a more dramatic effect on cPKC but not Akt expression levels? The presence of the priming site phosphorylation in PKC is required for kinase activation and subcellular localization [16, 56]. In contrast, PDK1 phosphorylation of the T-loop in Akt is less dependent on HM site docking of PDK1 [61]. Consistent with this interpretation, the T-loop phosphorylation of Akt is present whereas it is defective in cPKC in mTORC2-disrupted cells [19, 46]. Hence, despite the absence of both HM and TM site phosphorylation, a significant amount of Akt is stabilized by T-loop phosphorylation. It is also possible that cPKC expression may be attenuated at the level of transcription or translation in mTORC2-deficient cells.

PKC levels appear to be tightly controlled in the cell. The activation of many PKC isoforms is associated with increased degradation [62]. In cancer cells, elevated PKC levels have been observed. For example, increased PKCα expression in breast and ovarian cancer [63] and upregulated PKCβ in diffuse large B cell lymphoma have been reported [62]. In primary human glioblastoma tumors where EGFR becomes upregulated, increased PKCα but not Akt phosphorylation correlated with the abundant EGFR, suggesting that inhibiting PKC could be an important therapeutic target in malignant glioma [64]. It is not clear what led to increased PKCα phosphorylation although another study has shown that rictor levels and mTORC2 activity are elevated in glioma and glioblastoma cell lines [65]. Whether there is a correlation between mTORC2 component levels and cPKC expression remains to be determined.

D. SGK

The SGKs (serum- and glucocorticoid-induced kinase; SGK1, SGK2, SGK3) display about 55% identity in the kinase domain with Akt [66]. SGK1 is phosphorylated at the activation loop by PDK1 [67]. The HM site (Ser422) of SGK1 is phosphorylated by mTOR *in vitro* and *in vivo*. By over-expression of SGK1 in mTORC2-disrupted cells, the HM site phosphoryla-tion was found to be abolished, strongly suggesting that mTORC2 controls this phosphorylation site. This is supported by the finding that rapamycin does not abrogate phosphorylation of this site [68]. Phosphorylation of the previously identified SGK1 substrate, NDRG1, was also abolished in mTORC2-disrupted cells. In melanoma and MCF7 cells, HM phosphorylation is abolished upon rapamycin treatment implicating mTORC1 as the HM site regulator [69]. Activation of SGK1 correlated with p27 phosphorylation, which is also rapamycin sensitive. The discrepancy between the two studies may be explained by the use of different cell lines and methods of detecting SGK phosphorylation. p27 is also a target of Akt [70], which is not rapamycin sensitive. Notably, knockdown of rictor in melanoma cell lines appeared to more significantly diminish p27-Thr157 phosphorylation, in comparison to raptor knockdown wherein p27 phosphorylation was only minimally decreased and SGK1 phosphorylation was attenuated [69]. Another previous study has also found that the HM site phosphorylation is not rapamycin sensitive [71]. Whether SGK1 can indeed be targeted by mTORC1 under particular cellular conditions would need to be clarified.

SGK1/2 also contains conserved TM sites (Ser397 and Ser401 in SGK1). Phosphorylation of these sites is required for maximum SGK1 activity [72]. Ser401 is followed by a Pro and may putatively be an mTORC2 target based on similarity with the sites recognized in Akt and cPKC. It is also unclear if SGK1 protein expression could be attenuated in mTORC2-deficient cells. SGK expression levels are highly regulated in cells. Its N-terminus contains a hydrophobic degradation (HD)/PEST domain that mediates constitutive degradation via the ubiquitin–proteasome pathway [73, 74]. In recent stud-ies using mTORC2-disrupted cells, the phosphorylation state of SGK1 was assessed by overexpression of SGK1 with no apparent decrease in expres-sion of the exogenous SGK1 [68]. Similar to cPKC, the T-loop of SGK1 was also defectively phosphorylated, presumably due to the absence of HM site phosphorylation, which leads to inefficient PDK1 docking.

SGK1 forms a complex with Hsp90. Pharmacological inhibition of Hsp90 abolished phosphorylation of SGK1 [75]. Since the effect of Hsp90 inhibition on PDK1 levels occur at much longer time points than the acute dephos-phorylation of SGK1 [76, 77], it is likely that optimal SGK1 conformation is affected upon Hsp90 inhibition and prevents HM site phosphorylation.

VI. Phosphorylation of mTORCs by AGC Kinases

The phosphorylation of mTOR and its partners could serve as a mechanism to regulate mTORC activity [78]. Early studies have shown that mTOR is phosphorylated at three sites (Thr2446, Ser2448, Ser2481) (Figure 7.1). Thr2446 and Ser2448 are part of a regulatory repressor domain [79]. Increased phosphorylation at Ser2448, a site that conforms to the Akt/S6K recognition motif (Table 7.2), correlates with growth factor stimulation. This site is phosphorylated by S6K1 [80, 81]. Ser2481 is an autophosphorylation site [82]. In several cancer cell lines treated with rapamycin, phosphorylation at Ser2481 correlated with an intact mTORC2 rather than Akt Ser473 phosphorylation (which either increased or remain unchanged) making it a better marker for mTORC2 activity [83]. The functional significance of the above phosphorylation sites is still a mystery. mTOR also undergoes phosphorylation at Ser1261 and this phosphorylation correlates with increased mTORC1 signaling [84]. The identity of the Ser1261 kinase remains to be investigated.

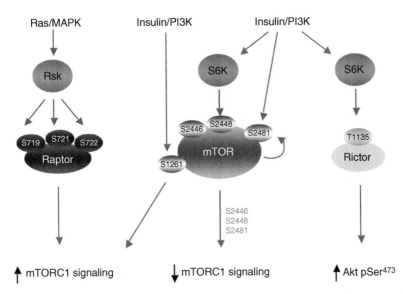

FIG. 7.1. AGC kinases phosphorylate mTOR complex components. mTOR and its partners raptor and rictor are phosphorylated by AGC kinases at sites that conform to the AGC kinase recognition motif. Phosphorylation by AGC kinases could serve to positively or negatively regulate mTORC signaling.

The mTOR partners are phosphoproteins and potentially phosphory-lated at multiple sites. Raptor is phosphorylated by Rsk1/2 both *in vivo* and *in vitro* [85]. In response to mitogens, Rsk phosphorylates raptor at three sites contained within a Rsk consensus motif. Phosphorylation *in vivo* occurs upon activation of the Ras/MAPK pathway. Mutation from Ser to Ala of the Rsk-dependent sites, Ser719, Ser721, Ser722, significantly reduced mTORC1 activity without affecting raptor interaction with mTOR or mTORC1 substrates. Thus, in addition to the Ras/MAPK/Rsk regulation of TSC2 [86], Rsk can directly signal to mTORC1 via phosphor-ylation of raptor. How phosphorylation of raptor may modify mTORC1 activity is unclear at the moment.

Numerous phosphorylation sites in rictor were also recently described [87, 88, 89]. Several of these sites are predicted to be targeted by AGC kinases based on the consensus recognition motif. Among these sites, Thr1135 was confirmed as an S6K1 target. Phosphorylation of Thr1135 is acutely sensitive to rapamycin but does not affect mTORC2 assembly nor *in vitro* kinase activity. Mutation of this residue to Ala led to a modest increase in Akt phosphorylation at Ser473, suggesting that phosphorylation of this site may modulate mTORC2. Interestingly, this mutation had no effect on other mTORC2 targets such as SGK1, cPKCα, and Akt-Thr450 phosphorylation. Thus, Thr1135 phosphorylation of rictor could play a more subtle or specific function in mTORC2 regulation. The phosphorylation of this site by S6K1 was proposed to serve as a feedback mechanism from mTORC1 signals to regulate the response of Akt to insulin. The above findings that demonstrate direct phosphorylation of mTOR partners by AGC kinases underscore the idea of feedback loop regulation between mTORC and AGC kinases that could also serve as a mechanism for cross talk between the two mTORCs.

VII. Phosphorylation of mTORC Regulators by AGC Kinases

Signaling molecules that control mTOR complex activity are also phos-phorylated by the AGC kinases (Figure 7.2). Growth factors signal to mTORC1 via a PI3K/Akt-dependent pathway leading to the phosphoryla-tion and inhibition of TSC2 and PRAS40 (Figure 7.3). The proline-rich Akt substrate of 40 kDa (PRAS40) was originally identified as a 14-3-3 binding partner and a major substrate of Akt [90]. Insulin stimulates Akt-mediated PRAS40 phosphorylation at Ser246 and consequently relieves the inhibi-tion of mTORC1 by PRAS40 [91, 92]. Insulin stimulated Akt activation can also promote mTORC1 activity via direct phosphorylation of TSC2 by Akt.

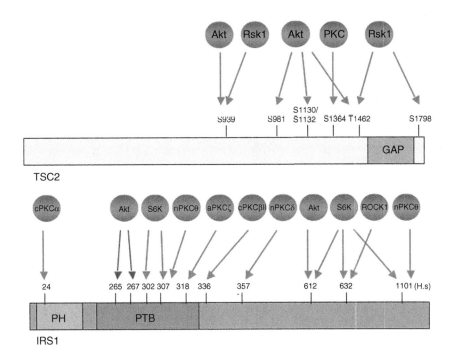

Fig. 7.2. AGC kinases phosphorylate mTORC regulators at multiple sites. Several AGC kinases phosphorylate TSC2 (tuberin) and the insulin receptor substrate-1 (IRS1) at multiple sites. Phosphorylation at Ser/Thr residues mainly serves to downregulate TSC2 and IRS1 activity. GTPase-activating protein domain (GAP); Pleckstrin homology domain (PH); phosphotyrosine-binding domain (PTB); human TSC2 numbering; mouse IRS1 numbering except for Ser1101 (*H. sapiens*); gray arrows indicate positive regulation while red arrows denote negative regulation. All numbered residues in IRS1 are Ser.

TSC2 (tuberin) forms a complex with TSC1 (hamartin), the products of tumor suppressor genes. Five phosphorylation sites in TSC2, which conform to the Akt recognition motif are Akt targets [93] (Figure 7.2). Mutations of different combination of these sites to Ala blocked Akt-mediated mTORC1 activation when these TSC2 mutants were overexpressed. What remains unsettled is how Akt inhibits the TSC1-TSC2 function to promote mTORC1 activation.

TSC2 is also phosphorylated by Rsk, an AGC kinase that is activated by the extracellular signal-regulated kinase (ERK). Rsk phosphorylates Ser1798 and also overlaps with Akt in the phosphorylation of Ser939 and Thr1462 [86]. Phosphorylation of TSC2 by Rsk at the above sites plus the ERK-targeted sites Ser540 and Ser664, contribute to ERK-mediated activation of mTORC1 signaling [94].

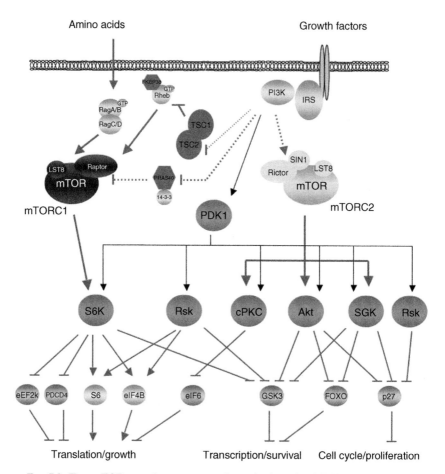

Fig. 7.3. Two mTOR complexes promote the activation of AGC kinases leading to the phosphorylation of several AGC kinase targets that mediate cell growth, survival, and proliferation. In response to amino acids and growth factors, mTORC1 and mTORC2 phosphorylate several AGC kinases at conserved motifs in their carboxyl-terminal tail. Phosphorylation at these sites, in conjunction with PDK1 phosphorylation of the AGC kinase activation loop, confers optimal activity to these kinases.

TSC2 is possibly phosphorylated by PKC at Ser1364 based on phosphorylation profiling studies and the induction of this phosphorylation by phorbol esters [95]. Its phosphorylation was sensitive to a selective PKC inhibitor but is insensitive to a MAPK inhibitor, implying that the kinase may be PKC instead of Rsk.

TSC1 is likely regulated by phosphorylation as well although it is not clear if there are AGC kinase-targeted sites. In *Drosophila*, Ser533 of dTSC1 is phosphorylated by dAkt. However, this site is not conserved in mammals and flies lacking dAkt phosphorylation sites in dTsc1 alone, or dTsc1 and dTsc2 are viable and are normal in size [96].

Another upstream regulator of mTORCs that is a "hotspot" for phosphorylation and regulation by various kinases including several AGC kinases is the insulin receptor substrate (IRS). IRS1, a member of the IRS family of adaptor molecules, is tyrosine phosphorylated in response to insulin, IGF-1 and cytokines [97]. Tyrosine phosphorylation recruits a number of SH2 containing proteins including PI3K. IRS1 is also phosphorylated potentially at over 70 Ser/Thr residues. Ser/Thr phosphorylation can positively or negatively regulate insulin signaling (Figure 7.2). Chronic hyperphosphorylation of Ser/Thr residues of IRS1 is linked to insulin resistance. It is suggested that serine phosphorylation may sterically inhibit the interactions between IRS1 and the insulin receptor (IR) making IRS1 a poorer substrate for the IR; alternatively phosphorylation could initiate IRS1 degradation.

Following insulin stimulation, IRS1 is phosphorylated within the phosphotyrosine-binding (PTB) domain by Akt at several sites. Phosphorylation at these sites could either enhance or inhibit insulin signaling [98]. Phosphorylation by most of the other Ser/Thr kinases downregulates IRS1 function. S6K can phosphorylate Ser302 (human Ser307) in vitro and knockdown of S6K inhibits this phosphorylation [99]. S6K (as well as mTOR) has been linked to the phosphorylation of Ser612 and Ser632, both proximal to tyrosine residues, thereby generating binding sites for PI3-kinase. In adipose tissues of various obese mice, activation of S6K correlates with elevated Ser632 phosphorylation. This phosphorylation is abrogated in S6K1 null mice [30]. S6K1 also directly phosphorylates Ser1101 in vitro. Phosphorylation of this site depends on insulin stimulation and is modulated by the presence of amino acids. Mutation of this residue to Ala led to increased insulin signaling to Akt despite the presence of phosphorylation at other insulin resistance-associated sites (Ser307, -312, -636/639) [100]. However, in TSC2-deficient MEFs, there was no apparent elevation in Ser1101 phosphorylation despite increased S6K activation. This is probably because of constitutive activation of mTOR/S6K, which then induce hyperphosphorylation of IRS1 on other serine residues and consequent IRS-1 degradation [101].

A number of different PKC isoforms contribute to insulin signaling and resistance. In the diabetic fat tissue the conventional cPKCβII plays a role in the phosphorylation of IRS1 at Ser336. cPKCβII expression and kinase activity is elevated in this tissue [102]. Knockout of cPKCβ indicates that this isoform is not essential for insulin-stimulated glucose transport and overall glucose homeostasis, suggesting that cPKCβ could play a negative regulatory role in insulin signaling [103]. Another conventional PKC isoform, cPKCα, may also downregulate IRS1. Knockout of PKCα led to enhanced insulin signaling in skeletal muscles and adipocytes [104]. A possible site of phosphorylation by PKCα is Ser24 based on bioinformatics and biochemical analyses [105]. Whether this site plays a role in insulin resistance remains to be investigated.

The novel PKCs have also been implicated in impaired insulin signaling. Inhibition of nPKCε prevents insulin resistance in the liver [106]. nPKCθ was also shown to inhibit insulin signaling via phosphorylation of IRS1 at Ser1101 [107]. Furthermore, nPKCθ activity is higher in muscles from obese diabetic patients. nPKCδ phosphorylates Ser357. Phosphorylation of this site and active nPKCδ are important for decreased insulin-induced Akt stimulation [108]. The atypical PKCs have been considered as positive modulators of insulin signaling, however, there is more recent evidence that they can also inhibit insulin signals. aPKCζ phosphorylates Ser318 and this phosphorylation decreases the tyrosine phosphorylation of IRS1 and interaction of IRS1 with the IR [109]. Yet another AGC kinase, ROCK1 was recently shown to play a role in insulin resistance. Disruption of ROCK1 in mice reveals that ROCK1 regulates glucose homeostasis and insulin sensitivity. *In vitro*, ROCK1 can phosphorylate IRS-1 at Ser632/635 [110]. The HM site of ROCK2 is required for substrate phosphorylation and kinase domain dimerization via interaction of the conserved HM site Thr405 with Asp39 of the N-terminal extension [111] but does not seem to be phosphorylated. Whether ROCK1 could be an mTORC target is, therefore, unlikely unless mTORC may indirectly facilitate the interaction of the HM with the N-terminus.

VIII.   mTORC Functions Mediated by AGC Kinases

The discovery of the recognition motif for AGC kinases and development of antibodies that recognize this motif have accelerated the identification of AGC kinase substrates. Phosphorylation of some of these AGC kinase substrates are emerging to be dependent on the mTOR complexes. Here, I will discuss some examples of defined mTORC cellular functions that involve AGC kinase activation and AGC-kinase-dependent substrate phosphorylation (Figure 7.3).

A. Translation/Cell Growth

mTORC1 functions in translation initiation largely because of its role in the phosphorylation and activation of S6K1. Early studies have shown that S6K1 (p70[S6K]/p85[S6K]) phosphorylates the 40S ribosomal protein S6 at several sites [112] (Table 7.2). The significance of this phosphorylation is currently controversial since neither S6K1 activity nor phosphorylation of S6 are required to drive 5′-terminal oligopyrimidine (TOP) translation [113]. While S6K1 may not be essential in the regulation of 5′-TOP translation, other translation-related functions for S6K1 have been reported. It phosphorylates and activates the eukaryotic initiation factor 4B (eIF4B) at Ser422, a site that is sensitive to PI3K and mTOR inhibitors [114, 115]. S6K1 also phosphorylates the elongation factor 2 kinase (eEF2k) at Ser366 [116]. Phosphorylation of eEF2k inactivates its kinase activity toward eEF2 and consequently enhances elongation.

S6K is also recruited to newly synthesized RNA by SKAR (S6K1/Aly/REF-like substrate) and promotes activation of the initiation complex through a series of phosphorylation events [117]. S6K1 phosphorylates and promotes degradation of the tumor suppressor, PDCD4, an inhibitor of the RNA helicase eIF4A [118]. Together with recruitment of eIF4B to eIF4A, the degradation of PDCD4 greatly enhances eIF4A helicase activity and facilitates 40S ribosomal subunit scanning to the initiation codon. Given these multiple roles of S6K in translation, why S6K null mice do not exhibit a more severe translation defect may seem puzzling [119]. However, Rsk, an AGC kinase that shares a number of substrates with S6K, may compensate for the loss of S6K. Rsk phosphorylates eIF4B on Ser422, the same residue that is targeted by S6K [120]. Rsk also phosphorylates S6 at Ser235/236 and promotes translation [121]. It is still unclear at this point if and how mTOR (as part of mTORC1 or mTORC2) may also regulate Rsk.

Other AGC kinases such as Akt and PKC play roles in translation as well but so far direct phosphorylation of proteins involved in the translation machinery have not been demonstrated and thus far their involvement is indirect. cPKC may regulate translation via interaction with the scaffold protein RACK1 (receptor for activated protein kinase C), which could serve to recruit PKC to the translation machinery. There is some evidence that cPKCβII modulates translation via phosphorylation of eIF6 [122]. eIF6 is proposed to regulate the joining of 40S and 60S subunits in the assembly of 80S ribosomes. The role of cPKCβII in the control of translation was shown to be independent of mTOR based on mTORC1 readouts. Thus, whether mTORC2 may play a role in the cPKC-mediated function and what the exact role of cPKC is during translation need to be further elucidated.

B. TRANSCRIPTION/CELL SURVIVAL

In addition to control of translation initiation, mTOR can promote transcription via the AGC kinases. The transcription factor FOXO1 is phosphorylated by Akt at Thr24, Ser256, and Ser319 [123]. Phosphorylation prevents FOXO-mediated transcription of genes that promote cell death. FOXO is also phosphorylated by SGK1 at the same sites targeted by Akt. In mTORC2-deficient cells including SIN1$^{-/-}$, rictor$^{-/-}$, or mLST8$^{-/-}$ MEFs, the phosphorylation of FOXO1/3a at the Akt target site Thr24/32 was greatly diminished [47, 49]. In line with this, SIN1$^{-/-}$ cells have decreased cell survival in the presence of stress-inducing agents [47]. Because SGK1 is also defectively phosphorylated and activated in mTORC2-disrupted cells whereas Akt had remaining activity despite defective HM phosphorylation, the loss of SGK activity could account for reduced phosphorylation of some AGC kinase substrates in mTORC2-deficient cells [68]. Another interpretation is that the partially active Akt in mTORC-2 deficient cells may have different substrate specificity. Given the overlapping substrate specificity of the AGC kinases, several mechanisms are likely to operate to determine how different AGC kinases specifically target their substrates *in vivo*.

In the presence of insulin, Akt phosphorylates and inhibits GSK3, a protein kinase that is involved in promoting apoptosis, among its many other cellular functions. The same sites in GSK3 (GSK3α at Ser21 and GSK3β at Ser9) are phosphorylated by Rsk in response to phorbol esters and growth factors [124] and by S6K upon induction by amino acids [125]. In cells that are TSC1 or TSC2 deficient, GSK3 becomes phosphorylated by S6K1 instead. This phosphorylation and inhibition of GSK3 contributes to proliferative defects in these TSC-deficient cells [126]. Thus, under conditions where Akt activity is attenuated and S6K activity is elevated, S6K becomes a predominant kinase for GSK3. In mTORC2-disrupted cells where Akt activity is also attenuated, GSK3 phosphorylation at the same sites remained normal [47, 48]. S6K activity does not appear to be significantly elevated in these cells, however. Thus, the suboptimal Akt activity may be sufficient to phosphorylate GSK3. Although SGK1 was reported to promote GSK3 phosphorylation as well [127], it is unlikely to contribute to GSK3 phosphorylation since it is highly defective in mTORC2-disrupted cells. Alternatively S6K or Rsk may contribute to its phosphorylation.

C. CELL CYCLE/CELL PROLIFERATION

mTOR coordinates cell growth with cell cycle progression [112]. More recent studies demonstrate that the AGC kinase-mediated phosphorylation of cell cycle regulators could be dependent on the mTOR complexes. In TSC2-deficient cells where mTORC1 is upregulated, the cyclin dependent kinase (CDK) inhibitor, p27^{Kip1}, becomes inactivated [128]. p27 blocks cyclin E-CDK2 activity during the G0 to G1 phase of the cell cycle. Akt phosphorylates p27 on Thr157. Both Akt and Rsk can also phosphorylate p27 on Thr198 [129]. Phosphorylation at these sites prevents p27 localization to the nucleus by sequestering it in the cytosol via 14-3-3 binding and thereby attenuates the cell-cycle inhibitory effects of p27. SGK1 phosphorylates p27 at the same residue *in vitro* and this phosphorylation was inhibited by rapamycin and SGK1 knockdown [69]. Since SGK1 is also an mTORC2 target [68], mTORC2 may thus regulate p27 via the AGC kinases Akt and SGK1.

IX. Conclusion

mTOR phosphorylates common motifs in the C-tail and allosterically activates the AGC kinases that include S6K, Akt/PKB, cPKC, and SGK. A few other AGC kinase family members are also critically regulated at these motifs, thus, it is possible that these are also mTOR targets. Phosphorylation at these motifs contributes to the stabilization and activation of the catalytic region. Given the multiple roles that these kinases play in response to different cellular inputs, modulation of their stability and activity would determine the cellular outcome. The HM and TM site phosphorylation of these kinases can also undergo self- or *trans*-phosphorylation by a kinase other than mTOR. Therefore, the exact role of mTOR complexes in the phosphorylation of these sites needs to be further elucidated. Could mTOR provide the priming phosphorylation or is this phosphorylation cell-type specific? Furthermore, can mTOR perform a function in AGC kinase regulation that requires its kinase activity but does not involve direct phosphorylation of these kinases? Phosphatases or AGC kinase-regulatory/scaffold proteins that are mTOR-dependent could mediate such regulation of AGC kinases by mTOR. Finally, potential cancer therapeutic agents that target the mTOR active site may be more effective but have more undesirable global cellular effects whereas drugs that can target the mTOR-dependent allosteric sites in AGC kinases may yield more specific effects.

ACKNOWLEDGMENTS

I thank the members of my laboratory for helpful discussions and comments on this manuscript. This work was supported in part by grants from the American Cancer Society RSG0721601TBE and NIH (GM079176).

References

1. Wullschleger, S., Loewith, R., and Hall, M.N. (2006). TOR signaling in growth and metabolism. *Cell* 124:471–484.
2. Polak, P., and Hall, M.N. (2009). mTOR and the control of whole body metabolism. *Curr Opin Cell Biol* 21:209–218.
3. Jacinto, E., and Hall, M.N. (2003). Tor signalling in bugs, brain and brawn. *Nat Rev Mol Cell Biol* 4:117–126.
4. Guertin, D.A., and Sabatini, D.M. (2009). The pharmacology of mTOR inhibition. *Sci Signal* 2:pe24.
5. Alessi, D.R., Pearce, L.R., and Garcia-Martinez, J.M. (2009). New insights into mTOR signaling: mTORC2 and beyond. *Sci Signal* 2:pe27.
6. Manning, G., Whyte, D.B., Martinez, R., Hunter, T., and Sudarsanam, S. (2002). The protein kinase complement of the human genome. *Science* 298:1912–1934.
7. Bosotti, R., Isacchi, A., and Sonnhammer, E.L. (2000). FAT: a novel domain in PIK-related kinases. *Trends Biochem Sci* 25:225–227.
8. Lovejoy, C.A., and Cortez, D. (2009). Common mechanisms of PIKK regulation. *DNA Repair (Amst)* 8:1004–1008.
9. Pearson, R.B., et al. (1995). The principal target of rapamycin-induced p70s6k inactivation is a novel phosphorylation site within a conserved hydrophobic domain. *EMBO J* 14:5279–5287.
10. Burnett, P.E., Barrow, R.K., Cohen, N.A., Snyder, S.H., and Sabatini, D.M. (1998). RAFT1 phosphorylation of the translational regulators p70 S6 kinase and 4E-BP1. *Proc Natl Acad Sci USA* 95:1432–1437.
11. Abraham, R.T. (2004). PI 3-kinase related kinases: "big" players in stress-induced signaling pathways *DNA Repair (Amst)* 3:883–887.
12. Gold, M.G., Barford, D., and Komander, D. (2006). Lining the pockets of kinases and phosphatases. *Curr Opin Struct Biol* 16:693–701.
13. Taylor, S.S., Kim, C., Cheng, C.Y., Brown, S.H., Wu, J., and Kannan, N. (2008). Signaling through cAMP and cAMP-dependent protein kinase: diverse strategies for drug design. *Biochim Biophys Acta* 1784:16–26.
14. Biondi, R.M., and Nebreda, A.R. (2003). Signalling specificity of Ser/Thr protein kinases through docking-site-mediated interactions. *Biochem J* 372:1–13.
15. Kannan, N., Haste, N., Taylor, S.S., and Neuwald, A.F. (2007). The hallmark of AGC kinase functional divergence is its C-terminal tail, a cis-acting regulatory module. *Proc Natl Acad Sci USA* 104:1272–1277.
16. Newton, A.C. (2003). Regulation of the ABC kinases by phosphorylation: protein kinase C as a paradigm. *Biochem J* 370:361–371.
17. Messerschmidt, A., et al. (2005). Crystal structure of the catalytic domain of human atypical protein kinase C-iota reveals interaction mode of phosphorylation site in turn motif. *J Mol Biol* 352:918–931.
18. Hauge, C., et al. (2007). Mechanism for activation of the growth factor-activated AGC kinases by turn motif phosphorylation. *EMBO J* 26:2251–2261.
19. Facchinetti, V., et al. (2008). The mammalian target of rapamycin complex 2 controls folding and stability of Akt and protein kinase C. *EMBO J* 27:1932–1943.
20. Jiang, T., and Qiu, Y. (2003). Interaction between Src and a C-terminal proline-rich motif of Akt is required for Akt activation. *J Biol Chem* 278:15789–15793.
21. Gould, C.M., Kannan, N., Taylor, S.S., and Newton, A.C. (2009). The chaperones Hsp90 and Cdc37 mediate the maturation and stabilization of protein kinase C through a conserved PXXP motif in the C-terminal tail. *J Biol Chem* 284:4921–4935.

22. Alessi, D.R., Caudwell, F.B., Andjelkovic, M., Hemmings, B.A., and Cohen, P. (1996). Molecular basis for the substrate specificity of protein kinase B; comparison with MAP-KAP kinase-1 and p70 S6 kinase. *FEBS Lett* 399:333–338.

23. Nishikawa, K., Toker, A., Johannes, F.J., Songyang, Z., and Cantley, L.C. (1997). Determination of the specific substrate sequence motifs of protein kinase C isozymes. *J Biol Chem* 272:952–960.

24. Jacinto, E., and Lorberg, A. (2008). TOR regulation of AGC kinases in yeast and mammals. *Biochem J* 410:19–37.

25. Schalm, S.S., and Blenis, J. (2002). Identification of a conserved motif required for mTOR signaling. *Curr Biol* 12:632–639.

26. Pullen, N., and Thomas, G. (1997). The modular phosphorylation and activation of p70s6k. *FEBS Lett* 410:78–82.

27. Kim, D.H., *et al.* (2002). mTOR interacts with raptor to form a nutrient-sensitive complex that signals to the cell growth machinery. *Cell* 110:163–175.

28. Polak, P., Cybulski, N., Feige, J.N., Auwerx, J., Ruegg, M.A., and Hall, M.N. (2008). Adipose-specific knockout of raptor results in lean mice with enhanced mitochondrial respiration. *Cell Metab* 8:399–410.

29. Bentzinger, C.F., *et al.* (2008). Skeletal muscle-specific ablation of raptor, but not of rictor, causes metabolic changes and results in muscle dystrophy. *Cell Metab* 8:411–424.

30. Um, S.H., *et al.* (2004). Absence of S6K1 protects against age- and diet-induced obesity while enhancing insulin sensitivity. *Nature* 431:200–205.

31. Saitoh, M., Pullen, N., Brennan, P., Cantrell, D., Dennis, P.B., and Thomas, G. (2002). Regulation of an activated s6 kinase 1 variant reveals a novel Mammalian target of rapamycin phosphorylation site. *J Biol Chem* 277:20104–20112.

32. Moser, B.A., *et al.* (1997). Dual requirement for a newly identified phosphorylation site in p70s6k. *Mol Cell Biol* 17:5648–5655.

33. Shah, O.J., and Hunter, T. (2004). Critical role of T-loop and H-motif phosphorylation in the regulation of S6 kinase 1 by the tuberous sclerosis complex. *J Biol Chem* 279:20816–20823.

34. Schalm, S.S., Tee, A.R., and Blenis, J. (2005). Characterization of a conserved C-terminal motif (RSPRR) in ribosomal protein S6 kinase 1 required for its mammalian target of rapamycin-dependent regulation. *J Biol Chem* 280:11101–11106.

35. Balendran, A., Currie, R., Armstrong, C.G., Avruch, J., and Alessi, D.R. (1999). Evidence that 3-phosphoinositide-dependent protein kinase-1 mediates phosphorylation of p70 S6 kinase in vivo at Thr-412 as well as Thr-252. *J Biol Chem* 274:37400–37406.

36. Belham, C., Comb, M.J., and Avruch, J. (2001). Identification of the NIMA family kinases NEK6/7 as regulators of the p70 ribosomal S6 kinase. *Curr Biol* 11:1155–1167.

37. Dennis, P.B., Pullen, N., Kozma, S.C., and Thomas, G. (1996). The principal rapamycin-sensitive p70(s6k) phosphorylation sites, T-229 and T-389, are differentially regulated by rapamycin-insensitive kinase kinases. *Mol Cell Biol* 16:6242–6251.

38. Cheatham, L., Monfar, M., Chou, M.M., and Blenis, J. (1995). Structural and functional analysis of pp 70S6k. *Proc Natl Acad Sci USA* 92:11696–11700.

39. Weng, Q.P., Andrabi, K., Kozlowski, M.T., Grove, J.R., and Avruch, J. (1995). Multiple independent inputs are required for activation of the p70 S6 kinase. *Mol Cell Biol* 15:2333–2340.

40. Ali, S.M., and Sabatini, D.M. (2005). Structure of S6 kinase 1 determines whether raptor-mTOR or rictor-mTOR phosphorylates its hydrophobic motif site. *J Biol Chem* 280:19445–19448.

41. Balendran, A., *et al.* (1999). PDK1 acquires PDK2 activity in the presence of a synthetic peptide derived from the carboxyl terminus of PRK2. *Curr Biol* 9:393–404.

42. Romanelli, A., Dreisbach, V.C., and Blenis, J. (2002). Characterization of phosphatidy-linositol 3-kinase-dependent phosphorylation of the hydrophobic motif site Thr(389) in p70 S6 kinase 1. *J Biol Chem* 277:40281–40289.

43. Ananthanarayanan, B., Fosbrink, M., Rahdar, M., and Zhang, J. (2007). Live-cell molecular analysis of Akt activation reveals roles for activation loop phosphorylation. *J Biol Chem* 282:36634–36641.

44. Yang, J., *et al.* (2002). Molecular mechanism for the regulation of protein kinase B/Akt by hydrophobic motif phosphorylation. *Mol Cell* 9:1227–1240.

45. Sarbassov, D.D., Guertin, D.A., Ali, S.M., and Sabatini, D.M. (2005). Phosphorylation and regulation of Akt/PKB by the rictor-mTOR complex. *Science* 307:1098–1101.

46. Ikenoue, T., Inoki, K., Yang, Q., Zhou, X., and Guan, K.L. (2008). Essential function of TORC2 in PKC and Akt turn motif phosphorylation, maturation and signalling. *EMBO J* 27:1919–1931.

47. Jacinto, E., *et al.* (2006). SIN1/MIP1 maintains rictor-mTOR complex integrity and regulates Akt phosphorylation and substrate specificity. *Cell* 127:125–137.

48. Shiota, C., Woo, J.T., Lindner, J., Shelton, K.D., and Magnuson, M.A. (2006). Multiallelic disruption of the rictor gene in mice reveals that mTOR complex 2 is essential for fetal growth and viability. *Dev Cell* 11:583–589.

49. Guertin, D.A., *et al.* (2006). Ablation in mice of the mTORC components raptor, rictor, or mLST8 reveals that mTORC2 is required for signaling to Akt-FOXO and PKC alpha, but not S6K1. *Dev Cell* 11:859–871.

50. Cybulski, N., Polak, P., Auwerx, J., Ruegg, M.A., and Hall, M.N. (2009). mTOR complex 2 in adipose tissue negatively controls whole-body growth. *Proc Natl Acad Sci USA* 106:9902–9907.

51. Williams, M.R., *et al.* (2000). The role of 3-phosphoinositide-dependent protein kinase 1 in activating AGC kinases defined in embryonic stem cells. *Curr Biol* 10:439–448.

52. Alessi, D.R., *et al.* (1996). Mechanism of activation of protein kinase B by insulin and IGF-1. *EMBO J* 15:6541–6551.

53. Feldman, M.E., *et al.* (2009). Active-site inhibitors of mTOR target rapamycin-resistant outputs of mTORC1 and mTORC2. *PLoS Biol* 7:e38.

54. Surucu, B., Bozulic, L., Hynx, D., Parcellier, A., and Hemmings, B.A. (2008). In vivo analysis of protein kinase B (PKB)/Akt regulation in DNA-PKCs-null mice reveals a role for PKB/Akt in DNA damage response and tumorigenesis. *J Biol Chem* 283:30025–30033.

55. Bozulic, L., Surucu, B., Hynx, D., and Hemmings, B.A. (2008). PKBalpha/Akt1 acts downstream of DNA-PK in the DNA double-strand break response and promotes survival. *Mol Cell* 30:203–213.

56. Cameron, A.J., *et al.* (2007). Protein kinases, from B to C. *Biochem Soc Trans* 35:1013–1017.

57. Sonnenburg, E.D., Gao, T., and Newton, A.C. (2001). The phosphoinositide-dependent kinase, PDK-1, phosphorylates conventional protein kinase C isozymes by a mechanism that is independent of phosphoinositide 3-kinase. *J Biol Chem* 276:45289–45297.

58. Sarbassov, D., *et al.* (2004). Rictor, a novel binding partner of mTOR, defines a rapamycin-insensitive and raptor-independent pathway that regulates the cytoskeleton. *Curr Biol* 14:1296–1302.

59. Bornancin, F., and Parker, P.J. (1996). Phosphorylation of threonine 638 critically controls the dephosphorylation and inactivation of protein kinase C alpha. *Curr Biol* 6:1114–1123.

60. Gao, T., and Newton, A.C. (2002). The turn motif is a phosphorylation switch that regulates the binding of Hsp70 to protein kinase C. *J Biol Chem* 277:31585–31592.

61. Biondi, R.M., Kieloch, A., Currie, R.A., Deak, M., and Alessi, D.R. (2001). The PIF-binding pocket in PDK1 is essential for activation of S6K and SGK, but not PKB. *EMBO J* 20:4380–4390.
62. Roffey, J., Rosse, C., Linch, M., Hibbert, A., McDonald, N.Q., and Parker, P.J. (2009). Protein kinase C intervention: the state of play. *Curr Opin Cell Biol* 21:268–279.
63. Lahn, M., *et al.* (2004). Protein kinase C alpha expression in breast and ovarian cancer. *Oncology* 67:1–10.
64. Fan, Q.W., *et al.* (2009). EGFR signals to mTOR through PKC and independently of Akt in glioma. *Sci Signal* 2:ra4.
65. Masri, J., *et al.* (2007). mTORC2 activity is elevated in gliomas and promotes growth and cell motility via overexpression of rictor. *Cancer Res* 67:11712–11720.
66. Tessier, M., and Woodgett, J.R. (2006). Serum and glucocorticoid-regulated protein kinases: variations on a theme. *J Cell Biochem* 98:1391–1407.
67. Kobayashi, T., and Cohen, P. (1999). Activation of serum- and glucocorticoid-regulated protein kinase by agonists that activate phosphatidylinositide 3-kinase is mediated by 3-phosphoinositide-dependent protein kinase-1 (PDK1) and PDK2. *Biochem J* 339(Pt 2):319–328.
68. Garcia-Martinez, J.M., and Alessi, D.R. (2008). mTOR complex 2 (mTORC2) controls hydrophobic motif phosphorylation and activation of serum- and glucocorticoid-induced protein kinase 1 (SGK1). *Biochem J* 416:375–385.
69. Hong, F., Larrea, M.D., Doughty, C., Kwiatkowski, D.J., Squillace, R., and Slingerland, J.M. (2008). mTOR-raptor binds and activates SGK1 to regulate p27 phosphorylation. *Mol Cell* 30:701–711.
70. Liang, J., *et al.* (2002). PKB/Akt phosphorylates p27, impairs nuclear import of p27 and opposes p27-mediated G1 arrest. *Nat Med* 8:1153–1160.
71. Park, J., Leong, M.L., Buse, P., Maiyar, A.C., Firestone, G.L., and Hemmings, B.A. (1999). Serum and glucocorticoid-inducible kinase (SGK) is a target of the PI 3-kinase-stimulated signaling pathway. *EMBO J* 18:3024–3033.
72. Chen, W., *et al.* (2009). Regulation of a third conserved phosphorylation site in SGK1. *J Biol Chem* 284:3453–3460.
73. Kobayashi, T., Deak, M., Morrice, N., and Cohen, P. (1999). Characterization of the structure and regulation of two novel isoforms of serum- and glucocorticoid-induced protein kinase. *Biochem J* 344 (Pt 1):189–197.
74. Bogusz, A.M., Brickley, D.R., Pew, T., and Conzen, S.D. (2006). A novel N-terminal hydrophobic motif mediates constitutive degradation of serum- and glucocorticoid-induced kinase-1 by the ubiquitin-proteasome pathway. *FEBS J* 273:2913–2928.
75. Belova, L., Brickley, D.R., Ky, B., Sharma, S.K., and Conzen, S.D. (2008). Hsp90 regulates the phosphorylation and activity of serum- and glucocorticoid-regulated kinase-1. *J Biol Chem* 283:18821–18831.
76. Fujita, N., Sato, S., Ishida, A., and Tsuruo, T. (2002). Involvement of Hsp90 in signaling and stability of 3-phosphoinositide-dependent kinase-1. *J Biol Chem* 277:10346–10353.
77. Basso, A.D., Solit, D.B., Chiosis, G., Giri, B., Tsichlis, P., and Rosen, N. (2002). Akt forms an intracellular complex with heat shock protein 90 (Hsp90) and Cdc37 and is destabilized by inhibitors of Hsp90 function. *J Biol Chem* 277:39858–39866.
78. Jacinto, E. (2008). What controls TOR? *IUBMB Life* 60:483–496.
79. Sekulic, A., *et al.* (2000). A direct linkage between the phosphoinositide 3-kinase-AKT signaling pathway and the mammalian target of rapamycin in mitogen-stimulated and transformed cells. *Cancer Res* 60:3504–3513.
80. Holz, M.K., and Blenis, J. (2005). Identification of S6 kinase 1 as a novel mammalian target of rapamycin (mTOR)-phosphorylating kinase. *J Biol Chem* 280:26089–26093.

81. Chiang, G.G., and Abraham, R.T. (2005). Phosphorylation of mammalian target of rapamycin (mTOR) at Ser-2448 is mediated by p70S6 kinase. *J Biol Chem* 280:25485–25490.
82. Peterson, R.T., Beal, P.A., Comb, M.J., and Schreiber, S.L. (2000). FKBP12-rapamycin-associated protein (FRAP) autophosphorylates at serine 2481 under translationally repressive conditions. *J Biol Chem* 275:7416–7423.
83. Copp, J., Manning, G., and Hunter, T. (2009). TORC-specific phosphorylation of mammalian target of rapamycin (mTOR): phospho-Ser2481 is a marker for intact mTOR signaling complex 2. *Cancer Res* 69:1821–1827.
84. Acosta-Jaquez, H.A., *et al.* (2009). Site-specific mTOR phosphorylation promotes mTORC1-mediated signaling and cell growth. *Mol Cell Biol* 29:4308–4324.
85. Carriere, A., *et al.* (2008). Oncogenic MAPK signaling stimulates mTORC1 activity by promoting RSK-mediated raptor phosphorylation. *Curr Biol* 18:1269–1277.
86. Roux, P.P., Ballif, B.A., Anjum, R., Gygi, S.P., and Blenis, J. (2004). Tumor-promoting phorbol esters and activated Ras inactivate the tuberous sclerosis tumor suppressor complex via p90 ribosomal S6 kinase. *Proc Natl Acad Sci USA* 101:13489–13494.
87. Dibble, C.C., Asara, J.M., and Manning, B.D. (2009). Characterization of Rictor phosphorylation sites reveals direct regulation of mTOR complex 2 by S6K1. *Mol Cell Biol* 29:5657–5670.
88. Julien, L.A., Carriere, A., Moreau, J., and Roux, P.P. (2010). mTORC1-activated S6K1 phosphorylates Rictor on threonine 1135 and regulates mTORC2 signaling. *Mol Cell Biol* 30(4):908–921.
89. Treins, C., Warne, P.H., Magnuson, M.A., Pende, M., and Downward, J. (2009). Rictor is a novel target of p70 S6 kinase-1. *Oncogene.* Epub.
90. Kovacina, K.S., *et al.* (2003). Identification of a proline-rich Akt substrate as a 14-3-3 binding partner. *J Biol Chem* 278:10189–10194.
91. Sancak, Y., *et al.* (2007). PRAS40 is an insulin-regulated inhibitor of the mTORC1 protein kinase. *Mol Cell* 25:903–915.
92. Vander Haar, E., Lee, S.I., Bandhakavi, S., Griffin, T.J., and Kim, D.H. (2007). Insulin signalling to mTOR mediated by the Akt/PKB substrate PRAS40. *Nat Cell Biol* 9:316–323.
93. Huang, J., and Manning, B.D. (2008). The TSC1-TSC2 complex: a molecular switchboard controlling cell growth. *Biochem J* 412:179–190.
94. Ma, L., Chen, Z., Erdjument-Bromage, H., Tempst, P., and Pandolfi, P.P. (2005). Phosphorylation and functional inactivation of TSC2 by Erk implications for tuberous sclerosis and cancer pathogenesis. *Cell* 121:179–193.
95. Ballif, B.A., Roux, P.P., Gerber, S.A., MacKeigan, J.P., Blenis, J., and Gygi, S.P. (2005). Quantitative phosphorylation profiling of the ERK/p90 ribosomal S6 kinase-signaling cassette and its targets, the tuberous sclerosis tumor suppressors. *Proc Natl Acad Sci USA* 102:667–672.
96. Schleich, S., and Teleman, A.A. (2009). Akt phosphorylates both Tsc1 and Tsc2 in *Drosophila*, but neither phosphorylation is required for normal animal growth. *PLoS ONE* 4:e6305.
97. Taniguchi, C.M., Emanuelli, B., and Kahn, C.R. (2006). Critical nodes in signalling pathways: insights into insulin action. *Nat Rev Mol Cell Biol* 7:85–96.
98. Paz, K., *et al.* (1999). Phosphorylation of insulin receptor substrate-1 (IRS-1) by protein kinase B positively regulates IRS-1 function. *J Biol Chem* 274:28816–28822.
99. Harrington, L.S., *et al.* (2004). The TSC1-2 tumor suppressor controls insulin-PI3K signaling via regulation of IRS proteins. *J Cell Biol* 166:213–223.

100. Tremblay, F., *et al.* (2007). Identification of IRS-1 Ser-1101 as a target of S6K1 in nutrient- and obesity-induced insulin resistance. *Proc Natl Acad Sci USA* 104:14056–14061.

101. Shah, O.J., and Hunter, T. (2006). Turnover of the active fraction of IRS1 involves raptor- mTOR- and S6K1-dependent serine phosphorylation in cell culture models of tuberous sclerosis. *Mol Cell Biol* 26:6425–6434.

102. Liberman, Z., Plotkin, B., Tennenbaum, T., and Eldar-Finkelman, H. (2008). Coordi- nated phosphorylation of insulin receptor substrate-1 by glycogen synthase kinase-3 and protein kinase C betaII in the diabetic fat tissue. *Am J Physiol Endocrinol Metab* 294: E1169–E1177.

103. Standaert, M.L., *et al.* (1999). Effects of knockout of the protein kinase C beta gene on glucose transport and glucose homeostasis. *Endocrinology* 140:4470–4477.

104. Leitges, M., *et al.* (2002). Knockout of PKC alpha enhances insulin signaling through PI3K. *Mol Endocrinol* 16:847–858.

105. Nawaratne, R., Gray, A., Jorgensen, C.H., Downes, C.P., Siddle, K., and Sethi, J.K. (2006). Regulation of insulin receptor substrate 1 pleckstrin homology domain by protein kinase C: role of serine 24 phosphorylation. *Mol Endocrinol* 20:1838–1852.

106. Samuel, V.T., *et al.* (2007). Inhibition of protein kinase Cepsilon prevents hepatic insulin resistance in nonalcoholic fatty liver disease. *J Clin Invest* 117:739–745.

107. Li, Y., *et al.* (2004). Protein kinase C Theta inhibits insulin signaling by phosphorylating IRS1 at Ser(1101). *J Biol Chem* 279:45304–45307.

108. Waraich, R.S., *et al.* (2008). Phosphorylation of Ser357 of rat insulin receptor substrate-1 mediates adverse effects of protein kinase C-delta on insulin action in skeletal muscle cells. *J Biol Chem* 283:11226–11233.

109. Moeschel, K., *et al.* (2004). Protein kinase C-zeta-induced phosphorylation of Ser318 in insulin receptor substrate-1 (IRS-1) attenuates the interaction with the insulin receptor and the tyrosine phosphorylation of IRS-1. *J Biol Chem* 279:25157–25163.

110. Lee, D.H., *et al.* (2009). Targeted disruption of ROCK1 causes insulin resistance in vivo. *J Biol Chem* 284:11776–11780.

111. Couzens, A.L., Saridakis, V., and Scheid, M.P. (2009). The hydrophobic motif of ROCK2 requires association with the N-terminal extension for kinase activity. *Biochem J* 419:141–148.

112. Fingar, D.C., and Blenis, J. (2004). Target of rapamycin (TOR): an integrator of nutrient and growth factor signals and coordinator of cell growth and cell cycle progression. *Oncogene* 23:3151–3171.

113. Ruvinsky, I., and Meyuhas, O. (2006). Ribosomal protein S6 phosphorylation: from protein synthesis to cell size. *Trends Biochem Sci* 31:342–348.

114. Raught, B., *et al.* (2004). Phosphorylation of eucaryotic translation initiation factor 4B Ser422 is modulated by S6 kinases. *EMBO J* 23:1761–1769.

115. Proud, C.G. (2007). Signalling to translation: how signal transduction pathways control the protein synthetic machinery. *Biochem J* 403:217–234.

116. Wang, X., Li, W., Williams, M., Terada, N., Alessi, D.R., and Proud, C.G. (2001). Regulation of elongation factor 2 kinase by p90(RSK1) and p70 S6 kinase. *EMBO J* 20:4370–4379.

117. Ma, X.M., Yoon, S.O., Richardson, C.J., Julich, K., and Blenis, J. (2008). SKAR links pre- mRNA splicing to mTOR/S6K1-mediated enhanced translation efficiency of spliced mRNAs. *Cell* 133:303–313.

118. Dorrello, N.V., Peschiaroli, A., Guardavaccaro, D., Colburn, N.H., Sherman, N.E., and Pagano, M. (2006). S6K1- and betaTRCP-mediated degradation of PDCD4 promotes protein translation and cell growth. *Science* 314:467–471.

119. Pende, M., *et al.* (2004). S6K1(−/−)/S6K2(−/−) mice exhibit perinatal lethality and rapamycin-sensitive 5'-terminal oligopyrimidine mRNA translation and reveal a mitogen-activated protein kinase-dependent S6 kinase pathway. *Mol Cell Biol* 24:3112–3124.

120. Shahbazian, D., *et al.* (2006). The mTOR/PI3K and MAPK pathways converge on eIF4B to control its phosphorylation and activity. *EMBO J* 25:2781–2791.

121. Roux, P.P., *et al.* (2007). RAS/ERK signaling promotes site-specific ribosomal protein S6 phosphorylation via RSK and stimulates cap-dependent translation. *J Biol Chem* 282:14056–14064.

122. Grosso, S., *et al.* (2008). PKCbetaII modulates translation independently from mTOR and through RACK1. *Biochem J* 415:77–85.

123. Manning, B.D., and Cantley, L.C. (2007). AKT/PKB signaling: navigating downstream. *Cell* 129:1261–1274.

124. Frame, S., and Cohen, P. (2001). GSK3 takes centre stage more than 20 years after its discovery. *Biochem J* 359:1–16.

125. Armstrong, J.L., Bonavaud, S.M., Toole, B.J., and Yeaman, S.J. (2001). Regulation of glycogen synthesis by amino acids in cultured human muscle cells. *J Biol Chem* 276:952–956.

126. Zhang, H.H., Lipovsky, A.I., Dibble, C.C., Sahin, M., and Manning, B.D. (2006). S6K1 regulates GSK3 under conditions of mTOR-dependent feedback inhibition of Akt. *Mol Cell* 24:185–197.

127. Sakoda, H., *et al.* (2003). Differing roles of Akt and serum- and glucocorticoid-regulated kinase in glucose metabolism, DNA synthesis, and oncogenic activity. *J Biol Chem* 278:25802–25807.

128. Rosner, M., Freilinger, A., and Hengstschlager, M. (2006). The tuberous sclerosis genes and regulation of the cyclin-dependent kinase inhibitor p27. *Mutat Res* 613:10–16.

129. Besson, A., Dowdy, S.F., and Roberts, J.M. (2008). CDK inhibitors: cell cycle regulators and beyond. *Dev Cell* 14:159–169.

8

mTORC1 and Cell Cycle Control

CHRISTOPHER G. PROUD

School of Biological Sciences
University of Southampton
Boldrewood Campus
United Kingdom

I. Abstract

Target of rapamycin (TOR) proteins play key roles in regulating cellular functions in eukaryotes. They form signaling complexes with partner proteins and respond to nutritional, hormonal, and other cues. It is now clear that TOR proteins can control several stages of the cell cycle, although the mechanisms by which they do this appear to differ considerably between different organisms. In budding yeast, TOR complex 1 (TORC1) promotes exit of cells from a quiescent state known as G_0. In mammals, and likely in many other species, (m)TORC1 promotes progression of cells from G1 into S-phase, for example, by regulating the levels of specific cyclins and thus the activity of cyclin-dependent kinases. mTORC1 may also influence this step of the cell cycle through its control of mitochondrial function and/or ribosome biogenesis. Recent data from diverse organisms also identify roles for TOR proteins in G2 progression and/or M-phase entry, although in different systems TOR signaling may promote or retard M-phase entry.

These effects on cell cycle underlie the utility of rapamycin (and related compounds) which inhibit mTORC1 as immunosuppressants and the potential value of such reagents as anticancer drugs. Existing data are stimulating further studies to elucidate the intricate web of regulatory processes through which TOR proteins control cell size and cell proliferation and may lead to further developments in the use of TOR inhibitors as therapeutic agents.

THE ENZYMES, Vol. XXVII   129   ISSN NO: 1874-6047
 DOI: 10.1016/S1874-6047(10)27008-7

II.   Introduction

It is now abundantly clear that the target of rapamycin (TOR) proteins, including the mammalian target of rapamycin, mTOR, regulate a multiplicity of cellular functions. These include events linked to cell cycle progression. As described in detail elsewhere in this volume, (m)TOR forms two main types of multiprotein complexes, mTORC1 and mTORC2. Certain functions of mTORC1 (but not all) are inhibited by the macrolide rapamycin whereas mTORC2 signaling is not inhibited by rapamycin, at least in the short term. This review focuses on (m)TOR signaling and the control of the cell cycle.

mTORC1 and the corresponding complexes in other species respond to many types of signals, including environmental/nutrient cues, as well as hormone and growth factor signaling. They are thus ideally placed to "integrate" this information and control the key events that are critical to cell cycle progression.

A.   RAPAMYCIN INHIBITS CELL CYCLE PROGRESSION

The first clinical application of the mTORC1 inhibitor rapamycin was as an immunosuppressant, to block transplant (graft) rejection. It blocks T-cell activation at the G_1/S boundary. They are still widely used for this purpose.

Cell cycle progression involves a complex interplay between regulatory components and specific "checkpoints," notably those at the transitions between G1 and S-phase and between G2 and M-phase. Progression is regulated by protein kinases whose activity is dependent upon cell cycle regulator proteins termed cyclins; they are therefore termed cyclin-dependent kinases or CDKs. They are also sometimes named as the products of "cell division control" (cdc) genes. Thus the mammalian kinase cdc2 is also termed CDK1 and can be activated by association with cyclin B, in which state it promotes the entry of cells into mitosis.

B.   MTORC1 CONTROLS THE CELL CYCLE

Conventionally, the cell cycle is divided into four stages: G1 (Gap 1), where cells synthesize proteins, make ribosomes, and increase in mass; S, where DNA synthesis occurs; G2, where cells make additional proteins and prepare for division; and M, mitosis, where the cell divides (Figure 8.1). The transitions between G1 and S phases, and also between G2 and M, must be tightly regulated, and sophisticated mechanisms operate to ensure this.

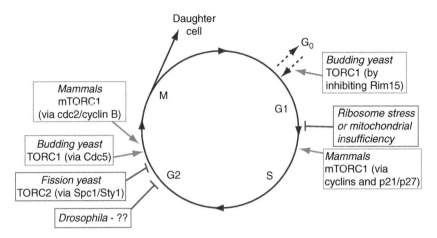

Fig. 8.1. Overview of the cell cycle and its control by (m)TOR signaling. This schematic depiction of the cell cycle also includes selected inputs from TOR signaling to the control of the cell cycle.

Cells may also exist in a fifth state, termed G_0. This represents a nondividing "resting" or quiescent condition, which may be because the cells are terminally differentiated or because (like naïve T cells) they await signals to stimulate them to enter the main cell cycle and divide.

Almost 20 years ago, it was discovered that rapamycin blocks T-cell activation by inhibiting entry of stimulated T-cells into S-phase (reviewed in Ref. [1]; Figure 8.1), an effect that underlies the application of rapamycin as an immunosuppressant (e.g., to prevent organ graft rejection [2]). The ability of rapamycin ability to block G1/S progression is probably also important for the antiproliferative effects of rapamycin and related compounds on tumor cells and for its use to prevent restenosis (a complication of cardiac surgery [3]).

Before examining in more detail how (m)TOR signaling affects the cell cycle, it is important to mention some of the major cellular targets for control by mTORC1, as they may be involved in the control of the cell cycle by this complex.

C.   SIGNALING DOWNSTREAM OF MTORC1: THE SHORT VERSION

The process whose control by mTORC1 is best understood is protein synthesis. Two sets of proteins linked to mRNA translation are direct targets for phosphorylation by mTORC1, the eukaryotic initiation factor

4E-binding proteins (three isoforms; e.g., 4E-BP1) and the ribosomal protein S6 kinases (S6K1 and S6K2) (Figure 8.2).

Briefly, 4E-BPs bind to and inhibit the function of eIF4E, the protein that interacts with the cap structure of cytoplasmic mRNAs and also with the scaffold protein eIF4G, which in turn recruits other components involved in efficient translation initiation (i.e., the binding of ribosomes to the 5′-end of the mRNA). The 4E-BPs undergo phosphorylation at multiple sites. The association of 4E-BPs with eIF4E is disrupted upon their extensive phosphorylation by mTORC1 (Figure 8.2), liberating eIF4E and allowing it to bind eIF4G and its partners. These include the RNA helicase eIF4A, which unwinds potentially inhibitory stem-loops (secondary structure) in the 5′-UTR of mRNAs, thereby facilitating their translation. Some of the phosphorylation sites in 4E-BP1—and in certain cell types, most, if not all of them—are resistant to treatment of cells with rapamycin. However, they are dephosphorylated when cells are treated with inhibitors of the kinase activity of mTOR, indicating that their control represents a

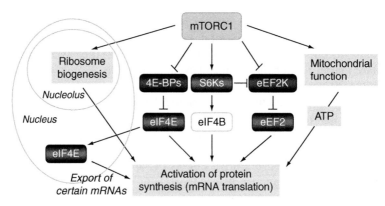

Fig. 8.2. Signaling downstream of mTORC1—the short version. mTORC1 (and likely the orthologous complexes in other organisms) positively regulates mRNA translation (protein synthesis) and other processes, including mitochondrial function and ribosome biogenesis (which augments cellular capacity for protein synthesis). Ribosome biogenesis involves processes that occur in the nucleolus (rRNA transcription, ribosome assembly) and the cytoplasm (ribosomal protein synthesis), as well as the nucleoplasm (transcription of genes for ribosomal proteins). mTORC1 positively regulates ribosome biogenesis, and, potentially, the production of ATP (and GTP) which are consumed by protein synthesis. Translation components positively/negatively regulated by mTORC1 are shown in green/red. S6K phosphorylates eIF4B, although the importance of this for the control of the translational machinery is incompletely understood. Note that 4E-BP1 appears to inhibit export of eIF4E from the nucleus, thereby promoting eIF4E's nuclear accumulation. The dotted line indicates that the control of eEF2K by mTORC1 is indirect.

rapamycin-insensitive output from mTOR, almost certainly from mTORC1 [4–6]]. There is a surprising paucity of data about the control of translation factors during the cell cycle. Data concerning the phosphorylation of 4E-BP1 during mitosis appear contradictory: for example, it has been reported to be hypophosphorylated (which would impair eIF4E function and translation of certain mRNAs [7]) or hyperphosphorylated [8].

The function(s) of the S6 kinases is more enigmatic: they were first identified for their ability to phosphorylate a component of the small ribosomal subunit, S6, but the biological function of this modification is unclear [9]. S6Ks have several other substrates including eIF4B (an auxiliary factor for eIF4A; [10]) and the kinase that phosphorylates and inhibits eukaryotic elongation factor (eEF) 2 [11]. This enzyme, eEF2 kinase (eEF2K; Figure 8.2), is inactivated by signaling through mTORC1, although it is not a direct substrate for mTORC1 [12]. One report indicates that S6K1 and S6K2 are both activated during mitosis [13].

mTOR and its orthologs in lower eukaryotes also regulate gene transcription and ribosome biogenesis, as well as diverse other processes including mitochondrial biogenesis.

III. TOR Signaling and G_0

In *Saccharomyces cerevisiae*, inhibition of TORC1 results in cells entering G_0, (Figures 8.1 and 8.2). This involves protein kinase called Rim15, which is maintained in a phosphorylated state by TORC1, probably via phosphorylation by Sch9 [14], an S6K ortholog [15]. Phosphorylated Rim1 associates with 14-3-3 proteins and this keeps Rim15 in the cytoplasm (Figure 8.3). Inactivation of TORC1 leads to the inhibition of Sch9 and thus to the dephosphorylation of Rim15, allowing it to enter the nucleus and induces the G_0 program. Rim15 is also regulated by additional pathways which sense nutrient availability. These include PKA (cAMP-dependent protein kinase) and Pho80 (a cyclin-dependent kinase, CDK, which is involved in sensing phosphate availability [16]).

IV. Control of G1/S-Phase Progression by (m)TORC1

Although it is almost two decades since rapamycin was found to block lymphocyte activation, it is still not fully clear how mTORC1 controls the cell cycle. Rapamycin binds to FKBP12 (FK506-binding protein, 12 kDa) and this complex binds to a region of (m)TOR that is close to its kinase domain. FKBP12 belongs to a family of proteins termed "immunophilins."

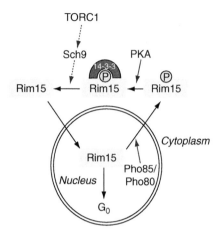

FIG. 8.3. TOR and the control of G_0 in budding yeast. TORC1 in *S. cerevisiae* positively regulates the kinase Sch9, which in turn phosphorylates Rim15, resulting in its binding to 14-3-3 and cytoplasmic localization. Inhibition of TORC1 and Sch9 (dotted line) results in dephosphorylation of Rim15, and its entry to the nucleus where it initiates a G_0 program.

Some other small molecules that bind to immunophilins also block T-cell activation, but do so differently from rapamycin—by blocking entry of cells from G_0 into G_1, while rapamycin blocks late in G_1, preventing cells from entering S-phase.

The CDK/cyclin complexes required for S-phase progression can be inhibited by specific CDK-inhibitor proteins (CDKIs) [17], including p21 (also called WAF1 or CIP1) and p27 (or Kip1) (see Figure 8.4). The expression of both p27 and p21 is regulated at transcriptional and posttranscriptional levels. Active CDKs (cyclin D/CDK4,6 and also cyclin E/CDK2) can phosphorylate the retinoblastoma gene product (Rb): this removes its inhibitory effect on E2F family transcription factors. Activated E2F factors turn on the transcription of genes whose products are required for S-phase entry and/or progression, thereby promoting progression of cells through the cell cycle. These genes include those encoding G1/S and S-phase cyclins, as well as the gene for E2F itself.

p27^{Kip1} inhibits both cyclin E-CDK2 and cyclin D-CDK4 complexes. The expression of p27^{Kip1} is regulated by signaling through mTORC1 (Figure 8.4). p27 levels are elevated in quiescent cells, while mitogenic stimulation causes a fall in p27 levels. For example, interleukin 2 promotes a decrease in p27^{Kip1} levels in T-cells, favoring S-phase progression; in contrast, transforming growth factor β can elicit an increase in p27^{Kip1}, inhibiting proliferation. Rapamycin blocks the ability of mitogenic stimuli

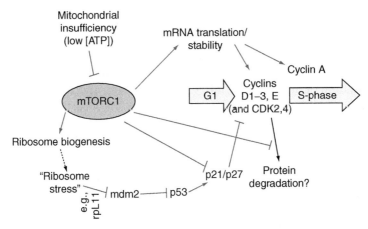

FIG. 8.4. (m)TOR and G1 → S progression. In mammalian cells, mTORC1 regulates G1/S progression by, among other things, controlling the levels of cyclins as well as the Cdk inhibitor p27, by affecting their rates of synthesis and/or degradation. Low mitochondrial activity may impair mTORC1 signaling (due to decreased ATP levels?). mTORC1 promotes ribosome biogenesis: defects in this process (dotted black arrow) give rise to ribosome stress which, via proteins such as rpL11, impairs mdm2, stabilizes p53, and leads to induction of the cell cycle inhibitor proteins p21 and p27.

to downregulate p27[Kip1]. Furthermore, expression of a so-called "constitutively active" form of 4E-BP1 can enhance the expression of p27[Kip1] [18].

Although data from p27[Kip1−/−] cells indicate that the downregulation of this protein plays an important role in the antiproliferative effects of rapamycin [19], other effects also play a role since 27[−/−] T-cells still showed residual sensitivity to rapamycin. This implies that mTORC1 signaling regulates additional components that are involved in the G1/S transition or S-phase progression (Figure 8.4). These include the G1/S cyclins D1 [20], D2, D3 [21], and E [21], as well as cyclin A, which is required for the initiation of DNA replication during S-phase [22] (and likely also plays a role in S-phase entry). In T-cells, rapamycin inhibits the expression of cyclins D2 and D3 at a posttranscriptional level [21, 23]. Indeed, this drug appears to inhibit the translation of the mRNAs for cyclins D1 [24] and D3 [21], thereby promoting a decrease in their cellular concentrations. Inhibitory effects of rapamycin on cyclin D3 levels have been noted in various cell types, including mantle-cell lymphomas and HER2-positive breast cancer cells [25, 26].

mTORC1 signaling can also regulate the expression of cyclin D1 at additional levels, including, under at least some conditions, the transcription and the stability of the cyclin D1 mRNA and of the cyclin D1 protein, likely through proteasome-dependent degradation [27] (Figure 8.4). Treating cells already growing in serum with rapamycin for 2–4 h causes a decrease in the levels of cyclin D1: this effect is not due to decreased cytoplasmic levels of cyclin D1 mRNA and is abolished by knocking down 4E-BP1 [24]. This is consistent with the idea that rapamycin inhibits thetranslation of the cyclin D1 mRNA by activating 4E-BP1 and inhibiting eIF4E availability for formation of active eIF4F initiation complexes. Since cyclin D1 can sequester p27^{Kip1}, decreased levels of cyclin D1 may cause inhibition of cyclin E-Cdk2 in addition to decreasing the activity of Cdk4, a partner for cyclin D1.

A number of other studies have indicated that eIF4E can also modulate expression of cyclin D1 (and other proteins, including cyclins A2, B1, and E1 [28]) by promoting the export of specific mRNAs from the nucleus to the cytoplasm (Figure 8.2, see also Figure 8.4). This effect was initially noted in the mid-1990s [29, 30] in studies in which eIF4E was artificially overexpressed: since then, Borden and colleagues have explored the role of eIF4E in mRNA export in considerable detail, identifying regulatory features of eIF4E (including its phosphorylation [31]); elements of the 3'-UTR of the cyclin D1 mRNA [32] and ancillary proteins [33] which are involved in this process. In fact, this process of eIF4E-mediated export may apply to many mRNAs involved in cell cycle control and cell transformation [34, 35]. Recent work has shown that a fraction of the 4E-BP1 is found in the nucleus where it affects the localization of eIF4E: when cells are starved of serum or treated with rapamycin (conditions which activate 4E-BP1), the proportion of eIF4E seen in the nucleus increases [36]. This finding is surprising given that 4E-BP1 should compete with the proposed nuclear import protein for eIF4E, 4E-T [37], for binding to eIF4E (suggesting that rapamycin should decrease nuclear eIF4E levels). Nonetheless, this observation and the finding that eIF4E mediates nuclear export of some mRNAs lead to a model where mTORC1 signaling could control the export of certain mRNAs to the cytoplasm by impairing the exit of eIF4E from the nucleus (Figure 8.2). Overexpression of eIF4E, either artificially [29, 38] or, as often occurs in tumors [39, 40], may promote the export of eIF4E-dependent mRNAs, by increasing nuclear levels of eIF4E, perhaps to levels that exceed those of nuclear 4E-BP1, thus evading control by mTORC1. Further studies are needed to investigate these ideas.

Recent work has suggested [41] that mTORC1 activates SGK1, a protein kinase closely related to PKB/Akt, and that SGK1 phosphorylates p27^{Kip1} causing it to be retained in the cytoplasm (i.e., where it is unable to inhibit cyclin E-Cdk2 complexes). This effect would help promote S-phase progression, potentially providing a link mTORC1 and S-phase entry. However, other work indicates SGK1 is actually phosphorylated by the rapamycin-insensitive mTORC2 complex (which also phosphorylates PKB/ Akt) SGK1 [42]. Thus, additional studies are needed to unravel the links between mTORC1 and/or mTORC2 and the control of the localization and activity of p27^{Kip1}.

V. Control of Mitotic Entry by TORCs

Compared to G1/S progression, far less is known about the potential role of (m)TOR in controlling later stages in the cell cycle. However, several recent studies, from eukaryotes as evolutionarily diverse as yeast and mammals, reveal that (m)TOR signaling also regulates events linked to mitotic entry.

Nakashima *et al.* [43] showed that temperature-sensitive (ts) mutants in KOG1 (which encodes the TORC1 component raptor in *S. cerevisiae*) experience delayed mitotic entry at nonpermissive temperatures, G2-phase being prolonged under these conditions. Rapamycin exerts a similar effect, indicating that TORC1 is required for G2→M progression. Two main lines of evidence indicate that Cdc5, a "polo-like" kinase, plays a key role in this: first, overexpressing Cdc5 (but not an inactive mutant) overcomes the defective mitotic entry seen in KOG1^{ts} cells and, second, in KOG1^{ts} cells Cdc5 fails to undergo normal activation as the cell cycle proceeds. In addition, Cdc5 normally localizes to the nucleus, but does not do so in KOG1 mutant cells. Further data from this study implicate the protein phosphatase PP2A, whose activity can be controlled by TORC1 [44], as being involved in the nuclear localization and activation of Cdc5 (Figure 8.1). Yeast Cdc5 is also required for exit from mitosis but this function is not perturbed by rapamycin or the KOG1^{ts} mutant.

As already mentioned (see Figure 8.2), eEF2K is regulated by mTORC1. However, this effect is indirect as eEF2K is not phosphorylated by mTORC1 [12]. mTORC1 signaling acts to inhibit eEF2K by regulating its phosphorylation at three inhibitory sites, Ser78, 359, and 366 [45]. Ser366 is phosphorylated by the S6 kinases. Smith and Proud [12] used a biochemical approach to isolate and identify the Ser359 kinase, revealing it to be cdc2/ cyclin B (cdc2 is also termed CDK1). Treating cells with roscovitine, an inhibitor of cdc2, blocked the phosphorylation of eEF2K at Ser359 and increased the phosphorylation of eEF2, consistent with cdc2 being an important eEF2K kinase in living cells. Although treating cells with rapamycin did not markedly affect cdc2 activity, two other manipulations that, respectively, inhibit or activate mTORC1 signaling (amino acid starvation and ablation of TSC2) did affect cdc2 activity in ways consistent with cdc2 being positively regulated by mTORC1 [12].

The phosphorylation of eEF2 is low in mitotic cells [12], consistent with the ability of cdc2/cyclin B to phosphorylate and inactivate eEF2K. This may serve to maintain the translation elongation machinery in an active state during mitosis, when certain proteins continue to be made [46]. Some reports indicate that 4E-BP1 is dephosphorylated in G2/M-phase, leading to suppression of cap-dependent translation (and implying that mTORC1 signaling is inhibited) [7, 46]. However, other data indicate that 4E-BP1 phosphorylation is high during mitosis and that cdc2 phosphorylates Thr70, a regulatory site in 4E-BP1 [8]. Thus, cdc2 may help to maintain active translation initiation and elongation.

It is not known how amino acids/mTORC1 regulate cdc2/cyclin B activity against eEF2K or whether its activity against other substrates is regulated in a similar way. The finding that leucine starvation delayed of mitotic entry of HeLa cells [12] suggests that the mitosis-promoting function of cdc2/cyclin B is also affected by mTORC1 signaling. This implies that mTORC1 promotes entry into S-phase and into mitosis (Figure 8.1).

Studies in *Schizosaccharomyces pombe* [47] and in *Drosophila* [48] have revealed quite different situations. In *Drosophila* S2 cells, insulin delays the progression from S-phase through G2/M, an effect that is countered by rapamycin [48], implying that here dTORC1 signaling inhibits the progress of cells through this stage of the cell cycle. Indeed, modestly impairing dTORC1 signaling (but not dTORC2) in S2 cells can actually speed up cell division and increase cell number. For example, treating S2 cells with rapamycin caused cells to pass more rapidly from S-phase through G2/M. This contrasts with the positive role played by dTORC1 in G1 and indicates that, in this system, insulin and dTORC1 have opposing effects on cell cycle progression. Consistent with this, treating fly larvae with low concentrations of rapamycin actually increased the numbers of cells in the wings of the resulting adult animals. The biochemical mechanisms underlying these very interesting effects remain to be clarified. The authors speculate that this effect may allow nutrient-deficient cells to pass rapidly through mitosis; interestingly, starving *S. pombe* (fission yeast) cells also accelerates their rate of division [49], resulting in decreased cell size.

Treating *S. pombe* cells with rapamycin also advances mitosis [47] (and decreases cell size) suggesting that TORC1 controls mitotic progression here. However, rapamycin apparently interferes with the functions of TORC1 and TORC2 in *S. pombe* [47]; furthermore, both complexes also appear to be regulated by amino acid (nitrogen) availability in this species. The data indicate that TORC2 controls mitotic entry in this organism, which should prompt studies of the role of this (usually rapamycin-insensitive) complex in M-phase progression in other systems. A further difference between *S. pombe* and many other systems is that cell size control is mainly exerted during G2 rather than G1.

What links nutrient control of TOR signaling to M-phase progression in *S. pombe*? Petersen and Nurse [47] show that TOR negatively regulates Sty1, a MAP kinase that normally promotes mitotic entry by activating Cdc2 through a Polo kinase termed Plo1 [50]. This regulation involves a nutrient-sensitive kinase, Gcn2, but is apparently independent of its only characterized substrate, the translation initiation factor eIF2. The elegant work of Petersen and Nurse thus provides important new insights into the connections between nutrients, TOR complexes and cell cycle control. It also raises many new and important questions.

(m)TOR therefore regulates mitotic entry in organisms as diverse as budding or fission yeast and mammalian cells, albeit in contrasting ways. There are important differences between these systems: first, in the fission yeast experiments, the quality of the nitrogen available to an organism that can make its own amino acids is being altered. Here, it may thus be appropriate for cells to continue to divide but at smaller cell size. In mammals, inhibiting mTORC1 signaling mimics the effect of starving cells for essential amino acids: in such auxotrophic organisms, stopping division seems the logical response to this situation.

VI.   A Link Between Mitochondrial Function, mTORC1, and Cell Cycle Progression?

Cell cycle progression—especially the biosynthetic processes required for this—makes high demands in terms of energy and nutrients, it would thus be logical for there to be a "metabolic checkpoint" that modulates cell cycle progression. Several disparate lines of evidence suggest that this may be so. For example, depletion of ATP, due to nutrient limitation, can trigger G1-arrest via a mechanism involving the activation of the AMP-activated protein kinase (AMPK; a sensor of cellular energy [ATP/AMP] levels [51]). A link to mTORC1 signaling is suggested by work on the fluctuations in mitochondrial membrane potential that occur during G1 [52]. The mitochondrial membrane potential ($\Delta\Psi_m$) is greater in Jurkat cells in late G1 than in cells early in G1. Cells in S and G2/M (obtained by cell sorting) also exhibit higher $\Delta\Psi_m$ values. Given that mTORC1 signaling can be controlled by AMPK [53, 54], these investigators then analyzed mTORC1 complexes in cells showing high or low values for $\Delta\Psi_m$. They found that cells with higher $\Delta\Psi_m$ also showed higher levels of mTOR-raptor association (i.e., mTORC1), although there was no apparent change in the phosphorylation of the mTORC1 target S6K1 [52]. The underlying mechanism and significance of these observations is unclear: however, these studies do point to a further connection between mTORC1 and mitochondrial (oxidative) function inasmuch as rapamycin inhibited the increase in O_2 consumption induced by serum. These data are consistent with the emerging data showing that mTORC1 signaling promotes mitochondrial function in mammalian cells [55, 56], at least in part through the transcriptional regulators PGC-1α and YY1.

It is possible that there exists a "metabolic checkpoint" late in G1 which may integrate a variety of signals including ones emanating from mTORC1 [52].

VII.   mTORC1, Ribosome Biogenesis, and Cell Cycle Control

mTORC1 also positively regulates ribosome biogenesis [57]; since ribosome biogenesis is required for continued protein synthesis, prolonged inhibition of mTORC1 will impair protein synthesis by reducing the cellular capacity for mRNA translation as well as through the shorter term (minutes) effects on the phosphorylation and/or activity of translation factors (described above). In fact, ribosome biogenesis is crucial for cell proliferation, as cells must double their number of ribosomes if they are to maintain their capacity for protein synthesis following division. A number of recent studies point to the existence of a "ribosome checkpoint," involving p53, which couples defects in ribosome production to cell cycle progression at the G1/S boundary.

Producing ribosomes is a complicated process, partly because of the large numbers of components which must be made in stoichiometric proportions (four types of rRNA and >80 different proteins). Insufficient synthesis of some components may lead to an unwanted excess of others. mTORC1 and TOR orthologs in other organisms are important in promoting rRNA synthesis [57]. In mammals, the mRNAs encoding ribosomal proteins (termed "5'-TOP mRNAs" because they contain a regulatory element tract rich in pyrimidines at their 5'-ends) are subject to translational control by mTORC1 signaling, which stimulates ribosomal protein production. The mechanisms underlying the activation of 5'-TOP mRNA translation by mTORC1 signaling are unclear [58, 59]. Interfering with ribosome biogenesis (by impairing rRNA processing [60] or via an imbalance in the output of ribosomal proteins, discussed in Ref. [61]) can activate p53 and thereby lead to cell cycle arrest in G1.

For example, a truncation version of the protein Bop1 ("Bop1Δ"), which plays a key role in processing rRNA [62], inhibits the maturation of 28S and 5.8S rRNA and thus interferes with production of 60S ribosomes.

Expressing Bop1Δ arrests cells in G1, leading to inhibition of Cdk4 and Cdk2 and hyperphosphorylation of Rb [60]. p53 is required for these effects indicating that they are not caused by 60S deficiency *per se*. Consistent with this, Bop1Δ causes increased expression of p21 and p27, which are transcriptional targets of p53.

How could perturbations in 60S subunit biogenesis regulate p53? Certain ribosomal proteins (e.g., rpL11, a component of the large subunit) can bind to mdm2 (or its human ortholog, hdm2) [63, 64]. Other 60S proteins (rpL5, rpL23) can also bind to Hdm2. Deficient production of rRNAs of the 60S subunit is expected to cause accumulation of free 60S proteins, such as rpL11, which would then interact with mdm2/hdm2. The disruption of the nucleolus, and concomitant release of rp's into the cytoplasm (where they encounter mdm2) likely plays a role at least under some conditions where ribosome biogenesis is perturbed [65]. Mdm2/hdm2 can bind to p53, acting as a ubiquitin ligase to bring about its destruction. Association of mdm2/hdm2 with rpL11 leads to the stabilization of p53. p53 levels therefore increase, promoting the transcription of specific genes (including the cell cycle inhibitors p27 and p14). This response is logical, since it would prevent cell cycle progress under conditions where insufficient 60S ribosomal subunits are being produced: it may be termed a "ribosomal stress checkpoint." Data have been presented that the unfolded protein response brings about cell cycle arrest via this mechanism [64]. As mTORC1 signaling affects the transcription of rRNAs [57, 66, 67], this mechanism may provide an additional link between the mTORC1 pathway and cell cycle progression.

Regulation of p53 is not restricted to proteins of the 60S subunit. For example, rpS7 (of the 40S subunit) also binds to mdm2 causing stabilization of p53 [68]. Disruption of the rpS6 gene in mice, or knockdown of rpS6 by siRNA, also causes activation of p53 [69], even in the absence of nucleolar disruption. It is now clear that hypomorphic mutations in genes for ribosomal proteins underlie a number of disease states including Diamond Blackfan anemia (which, in addition to low red cell counts, is associated with congenital abnormalities and increased risk of leukemia, reviewed in Ref. [70]). It remains to be established whether ribosome stress-induced of p53 plays a role in the etiology of such conditions.

VIII.   Conclusions and Perspective

Recent data demonstrate that TOR proteins control several stages of the cell cycle in organisms as distantly related as budding and fission yeasts and mammalian cells. This is consistent with the central role that TOR signaling plays in coordinating diverse responses to a wide range of stimuli. Further work is required to delineate the underlying mechanisms, especially in higher eukaryotes. mTOR signaling has emerged as a strong therapeutic candidate in a number of settings, and further understanding of its role in cell cycle control will aid the development of drugs targeting (m)TOR as potential treatments for hyperplastic disorders including cancers.

REFERENCES

1. Kay, J.E., Kromwel, L., Doe, S.E., and Denyer, M. (1991). Inhibition of T and B lymphocyte proliferation by rapamycin. *Immunology* 72:544–549.
2. Scherer, M.N., Banas, B., Mantouvalou, K., Schnitzbauer, A., Obed, A., Kramer, B.K., and Schlitt, H.J. (2007). Current concepts and perspectives of immunosuppression in organ transplantation. *Langenbecks Arch Surg* 392:511–523.
3. Butt, M., Connolly, D., and Lip, G.Y. (2009). Drug-eluting stents: a comprehensive appraisal. *Future Cardiol* 5:141–157.
4. Feldman, M.E., Apsel, B., Uotila, A., Loewith, R., Knight, Z.A., Ruggero, D., and Shokat, K.M. (2009). Active-site inhibitors of mTOR target rapamycin-resistant outputs of mTORC1 and mTORC2. *PLoS Biol* 7:e38.
5. Thoreen, C.C., Kang, S.A., Chang, J.W., Liu, Q., Zhang, J., Gao, Y., Reichling, L.J., Sim, T., Sabatini, D.M., and Gray, N.S. (2009). An ATP-competitive mTOR inhibitor reveals rapamycin-insensitive functions of mTORC1. *J Biol Chem* 284:8023–8032.
6. Wang, X., Beugnet, A., Murakami, M., Yamanaka, S., and Proud, C.G. (2005). Distinct signaling events downstream of mTOR cooperate to mediate the effects of amino acids and insulin on initiation factor 4E-binding proteins. *Mol Cell Biol* 25:2558–2572.
7. Pyronnet, S., Dostie, J., and Sonenberg, N. (2001). Suppression of cap-dependent translation in mitosis. *Genes Dev* 15:2083–2093.
8. Heesom, K., Gampel, A., Mellor, H., and Denton, R.M. (2001). Cell cycle-dependent phosphorylation of the translational repressor eIF4E binding protein-1 (4E-BP1). *Curr Biol* 11:1374–1379.
9. Ruvinsky, I., and Meyuhas, O. (2006). Ribosomal protein S6 phosphorylation: from protein synthesis to cell size. *Trends Biochem Sci* 31:342–348.
10. Shahbazian, D., Roux, P.P., Mieulet, V., Cohen, M.S., Raught, B., Taunton, J., Hershey, J. W., Blenis, J., Pende, M., and Sonenberg, N. (2006). The mTOR/PI3K and MAPK pathways converge on eIF4B to control its phosphorylation and activity. *EMBO J* 25:2781–2791.
11. Wang, X., Li, W., Williams, M., Terada, N., Alessi, D.R., and Proud, C.G. (2001). Regulation of elongation factor 2 kinase by p90[RSK1] and p70 S6 kinase. *EMBO J* 20:4370–4379.
12. Smith, E.M., and Proud, C.G. (2008). cdc2-cyclin B regulates eEF2 kinase activity in a cell cycle- and amino acid-dependent manner. *EMBO J* 27:1005–1016.
13. Boyer, D., Quintanilla, R., and Lee-Fruman, K.K. (2008). Regulation of catalytic activity of S6 kinase 2 during cell cycle. *Mol Cell Biochem* 307:59–64.
14. Wanke, V., Cameroni, E., Uotila, A., Piccolis, M., Urban, J., Loewith, R., and De, V.C. (2008). Caffeine extends yeast lifespan by targeting TORC1. *Mol Microbiol* 69:277–285.
15. Urban, J., Soulard, A., Huber, A., Lippman, S., Mukhopadhyay, D., Deloche, O., Wanke, V., Anrather, D., Ammerer, G., Riezman, H., Broach, J.R., De, V.C., Hall, M.N., and

Loewith, R. (2007). Sch9 is a major target of TORC1 in *Saccharomyces cerevisiae*. *Mol Cell* 26:663–674.

16. Wanke, V., Pedruzzi, I., Cameroni, E., Dubouloz, F., and De, V.C. (2005). Regulation of G_0 entry by the Pho80-Pho85 cyclin-CDK complex. *EMBO J* 24:4271–4278.

17. Sherr, C.J., and Roberts, J.M. (1999). CDK inhibitors: positive and negative regulators of G1-phase progression. *Genes Dev* 13:1501–1512.

18. Jiang, H., Coleman, J., Miskimins, R., and Miskimins, W.K. (2003). Expression of constitutively active 4EBP-1 enhances p27Kip1 expression and inhibits proliferation of MCF7 breast cancer cells. *Cancer Cell Int* 3:2.

19. Luo, Y., Marx, S.O., Kiyokawa, H., Koff, A., Massague, J., and Marks, A.R. (1996). Rapamycin resistance tied to defective regulation of $p27^{kip1}$. *Mol Cell Biol* 16:6744–6751.

20. Takuwa, N., Fukui, Y., and Takuwa, Y. (1999). Cyclin D1 expression mediated by phosphatidylinositol 3-kinase through mTOR-p70(S6K)-independent signaling in growth factor-stimulated NIH 3T3 fibroblasts. *Mol Cell Biol* 19:1346–1358.

21. Hleb, M., Murphy, S., Wagner, E.F., Hanna, N.N., Sharma, N., Park, J., Li, X.C., Strom, T.B., Padbury, J.F., Tseng, Y.T., and Sharma, S. (2004). Evidence for cyclin D3 as a novel target of rapamycin in human T lymphocytes. *J Biol Chem* 279:31948–31955.

22. Decker, T., Hipp, S., Ringshausen, I., Bogner, C., Oelsner, M., Schneller, F., and Peschel, C. (2003). Rapamycin-induced G1 arrest in cycling B-CLL cells is associated with reduced expression of cyclin D3, cyclin E, cyclin A, and survivin. *Blood* 101:278–285.

23. Breslin, E.M., White, P.C., Shore, A.M., Clement, M., and Brennan, P. (2005). LY294002 and rapamycin co-operate to inhibit T-cell proliferation. *Br J Pharmacol* 144:791–800.

24. Averous, J., Fonseca, B.D., and Proud, C.G. (2007). Regulation of cyclin D1 expression by mTORC1 signaling requires eukaryotic initiation factor 4E-binding protein 1. *Oncogene* 27:1106–1113.

25. Garcia-Morales, P., Hernando, E., Carrasco-Garcia, E., Menendez-Gutierrez, M.P., Saceda, M., and Martinez-Lacaci, I. (2006). Cyclin D3 is down-regulated by rapamycin in HER-2-overexpressing breast cancer cells. *Mol Cancer Ther* 5:2172–2181.

26. Hipp, S., Ringshausen, I., Oelsner, M., Bogner, C., Peschel, C., and Decker, T. (2005). Inhibition of the mammalian target of rapamycin and the induction of cell cycle arrest in mantle cell lymphoma cells. *Haematologica* 90:1433–1434.

27. Hashemolhosseini, S., Nagamine, Y., Morley, S.J., Desrivieres, S., Mercep, L., and Ferrari, S. (1998). Rapamycin inhibition of the G1 to S transition is mediated by effects on cyclin D1 mRNA and protein stability. *J Biol Chem* 273:14424–14429.

28. Culjkovic, B., Topisirovic, I., Skrabanek, L., Ruiz-Gutierrez, M., and Borden, K.L. (2006). eIF4E is a central node of an RNA regulon that governs cellular proliferation. *J Cell Biol* 175:415–426.

29. Rosenwald, I.B., Kaspar, R., Rousseau, D., Gehrke, L., Leboulch, P., Chen, J.J., Schmidt, E.V., Sonenberg, N., and London, I.M. (1995). Eukaryotic translation initiation factor 4E regulates expression of cyclin D1 at transcriptional and post-transcriptional levels. *J Biol Chem* 270:21176–21180.

30. Rosenwald, I.B., Lazaris-Karatzas, A., Sonenberg, N., and Schmidt, E.V. (1993). Elevated levels of cyclin D1 protein in response to increased expression of eukaryotic initiation factor 4E. *Mol Cell Biol* 13:7358–7363.

31. Topisirovic, I., Ruiz-Gutierrez, M., and Borden, K.L. (2004). Phosphorylation of the eukaryotic translation initiation factor eIF4E contributes to its transformation and mRNA transport activities. *Cancer Res* 64:8639–8642.

32. Culjkovic, B., Topisirovic, I., Skrabanek, L., Ruiz-Gutierrez, M., and Borden, K.L. (2005). eIF4E promotes nuclear export of cyclin D1 mRNAs via an element in the 3′UTR. *J Cell Biol* 169:245–256.

33. Topisirovic, I., Siddiqui, N., Lapointe, V.L., Trost, M., Thibault, P., Bangeranye, C., Pinol-Roma, S., and Borden, K.L. (2009). Molecular dissection of the eukaryotic initiation factor 4E (eIF4E) export-competent RNP. *EMBO J* 28:1087–1098.
34. Culjkovic, B., Tan, K., Orolicki, S., Amri, A., Meloche, S., and Borden, K.L. (2008). The eIF4E RNA regulon promotes the Akt signaling pathway. *J Cell Biol* 181:51–63.
35. Culjkovic, B., Topisirovic, I., and Borden, K.L. (2007). Controlling gene expression through RNA regulons: the role of the eukaryotic translation initiation factor eIF4E. *Cell Cycle* 6:65–69.
36. Rong, L., Livingstone, M., Sukarieh, R., Petroulakis, E., Gingras, A.C., Crosby, K., Smith, B., Polakiewicz, R.D., Pelletier, J., Ferraiuolo, M.A., and Sonenberg, N. (2008). Control of eIF4E cellular localization by eIF4E-binding proteins, 4E-BPs. *RNA* 14:1318–1327.
37. Dostie, J., Ferraiuolo, M., Pause, A., Adam, S.A., and Sonenberg, N. (2000). A novel shuttling protein, 4E-T, mediates the nuclear import of the mRNA 5′ cap-binding protein, eIF4E. *EMBO J* 19:3142–3156.
38. Rousseau, D., Kaspar, R., Rosenwald, I., Gehrke, L., and Sonenberg, N. (1996). Translation initiation of ornithine decarboxylase and nucleocytoplasmic transport of cyclin D1 messenger-RNA are increased in cells overexpressing eukaryotic initiation factor 4E. *Proc Natl Acad Sci USA* 93:1065–1070.
39. De Benedetti, A., and Graff, J.R. (2004). eIF-4E expression and its role in malignancies and metastases. *Oncogene* 23:3189–3199.
40. Graff, J.R., Konicek, B.W., Carter, J.H., and Marcusson, E.G. (2008). Targeting the eukaryotic translation initiation factor 4E for cancer therapy. *Cancer Res* 68:631–634.
41. Hong, F., Larrea, M.D., Doughty, C., Kwiatkowski, D.J., Squillace, R., and Slingerland, J.M. (2008). mTOR-raptor binds and activates SGK1 to regulate p27 phosphorylation. *Mol Cell* 30:701–711.
42. Garcia-Martinez, J.M., and Alessi, D.R. (2008). mTOR complex 2 (mTORC2) controls hydrophobic motif phosphorylation and activation of serum- and glucocorticoid-induced protein kinase 1 (SGK1). *Biochem J* 416:375–385.
43. Nakashima, A., Maruki, Y., Imamura, Y., Kondo, C., Kawamata, T., Kawanishi, I., Takata, H., Matsuura, A., Lee, K.S., Kikkawa, U., Ohsumi, Y., Yonezawa, K., and Kamada, Y. (2008). The yeast Tor signaling pathway is involved in G2/M transition via polo-kinase. *PLoS ONE* 3:e2223.
44. Di Como, C.J., and Arndt, K.T. (1996). Nutrients, via the TOR proteins, stimulate the association of Tap42 with Type phosphatases. *Genes Dev* 10:1904–1916.
45. Proud, C.G. (2007). Signalling to translation: how signal transduction pathways control the protein synthetic machinery. *Biochem J* 403:217–234.
46. Pyronnet, S., and Sonenberg, N. (2001). Cell-cycle-dependent translational control. *Curr Opin Gen Dev* 11:13–18.
47. Petersen, J., and Nurse, P. (2007). TOR signalling regulates mitotic commitment through the stress MAP kinase pathway and the Polo and Cdc2 kinases. *Nat Cell Biol* 9:1263–1272.
48. Wu, M.Y., Cully, M., Andersen, D., and Leevers, S.J. (2007). Insulin delays the progression of Drosophila cells through G2/M by activating the dTOR/dRaptor complex. *EMBO J* 26:371–379.
49. Fantes, P.A., and Nurse, P. (1978). Control of the timing of cell division in fission yeast. Cell size mutants reveal a second control pathway. *Exp. Cell Res.* 115:317–329.
50. Petersen, J., and Hagan, I.M. (2005). Polo kinase links the stress pathway to cell cycle control and tip growth in fission yeast. *Nature* 435:507–512.
51. Hardie, D.G. (2007). AMP-activated/SNF1 protein kinases: conserved guardians of cellular energy. *Nat Rev Mol Cell Biol* 8:774–785.

52. Schieke, S.M., McCoy, J.P., Jr., and Finkel, T. (2008). Coordination of mitochondrial bioenergetics with G1 phase cell cycle progression. *Cell Cycle* 7:1782–1787.
53. Inoki, K., Ouyang, H., Zhu, T., Lindvall, C., Wang, Y., Zhang, X., Yang, Q., Bennett, C., Harada, Y., Stankunas, K., Wang, C.Y., He, X., MacDougald, O.A., You, M., Williams, B. O., and Guan, K.L. (2006). TSC2 integrates Wnt and energy signals via a coordinated phosphorylation by AMPK and GSK3 to regulate cell growth. *Cell* 126:955–968.
54. Inoki, K., Zhu, T., and Guan, K.L. (2003). TSC2 mediates cellular energy response to control cell growth and survival. *Cell* 115:577–590.
55. Cunningham, J.T., Rodgers, J.T., Arlow, D.H., Vazquez, F., Mootha, V.K., and Puigserver, P. (2007). mTOR controls mitochondrial oxidative function through a YY1-PGC-1alpha transcriptional complex. *Nature* 450:736–740.
56. Schieke, S.M., Phillips, D., McCoy, J.P., Jr., Aponte, A.M., Shen, R.F., Balaban, R.S., and Finkel, T. (2006). The mammalian target of rapamycin (mTOR) pathway regulates mitochondrial oxygen consumption and oxidative capacity. *J Biol Chem* 281:27643–27652.
57. Mayer, C., and Grummt, I. (2006). Ribosome biogenesis and cell growth: mTOR coordinates transcription by all three classes of nuclear RNA polymerases. *Oncogene* 25:6384–6391.
58. Hamilton, T.L., Stoneley, M., Spriggs, K.A., and Bushell, M. (2006). TOPs and their regulation. *Biochem Soc Trans* 34:12–16.
59. Patursky-Polischuk, I., Stolovich-Rain, M., Hausner-Hanochi, M., Kasir, J., Cybulski, N., Avruch, J., Ruegg, M.A., Hall, M.N., and Meyuhas, O. (2009). The TSC-mTOR pathway mediates translational activation of TOP mRNAs by insulin largely in a raptor- or rictor-independent manner. *Mol Cell Biol* 29:640–649.
60. Pestov, D.G., Strezoska, Z., and Lau, L.F. (2001). Evidence of p53-dependent cross-talk between ribosome biogenesis and the cell cycle: effects of nucleolar protein Bop1 on G(1)/S transition. *Mol Cell Biol* 21:4246–4255.
61. Ferreira-Cerca, S., and Hurt, E. (2009). Cell biology: arrest by ribosome. *Nature* 459:46–47.
62. Strezoska, Z., Pestov, D.G., and Lau, L.F. (2000). Bop1 is a mouse WD40 repeat nucleolar protein involved in 28S and 5. 8S RRNA processing and 60S ribosome biogenesis. *Mol Cell Biol* 20:5516–5528.
63. Lohrum, M.A., Ludwig, R.L., Kubbutat, M.H., Hanlon, M., and Vousden, K.H. (2003). Regulation of HDM2 activity by the ribosomal protein L11. *Cancer Cell* 3:577–587.
64. Zhang, F., Hamanaka, R.B., Bobrovnikova-Marjon, E., Gordan, J.D., Dai, M.S., Lu, H., Simon, M.C., and Diehl, J.A. (2006). Ribosomal stress couples the unfolded protein response to p53-dependent cell cycle arrest. *J Biol Chem* 281:30036–30045.
65. Mayer, C., and Grummt, I. (2005). Cellular stress and nucleolar function. *Cell Cycle* 4:1036–1038.
66. Hannan, K.M., Brandenburger, Y., Jenkins, A., Sharkey, K., Cavanaugh, A., Rothblum, L., Moss, T., Poortinga, G., McArthur, G.A., Pearson, R.B., and Hannan, R.D. (2003). mTOR-dependent regulation of ribosomal gene transcription requires S6K1 and is mediated by phosphorylation of the carboxy-terminal activation domain of the nucleolar transcription factor UBF. *Mol Cell Biol* 23:8862–8877.
67. James, M.J., and Zomerdijk, J.C. (2004). Phosphatidylinositol 3-kinase and mTOR signaling pathways regulate RNA polymerase I transcription in response to IGF-1 and nutrients. *J Biol Chem* 279:8911–8918.
68. Chen, D., Zhang, Z., Li, M., Wang, W., Li, Y., Rayburn, E.R., Hill, D.L., Wang, H., and Zhang, R. (2007). Ribosomal protein S7 as a novel modulator of p53-MDM2 interaction: binding to MDM2, stabilization of p53 protein, and activation of p53 function. *Oncogene* 26:5029–5037.

69. Fumagalli, S., Di, C.A., Neb-Gulati, A., Natt, F., Schwemberger, S., Hall, J., Babcock, G.F., Bernardi, R., Pandolfi, P.P., and Thomas, G. (2009). Absence of nucleolar disruption after impairment of 40S ribosome biogenesis reveals an rpL11-translation-dependent mechanism of p53 induction. *Nat Cell Biol* 11:501–508.
70. Flygare, J., and Karlsson, S. (2007). Diamond-Blackfan anemia: erythropoiesis lost in translation. *Blood* 109:3152–3154.

9

TORC1 Signaling in Budding Yeast

ROBBIE LOEWITH

Department of Molecular Biology and NCCR Program "Frontiers in Genetics"
University of Geneva
Geneva, Switzerland

I. Abstract

Although small in size, the budding yeast *Saccharomyces cerevisiae* has contributed a lion's share of major advances to our understanding of target of rapamycin signaling. Indeed, rapamycin was first identified by virtue of its antifungal activity. Furthermore, both TOR (target of rapamycin) and the two protein complexes in which this conserved kinase resides were first described in yeast. In this chapter, I focus on the rapamycin-sensitive TOR complex 1 (TORC1) of *S. cerevisiae*. I highlight the recent discoveries of a novel upstream regulator and downstream effectors with an emphasis on the contributions made by system-wide studies such as the characterization of the rapamycin-sensitive phosphoproteome. The conservation of the pathways up- and downstream of *S. cerevisiae* TORC1 reconfirm that yeast remains a valuable system to elucidate how eukaryote cell growth is controlled.

II. The Discovery of TOR

Rapamycin, a secondary metabolite produced by the soil bacterium *Streptomyces hygroscopicus*, was first identified by virtue of its antifungal properties [1, 2]. In the nearly 35 years since this discovery, rapamycin has been found to have potent immunosuppressive, cytostatic, and antiaging

ISSN NO: 1874-6047
DOI: 10.1016/S1874-6047(10)27009-9

properties [3–5]. Clinical trials prompted by these observations have so far led to the approval of rapamycin and related compounds to treat renal cell carcinoma, to immunosuppress organ transplant recipients, and to prevent in-stent restenosis following angioplasty [6–8] but surely more clinical indications will be soon to follow.

What is this compound targeting to elicit such diverse effects on human biology? This was the question on many minds in the late 1980s; remarkably, it was a simple yet elegant genetic screen performed in budding yeast that provided the answer [9]. Characterization of spontaneous *Saccharomyces cerevisiae* mutants that were able to form colonies in the presence of an otherwise cytostatic concentration of rapamycin demonstrated the involvement of three genetic loci. The locus most frequently hit was the highly conserved *FPR1* gene. The FPR1 protein, also known as FKBP12, is a small proline isomerase normally involved in protein folding. *FPR1* is not an essential gene and thus FKBP12 is not the cytostatic target of rapamycin; rather it was determined to be a cofactor required for the toxicity of rapamycin [9]. The cytostatic targets of rapamycin (actually the rapamycin·FKBP12 dimer) were encoded by the two other loci identified in this screen which were thus named *TOR1* and *TOR2*.

The rapamycin-resistant *TOR1* and *TOR2* alleles were eventually cloned, sequenced and found to encode paralogs that structurally resemble phophatidylinositol (PI) kinases [10–12]. These rare alleles contained single missense mutations, Ser1972Arg and Ser1975Ile, respectively, which prevent the binding of rapamycin·FKBP12 to the encoded proteins and thus confer dominant resistance to rapamycin [10, 13]. Subsequently, *TOR* genes have been found in all eukaryote genomes examined, but unlike yeast, which possess two *TOR*s most other eukaryotes possess only a single *TOR* gene [14]. This quirk of yeast biology turned out to be quite providential as described below. Curiously, although they resemble lipid kinases, TOR proteins are the founding members of a family of related Ser/Thr protein kinases known as phosphatidylinositol kinase-related kinases (PIKKs; [15]). The significance, if any, of the resemblance to lipid kinases is not known.

III.   The Discovery of TOR Complexes

Following their discovery, many groups using a variety of model systems (including yeasts, slime mould, worms, plants, flies, mammalian tissue culture cells, and eventually mice), launched studies aimed at understanding the function of TOR proteins. The take-home message from these studies is that TOR proteins function in signaling pathways that respond to environmental cues to coordinate cell and organismal growth [16]. Although most of these early studies relied on rapamycin to interrogate TOR function, genetic and biochemical dissection of TOR signaling in *S. cerevisiae* turned out to be especially informative.

In particular, the phenotypes of *tor1*, *tor2*, and *tor1 tor2* mutants were most revealing [10–12]. Loss of *TOR1* results in very mild phenotypes (slow growth and inability to grow at temperature extremes). In contrast, loss of *TOR2* is lethal, causing a random cell-cycle arrest and a pronounced depolarization of the actin cytoskeleton [17]. Simultaneous loss of both *TOR1* and *TOR2*, like exposure to rapamycin, is also lethal but cells arrest predominantly in G_1/G_0. These observations suggested that (a) TOR1 and TOR2 have redundant functions (e.g., transit through G_1); (b) these redundant functions are inhibited by rapamycin; and (c) TOR2 performs additional functions that TOR1 is unable to perform (e.g., maintenance of actin polarity). The observation that *TOR1-1* cells grow in the presence of rapamycin additionally suggested that (d) the TOR2-specific function is not sensitive to rapamycin [18].

As described in Chapter 1 of this volume, molecular support for these genetic predictions was provided with the biochemical purifications of TOR1 and TOR2 [19–22]. These studies demonstrated that TOR proteins function in two distinct multiprotein complexes named TOR complex 1 (TORC1) and TOR complex 2 (TORC2). Either TOR1 or TOR2 can function in TORC1 whereas only TOR2, and not TOR1, can assemble into TORC2. Furthermore, it was observed that rapamycin·FKBP12 can bind to either TOR1 or TOR2 but only in the context of TORC1 [20]. These observations satisfyingly explained why TOR2 has both a redundant function with TOR1 (i.e., as the catalytic component of TORC1) and a unique function (i.e., as the catalytic component of TORC2). The observation that TORC1 but not TORC2 is bound by rapamycin·FKBP12 explained why only the redundant TOR function is inhibited by rapamycin [20]. Like TOR, the TOR complexes are conserved from yeast to man [20, 23–25] (Chapter 2).

As eluded to above and discussed by others in this volume, TORC1 in mammals is implicated in pathologies ranging from cancer to aging. Consequently, an understanding of the signaling pathways in which this kinase complex operates is a major objective of many laboratories. Historically, studies in budding yeast have been vital in these efforts. As TOR signaling in yeast was comprehensively reviewed not long ago [26], in this chapter I aim to highlight the more recent advances specifically concerning the pathways up- and downstream of TORC1 in *S. cerevisiae* (referred to simply as "yeast" for the remainder of this chapter). Yeast TORC2 signaling will be addressed in the following chapter.

IV.   What is TORC1 ?

TORC1 is an ∼2 MDa complex consisting of LST8, KOG1, TCO89, and either TOR1 or TOR2 (Figure 9.1) [26]. With the exception of TCO89, all of these proteins have obvious mammalian orthologs (TOR1/2 = mTOR, LST8 = mLST8, KOG1 = raptor) [16]. Conversely, mammalian TORC1 (mTORC1) contains two proteins, PRAS40 and DEPTOR, not obviously conserved in yeast [27, 28] (Chapter 2). The TOR-associated proteins found in TORC1 appear to function positively with TOR in terms of signal propagation, though, specific functions of these proteins are not known. TORC1 is likely oligomeric (drawn as dimer in Figure 9.1) built on a TOR–TOR backbone; this oligomerization is not obviously altered by rapamycin or environmental cues known to alter TORC1 activity [26].

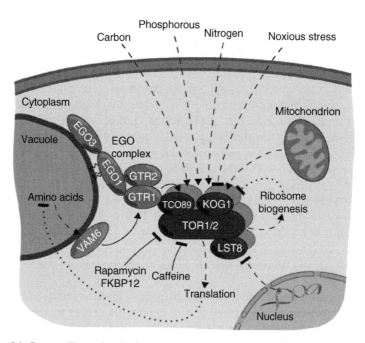

FIG. 9.1 Cartoon illustrating the known physiological cues that influence TORC1 activity. As illustrated, many environmental cues are known to influence TORC1 activity. Understanding the mechanisms by which these cues are sensed and how this information is communicated to TORC1 remains a major challenge in the field. Intravacuolar amino acids may influence the activity of VAM6, a guanine–nucleotide exchange factor for the GTR1 GTPase. As part of the EGO complex, GTP loaded GTR1 binds and somehow activates TORC1 kinase activity via the TORC1 component TCO89. TORC1 seems to be inhibited by translation and/or ribosome biogenesis rates in apparent feedback loops (dotted lines) (see text for details).

V. Where is TORC1?

Several independent studies using fluorescent-protein tagged TOR1 or other TORC1-specific components clearly show that TORC1 is concentrated on the limiting membrane of the vacuole [29–32]. These studies are largely congruent with previous localization studies employing alternative techniques [26]. However, it is not possible to exclude that a fraction of TORC1 resides elsewhere. Indeed the Zheng group controversially claims that TORC1 shuttles in and out of the nucleolus to regulate rDNA expression [33]. Like complex stability, the vacuolar localization of TORC1 is not obviously altered by environmental cues known to alter TORC1 activity [30]. The vacuolar localization of TORC1 is provocative in that nutrient levels appear to be an important determinant of TORC1 kinase activity (see Section VI) and the vacuole is a major nutrient reservoir in yeast cells [26].

VI. What Regulates TORC1?

Transcriptome profiling showed a remarkably similar response of yeast cells exposed to rapamycin, nutrient starvation, noxious stressors and, interestingly, caffeine [26, 34] strongly suggesting that TORC1 activity is inhibited by various forms of stress. The recent *bona fide* characterization of a TORC1 substrate, the AGC kinase SCH9 [32], allowed this hypothesis to be tested directly. Using SCH9 phosphorylation as a proxy for *in vivo* TORC1 activity confirmed that TORC1 is inhibited by a wide range of environmental stresses, including caffeine, nutrient starvation (carbon, nitrogen, phosphorous), hyperosmolarity (NaCl, KCl, sorbitol), oxidants (H_2O_2, diamide), ethanol, weak organic acids, heat shock, and SDS (Figure 9.1; [32, 35] and Loewith lab, unpublished observations).

TORC1 additionally responds to intracellular cues; DNA damage or mitochondrial dysfunction, for example, decrease TORC1 activity (A. Uotila, S. Kawai, and R. Loewith, unpublished). Intriguingly, signals from downstream effectors regulate TORC1 activity in apparent feedback loops: blocking translation with cycloheximide increases TORC1 activity, possibly by increasing the intracellular concentrations of free amino acids [30, 32, 36]; and deletion of *SCH9* or *SFP1* (which encodes a TORC1-regulated transcription factor—discussed below) hyperactivates TORC1, presumably as a consequence of reduced ribosome biogenesis (Figure 9.1; [37]).

Although many cues that affect TORC1 activity are now known, the mechanisms by which these cues are sensed and how this information is signaled to TORC1 is at present essentially unknown. Indeed, the identification of regulators upstream of TORC1 remains a major challenge, although some progress in this vein has been made recently.

A. THE EGO COMPLEX COUPLES AMINO ACID CUES TO TORC1

Exposure to rapamycin causes yeast to arrest in a G_0-like state. For some yeast strains this arrest is irreversible but for others, including BY4741, cells can eventually reenter the cell cycle subsequent to rapamycin removal. Interested in understanding how cells exit from G_0, the De Virgilio group probed the collection of viable deletion mutants (which is in the BY4741 background) to identify mutants that cannot exit from rapamycin-induced growth arrest [38]. Three of the eight *ego* mutants identified in this screen were characterized in more detail. EGO1 (aka MEH1 or GSE2), EGO3 (aka SLM4, NIR1, or GSE1), and GTR2 were found to form a stable complex—the EGO complex—that localizes, like TORC1, to the vacuolar membrane. Subsequently, a fourth component of this complex, GTR1, was identified (Figure 9.1; [39]. EGO1 and EGO3 appear to be yeast-specific proteins that anchor the complex to the vacuolar membrane. Conversely, GTR1 and GTR2 are small GTPases highly conserved in eukaryotes; GTR1 is the ortholog of mammalian RagA and RagB, and GTR2 is the ortholog of mammalian RagC and RagD.

The importance of the EGO complex in TORC1 signaling suggested by this initial yeast work became much clearer with work in fly and mammalian cell lines where the Rag GTPases were recently shown to couple amino acid cues to mammalian TORC1 [40]; and described in detail in Chapter 5. Subsequently, very similar observations were made in yeast (Figure 9.1; [30]). When amino acids are available, the EGO complex component GTR1 associates with TORC1, and SCH9 is phosphorylated; conversely, leucine starvation results in a reduced or weakened association and SCH9 dephosphorylation. Critically, expression of a GTR1 mutant unable to hydrolyze GTP partially blocks the leucine-starvation induced dissociation from TORC1 and subsequent SCH9 dephosphorylation.

The vacuolar localizations of neither the EGO complex nor TORC1 is obviously altered by leucine starvation [30]. This observation contrasts with the model in mammalian cells which proposes that amino acids, via the Rag GTPases, recruit mTORC1 to specific endomembranes for activation [41]. However, biochemical and genetic studies in yeast strongly imply that correct membrane localization is important for TORC1 function [42, 43]. At present it is not known how binding of the EGO complex alters TORC1 activity, but the yeast-specific TORC1 component TCO89 appears to play an important role as the EGO complex fails to regulate TORC1 activity in *tco89* mutants [30]. Given the conservation of the RAG GTPases and TORC1, one might speculate that a functional homolog of TCO89 remains to be indentified in metazoans. Curiously, the nucleotide loading of GTR1 and GTR2 appears to occur in a reciprocal fashion: the GTR1^{GTP}·GTR2^{GDP} form of the EGO complex stimulates TORC1 kinase activity toward SCH9, whereas, the GTR1^{GDP}·GTR2^{GTP} form antagonizes TORC1 activity.

It is tempting to hypothesize that the EGO complex responds to levels of intravacuolar amino acids, but how this would occur is not clear. VAM6, a vacuolar membrane protein and a conserved guanine–nucleotide exchange factor, binds to and regulates the nucleotide-binding status of GTR1, and thus could potentially be involved in coupling vacuolar amino acid signals to the EGO complex [30]. It will be very interesting to determine if mammalian orthologs of VAM6 function similarly.

VII.   What Does TORC1 Regulate?

Studies over the years with rapamycin have identified a number of growth-related processes under the control, or at least influence, of TORC1. In general terms, active TORC1 stimulates processes that promote the accumulation of mass and antagonizes processes involved in responding to stress (Figure 9.2A). The signaling pathways and distal effectors involved in these processes are known to varying extents as discussed in the following sections.

A.   TORC1 PROXIMAL EFFECTORS

There are two major effectors of TORC1 responsible for propagating the essential signals downstream of this complex. One of these is the AGC kinase SCH9. Introduced above, SCH9 is presently the best characterized TORC1 substrate [32, 44]. *In vivo* and *in vitro* data demonstrate that TORC1 directly phosphorylates five to six serine and threonine residues in the carboxy terminus of SCH9. Replacement of these residues with alanine yields a nonfunctional version of SCH9; cells expressing SCH9^{5A} like *sch9* cells are very nearly dead but quick growing "adapters" arise at a high frequency making it very difficult to work with these strains. Conversely, replacement with acidic amino acids yields a hypomorphic version (SCH9^{DE}) that unlike wt SCH9 retains activity following inhibition of TORC1 with rapamycin [32]; and Loewith lab, unpublished observations). SCH9 has been proposed to be the functional ortholog of the mTORC1 substrate S6K [32].

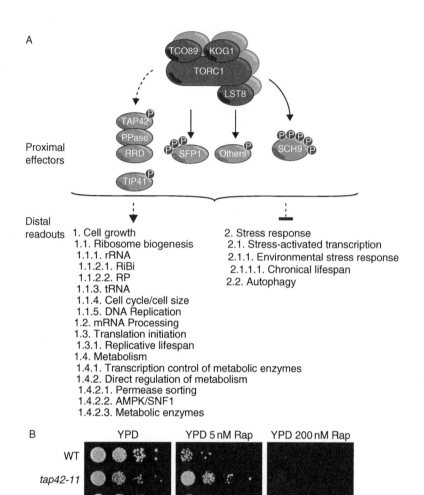

FIG. 9.2 TAP42 and SCH9 mediate the two major signaling outputs of TORC1. (A) Cartoon illustrating the major effectors and distal growth-related readouts downstream of TORC1. (B) TAP42 and SCH9 act in parallel downstream of TORC1. Tenfold serial dilutions of yeast cells expressing the indicated alleles of *TAP42* and *SCH9* were spotted onto rich medium (YPD) or YPD containing sublethal (5 nM) or lethal (200 nM) concentrations of rapamycin (rap) for 3–5 days at 25 °C. *tap42-11* encodes a temperature-sensitive allele of TAP42 which somehow renders TAP42-associated phosphatases insensitive to upstream signals from TORC1. In *SCH9^{DE}*, the codons encoding residues phosphorylated by TORC1 have been replaced with codons encoding aspartate or glutamate to mimic phosphorylation. This version of SCH9 retains activity in the absence of TORC1 function.

Type 2A (PP2A) and related protein phosphatases constitute the second major effector branch downstream of TORC1. Yeast cells encode three PP2A catalytic subunits (PPH21, PPH22, and PPH3; collectively PP2Ac) and several PP2A-related catalytic subunits, including SIT4. These enzymes are notoriously difficult to study due to the fact that they assemble into a number of different complexes (i.e., they have many different binding partners) and thus have pleiotropic functions, some of which are essential [26]. Analyses of SIT4 functions are additionally complicated by strain-dependent polymorphisms at the *SSD1* locus. Deletion of *SIT4* is lethal in *SSD1-d* backgrounds but viable in *SSD1-v* backgrounds. As SIT4 is an important mediator of TORC1 function, *SSD1* allele status will in some circumstances cause strain-specific responses to rapamycin [21].

A role for these phosphastases in TORC1 signaling was first uncovered by the Arndt lab [45]. In this work, a novel subset of PP2Ac and SIT4 complexes containing an essential regulatory subunit called TAP42 were described. Subsequently, two peptidyl-prolyl *cis/trans*-isomerases, RRD1 and RRD2, were characterized as components of these complexes (TAP42·PP2Ac·RRD2, and TAP42·SIT4·RRD1; [46]). The peptidyl-prolyl *cis/trans*-isomerase activity of the RRDs is required for activation of the associated phosphatase [47]. Presently, it is supposed that when TORC1 is active, TAP42 is phosphorylated and remains tightly bound to the phosphatase·RRD dimer [45, 46, 48]; upon TORC1 inactivation, TAP42 becomes dephosphorylated and dissociates from the phosphatase·RRD dimer which then presumably becomes active and/or has an altered access to substrates [49, 50]. How TORC1 regulates this proposed transition is not entirely clear. TORC1 may phosphorylate TAP42 directly [48], or TORC1 may dissociate TAP42 indirectly via another phosphoprotein known as TIP41 [51]. What is clear is that these complexes have an important role downstream of TORC1: *rrd1*, *rrd2*, and *tip41* deletions and temperature-sensitive alleles of *tap42*, *pp2ac*, and *sit4* all confer partial resistance to rapamycin [26]. As these genes appear to be conserved, it seems likely that the mammalian orthologs will have similar functions downstream of mTORC1. Indeed, α4, the mammalian ortholog of TAP42, was recently demonstrated to be an essential regulator of PP2A in mice [52]. However, the functional details elucidated in this study differ from the yeast model of TAP42 function highlighting the ambiguity in current models of PP2A regulation.

Depending on the yeast background employed, genetic uncoupling of SCH9 or TAP42·PPase function from TORC1 regulation yields weak to moderate resistance to rapamycin. In the TB50 background, $SCH9^{DE}$ cells and *tap42-11* (a ts allele of *TAP42*) cells grow only slightly better than isogenic wild-type cells in the presence of sublethal concentrations of rapamycin. Combining $SCH9^{DE}$ and *tap42-11* alleles, however, yields cells that grow, albeit slowly, in the presence of even very high concentrations of rapamycin [53]; Figure 9.2B). This demonstrates that SCH9 and TAP42 are the major effectors downstream of TORC1 and that both branches perform one or more functions essential for viability. The observation that growth of $SCH9^{DE}$ *tap42-11* cells is slowed by rapamycin predicts that TORC1 may have additional direct substrates.

B. THE RAPAMYCIN-SENSITIVE PHOSPHOPROTEOME

To find new TORC1 substrates, and to better understand the signaling pathways that couple TORC1 to its known distal readouts, label-free quantitative mass spectroscopy was recently employed to define the rapapmycin-sensitive phosphoproteome [53]. Although the coverage was far from complete—a total of 120 rapamycin-sensitive phosphoproteins were identified likely representing approximately one-third of the rapamycin-sensitive phosphoproteome—secondary assays indicate a low rate of false positives. The majority of these proteins (up to 75% depending on the cutoffs employed) have mammalian sequence homologs, reconfirming the importance of yeast studies in elucidating general properties of TOR signaling conserved in bigger eukaryotes. Excitingly, 109 of the 120 rapamycin-sensitive phosphoproteins were not previously implicated in TORC1 signaling. Repeating the mass spectroscopy analyses in $SCH9^{DE}$ and *tap42-11* cells allowed phosphorylation events to be partitioned between these effector branches.

In a complementary study, rapamycin-induced protein relocalization was assayed in 4159 GFP-tagged strains (representing $\sim$2/3 of all ORFs) with the presumption that a change in localization upon rapamycin treatment implies that the tagged protein functions downstream of TORC1 [53]. Indeed, a similar approach was used previously to identify TORC1-regulated transcription factors [55]. The data from these studies will be instrumental in defining how TORC1 signals to its distal readouts and are incorporated, where relevant, in the discussions below.

C. SIGNALING TO DISTAL READOUTS

In the presence of favorable environmental conditions, the primary functions of TORC1 are (1) to promote the accumulation of mass (i.e., growth) and (2) to inhibit stress responses that are incompatible with rapid growth (Figure 9.2A).

1. TORC1 Stimulates Cell Growth

1.1. Ribosome Biogenesis

Rapid growth requires robust protein synthesis. To support such robust protein synthesis, rapidly growing yeast cells need to produce incredible numbers of ribosomes and tRNAs which themselves require accessory factors (aka ribosome biogenesis or RiBi factors) for their processing and assembly. Expression of the genes encoding these factors requires the concerted activities of all three nuclear RNA polymerases (RNA Pols). In fact, approximately 95% of total transcription in rapidly growing yeast cells is dedicated to ribosome biogenesis. As this consumes an enormous amount of energy, yeast cells have developed tight control mechanisms to regulate and coordinate the expression of these genes [56, 57]. As illustrated in Figure 9.3, much of this regulation is mediated by TORC1-dependent signals.

1.1.1. Regulation of RNA Pol I, Figure 9.3A.   Production of rRNA is a limiting step in ribosome synthesis and is under the control of TORC1 as evidenced by the observation that rapamycin treatment elicits a dramatic reduction in the expression and processing of the RNA Pol I-transcribed 35S precursor rRNA [58, 59]. Rapamycin does not appear to alter the number of active genes within the rDNA repeats but rather decreases the association of RNA Pol I with active genes [59]. In this study, RRN3, a transcription factor required for RNA Pol I recruitment to the rDNA, was proposed to be the major target of regulation. Curiously, the ability of TORC1 to stimulate RNA Pol I activity appears to be largely mediated by SCH9 but the association of RRN3 with RNA Pol I does not appear to be the critical target of SCH9 in this regulation ([53]; Figure 9.3A).

1.1.2.1. Regulation of RNA Pol II—*RiBi* Genes, Figure 9.3B. The *RiBi* regulon consists of ~200 protein-encoding genes plus ~75 small nucleolar RNA genes that are coordinately expressed according to growth conditions. Collectively, these nonribosomal protein (RP) gene products serve to boost translation capacity by functioning in the expression, processing, assembly, nuclear import/export, etc. of the protein synthesis machinery (ribosomal subunits, tRNAs, etc.; [44, 56]. TORC1 regulates *RiBi* gene expression via SCH9 and SFP1 [32, 44].

RiBi genes characteristically have an *RRPE* and/or *PAC* element in their promoter. *RRPE* elements are bound by the SIN3-binding protein STB3 whereas *PAC* elements are bound by the myb-family transcription factors DOT6 and its homolog TOD6 [61–63]. These transcription factors appear to be repressors that function by recruiting the RPD3 histone deacetylase

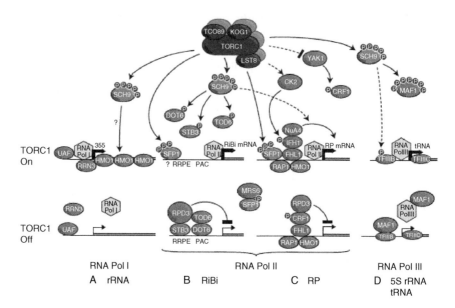

Fig. 9.3 Regulation of "ribosome biogenesis" by TORC1. (A) Regulation of rRNA expression. Yeast cells contain ~150 copies of the rDNA repeat, about 1/2 of which are active in rapidly growing cells. RNA Pol I is recruited to the promoters of these active genes via interactions with its essential initiation factor RRN3 and the upstream activation factor (UAF) and the transcribed region is bound by the high-mobility group protein HMO1. Addition of rapamycin (TORC1 off) dramatically reduces the occupancy of RNA Pol I, RRN3, and HMO1 but not UAF at active loci [60, 114]. SCH9 appears to be the major mediator of TORC1 signals to the rDNA [53]. SCH9 signals apparently do not influence the interaction between RNA Pol I and RRN3 but instead may target HMO1 [114–116]. (B) Regulation of *RiBi* gene expression. The *RRPE* and/or *PAC* elements found in *RiBi* gene promoters are bound by STB3 and TOD6/DOT6, respectively, but it is not known if these factors are bound constitutively. STB3 and DOT6/TOD6 appear to be direct substrates of SCH9 but how phosphorylation affects their activity is not known. These transcription factors likely function by recruiting the RPD3 histone deacetylase complex to repress transcription. TORC1 directly phosphorylates SFP1 which then activates *RiBi* gene expression in undefined ways. Hypophosphorylated SFP1, generated upon TORC1 inhibition, is sequestered in the cytoplasm by the Rab escort protein MRS6. (C) Regulatin of *RP* gene expression. When TORC1 is active, the forkhead-like transcription factor FHL1 binds the IFH1 coactivator which possibly leads to the corecruitment of the NuA4 histone acetyltransferase complex. TORC1 may regulate IFH1 phosphorylation via casein kinase 2 (CK2). The roles of SFP1, HMO1, and RAP1 are not clear although their presence seems to be necessary for stable FHL1 binding. Inactivation of TORC1 results in activation of the YAK1 kinase and phosphorylation of the CRF1 corepressor which subsequently displaces IFH1 from FHL1 and likely results in the recruitment of the RPD3 histone deacetylase complex. How SCH9 signals stimulate *RP* transcription is not clear. (D) Regulation of RNA Pol III activity. RNA Pol III regulation at a generic tDNA locus is depicted. RNA Pol III is recruited via the initiation factors TFIIIB and TFIIIC. In rapidly growing cells, MAF1 is heavily phosphorylated and predominantly cytoplasmic. Potentially a component of TFIIIB is also phosphorylated under these conditions. Upon TORC1 inhibition, MAF1 is dephosphorylated, accumulates in the nucleus and binds RNA Pol III and possibly the dephosphorylated TFIIIB. Binding of both RNA Pol III and TFIIIB appear to be essential to block tDNA transcription.

complex to silence the expression of adjacent genes [61, 64–67]. Interestingly, STB3 and DOT6/TOD6 are phosphorylated downstream of SCH9 [53], apparently directly (Loewith lab, unpublished) but the significance of these phosphorylation events has not yet been clarified. Thus, it is very likely that the regulation of *RiBi* genes by SCH9 observed previously is mediated by these transcription factors.

SFP1 is a putative transcription factor with a split zinc-finger at its carboxyl terminus. SFP1 binds to, and is phosphorylated by, TORC1 and thus, as predicted, SFP1 represents a third important effector of TORC1 in addition to SCH9 and TAP42 (Figure 9.2; [37]). In actively growing cells, SFP1 is mostly nuclear and stimulates *RiBi* gene expression [44, 68, 69]. SFP1 cannot be ChIPed at *RiBi* gene promoters and thus it is unclear how this factor can activate these genes [56]. Upon TORC1 inactivation with rapamycin, SFP1 is partially dephosphorylated [37] and anchored in the cytoplasm through interaction with a Rab escort protein known as MRS6 [70] and thus unable to activate *RiBi* genes.

1.1.2.1.1.   Regulation of RiBi proteins. Recently, it has been observed that the phosphorylation status and/or localization of a number of RiBi factors is altered upon brief exposure to rapamycin (Table 9.1; [53, 54]). These studies support previous reports [71] that beyond regulating the transcription of genes encoding RiBi factors, TORC1 also plays a much more direct (since exposure to rapamycin was for only a limited time— presumably before transcriptional effects become manifest) role in ribosome biogenesis.

1.1.2.2.   Regulation of RNA Pol II—Ribosomal Protein (*RP*) Genes, Figure 9.3C. TORC1 regulates the expression of *RP* genes in part via the fork head transcription factor FHL1 [26]. FHL1 binding to RP promoters appears to be constitutive and in some cases is facilitated by HMO1, a high-mobility group protein and RAP1, an essential protein with pleiotropic functions [26]. When TORC1 is active, FHL1 binds a coactivator IFH1; when TORC1 is inactive, IFH1 is displaced and replaced by a corepressor CRF1 although CRF1-independent mechanisms function in some strains [72]. Binding to these coregulators is mediated by their TORC1-dependent phosphorylation. Casein kinase 2 could potentially be the IFH1 kinase while YAK1 appears to be the CRF1 kinase ([73, 74]; J. Merwin and D. Shore, unpublished). How TORC1 regulates these phosphorylation events is not known. When active, these promoters typically are bound by the NuA4 histone acetyltransferase complex; when inactive, by the RPD3 histone deacetylase complex [65, 75, 76].

SCH9 and SFP1 also serve to couple TORC1 to *RP* gene expression. SFP1 can be ChIPed at *RP* promoters and sophisticated genetic interactions

TABLE 9.1

RIBOSOME BIOGENESIS FACTORS WHOSE PHOSPHORYLATION AND/OR LOCALIZATION IS
ALTERED BY RAPAMYCIN TREATMENT

Gene name	Function of encoded protein	Phosphorylation (P);localization (L) altered by rapamycin
FAP7	Small subunit synthesis	P
SRP40	Preribosome assembly/transport	P
LHP1	Maturation of tRNA and U6 snRNA precursors	P
RRP14	Constituent of 66S preribosomal particles	P
RRN3	RNA Pol I initiation factor	P, L
NOP9	90S preribosome/SSU processome	L
NHP2	rRNA processing	L
NSR1	rRNA processing	L
CBF5	90S preribosome/SSU processome	L
UTP4	90S preribosome/SSU processome	L
UTP5	90S preribosome/SSU processome	L
UTP18	90S preribosome/SSU processome	L
UTP22	90S preribosome/SSU processome	L
UTP25	90S preribosome/SSU processome	L
PWP2	90S preribosome/SSU processome	L
BFR2	90S preribosome/SSU processome	L
NOP58	90S preribosome/SSU processome	L
NOP56	90S preribosome/SSU processome	L
RRP5	90S preribosome/SSU processome	L
RPC10	Subunit of RNA Pols I, II, III	L
RPA12	Subunit of RNA Pol I	L
RPA135	Subunit of RNA Pol I	L
RPA34	Subunit of RNA Pol I	L
RPA43	Subunit of RNA Pol I	L
RPA49	Subunit of RNA Pol I	L
RPA190	Subunit of RNA Pol I	L
MRT4	Large subunit synthesis	L
NOG1	Small subunit synthesis	L
NAF1	rRNA processing	L
SHQ1	rRNA processing	L

Bolded genes have obvious mammalian homologs; YOGY and http://www.yeastgenome.org/.

studies suggest that SFP1 influences *RP* transcription via FHL1 and IFH1
while SCH9 appears to function in a parallel, but as of yet, uncharacterized
pathway [32, 44, 55].

1.1.3. Regulation of RNA Pol III, Figure 9.3D. RNA Pol III
transcribes the 5S, tRNA, and other stable nonconding RNAs in a
rapamycin-sensitive fashion [59]. Characterization of the rapamycin-
sensitive phosphoproteome demonstrated that TORC1 via SCH9

stimulates RNA Pol III activity by antagonizing the function of MAF1, a conserved RNA Pol III repressor [53]. Deletion of *MAF1* completely blocks the rapamycin-induced decrease in tRNA production. The mechanisms by which MAF1 inhibits RNA Pol III activity are not completely clear but likely involve binding to both RNA Pol III and the RNA Pol III initiation complex TFIIIB [77]. SCH9 phosphorylates seven residues in MAF1 and this is sufficient to prevent MAF1 from binding to RNA Pol III. Curiously, versions of MAF1 in which these seven residues have been replaced with either alanine or glutamate are still able to complement a *MAF1* deletion. This suggests that SCH9 phosphorylates an additional target involved in MAF1-dependent repression of RNA Pol III activity. Dephosphorylation of this hypothetical target—a component of TFIIIB would be a good guess— would subsequently recruit MAF1 to help repress RNA Pol III activity.

1.1.4. Regulation of the Cell Cycle/Cell Size. Yeast cells commit to a new round of cell division only after attaining a critical cell size, which varies according to growth conditions [78]. How yeast cells couple environmental cues to cell growth and cell cycle is a fascinating yet poorly defined process. To tackle this problem on a broad level the Tyers group performed a large-scale screen in yeast to identify mutants that appeared to be involved in size control. Amazingly, the two smallest mutants recovered in this screen were *sfp1* and *sch9* [44]. This observation, coupled with the fact that both the cell size threshold and the rate of ribosome biogenesis are dictated by environmental conditions, suggest the following cascade from environmental signals to G_1 to S phase transition (START) [78]:

Environment $\rightarrow$ $\rightarrow$ TORC1 $\rightarrow$ SFP1/SCH9 $\rightarrow$ ribosome biogenesis $\rightarrow$ $\rightarrow$ START

How the rate of ribosome biogenesis influences START is not clear. Potentially, downstream affects on translational output may contribute. The unstable G_1 cyclin Cln3, for example, requires a robust rate of translation to attain a critical abundance to drive cell-cycle progression [79]. Intriguingly, two START regulators, WHI5—a repressor of G_1 transcription [80] and MSA1—an activator of G_1 transcription [81], appear to be differentially phosphorylated upon rapamycin treatment. Although it is not immediately apparent how ribosome biogenesis might play a role in this regulation, it is tempting to think that the phosphorylation of these proteins could be involved in coupling growth cues to the G_1 to S phase transition.

An elegant study by the Kamada group recently demonstrated that TORC1, seemingly independently of ribosome biogenesis, also influences G_2/M transition [82]. This group found that cells expressing a temperature-sensitive allele of *kog1* (and also cells treated with rapamycin) have a mitotic delay due to a prolonged G_2 phase. A search for multicopy suppressors of the *kog1* allele led to the discovery that TORC1 via TAP42·P-P2Ac regulates the subcellular localization of the polo-like kinase CDC5. CDC5 activity is required to destabilize SWE1, a kinase that phosphorylates and thus inactivates the mitotic cyclin-dependent kinase CDC28. Consistently, the dysregulated localization of CDC5 in *kog1-105* cells led to the inappropriate stabilization of SWE1, inactivation of CDC28 and consequently prolonged G_2. Like budding yeast, *Schizosaccharomyces pombe* also appears to use TOR to regulate G_2/M transition, but, in this yeast, TORC2 (rather than TORC1) signals delay (rather than accelerate) G_2/M transit ([83]. It will of interest to see which yeast system will better model metazoan cell division.

1.1.5. Regulation of DNA Replication. Perhaps a consequence of the altered G_1 to S phase transition described above, rapamycin treatment was found to alter the phosphorylation of a number of proteins involved in DNA replication. Prominent in this class were factors involved in formation of the prereplication complex, MCM3, MCM10, and the DNA replication licensing factor TAH11 [53].

1.2. Regulation of mRNA Processing

In addition to regulating transcription, rapamycin treatment is known to alter the turnover of a large number of mRNAs likely via multiple mechanisms [84]. Interestingly, several factors involved in mRNA degradation/ processing were identified as rapamycin-sensitive phosphoproteins and/or their localization was altered upon rapamycin treatment (Table 9.2; [53, 54]. These data may help explain how TORC1 regulates mRNA stability.

1.3. Regulation of Translation Initiation

Although underappreciated at the time, a major breakthrough in TOR signaling came with a report from the Hall lab that rapamycin-sensitive TOR signaling regulates mRNA translation initiation [79]. This observation was key as it was the first indication that TOR functions to couple environmental cues to the cell growth machinery.

The best understood translation-related target of TORC1 is the Ser/Thr kinase GCN2. The kinase activity of GCN2 is stimulated by binding to uncharged tRNAs [85]. When active, GCN2 phosphorylates the translation initiation factor eIF2 and this phosphorylation event ultimately greatly reduces global translation initiation rates. TORC1 represses GCN2 activity, and thus promotes global translation, via both SCH9 and TAP42·PPase branches. GCN2 is negatively regulated by phosphorylation on Ser-577 which reduces tRNA binding and subsequently decreases kinase activity.

TABLE 9.2

mRNA Processing Factors Whose Phosphorylation and/or Localization is Altered by Rapamycin Treatment

Gene name	Function of encoded protein	Phosphorylation (P);localization (L) altered by rapamycin
CWC2	Pre-mRNA splicing	L
PRP9	Pre-mRNA splicing	L
SNU66	Pre-mRNA splicing	P
SKY1	SR protein kinase	P
GBP2	SR protein involved in mRNA transport	P
HRB1	SR protein involved in mRNA transport	P
PAT1	Deadenylation-dependent mRNA decay	P
CCR4	Deadenylation-dependent mRNA decay	P
NOT3	Deadenylation-dependent mRNA decay	P
MPP6	Degradation of cryptic noncoding mRNAs	P

Bolded genes have obvious mammalian homologs; YOGY and http://www.yeastgenome.org/.

Rapamycin treatment causes a rapid dephosphorylation of this residue in a TAP42·PPase-dependent fashion. Subsequently, GCN2 activity increases, eIF2α becomes phosphorylated and translation rates decline [86]. SCH9 also regulates eIF2α phosphorylation, seemingly in parallel to TAP42 [32]. SCH9 is not the GCN2-Ser-577 kinase (Loewith lab, unpublished).

In addition to eIF2α, TORC1 likely regulates other translation factors as well. Previous studies have suggested that eIF4G, and possibly the eIF4E-binding protein EAP1, are TORC1 effectors [26]. Furthermore, several translation-related targets were found to be differentially phosphorylated upon rapamycin treatment [53]. An interesting member of this list is the DEaD-box helicase DED1, an essential protein required for translation initiation of all yeast mRNAs [87]. Whether or not these are direct targets or secondary effects downstream of eIF2α phosphorylation remains to be determined.

TORC1 via SCH9 also regulates the phosphorylation of ribosomal protein S6 [26, 53]. Interestingly, the rapamycin-sensitive phosphorylation of S6 is conserved from yeast to man but its role remains mysterious [88, 89]. Additional ribosomal proteins also appear to be differentially phosphorylated upon rapamycin treatment [53]. It will be very interesting to determine the function of these phosphorylation events.

1.3.1.   Protein Synthesis and Replicative Lifespan.   Dietary restriction is, at present, the only protocol known generally to slow aging in eukaryota. Given the links between nutrient abundance and TORC1 activity described above, it is perhaps not surprising that epistasis studies in flies, worms, and yeast have confirmed that dietary restriction regulates longevity, at least in part, via TORC1 [90]. In yeast, aging is studied by two assays: *Replicative Lifespan* is defined by the number of daughter cells a single cell can produce prior to senescence; *Chronological Lifespan* is defined by the length of time a cell can survive in a quiescent state. TORC1 signaling figures prominently in both of these aging paradigms; but, strikingly, different TORC1 effector pathways are implicated. Specifically, TORC1 appears to influence Replicative Lifespan via its regulation of protein synthesis and Chronological Lifespan via its regulation of stress-response programs (consequently chronological lifespan is discussed in the "Stress Programs" section below).

The importance of TORC1 signaling in replicative lifespan was cemented in a painstaking study where 564 strains randomly picked from the collection of ~4800 viable yeast gene deletions were individually assayed by microdissection for increased replicative lifespan [91]. In addition to *tor1Δ* and *sch9Δ*, *rpl31aΔ* and *rpl6bΔ* cells were also found to be long-lived. A subsequent study determined that deletion of many but not all genes encoding 60S ribosomal protein subunits or assembly factors confer increased replicative lifespan [92]. These observations suggest that dietary restriction via TORC1 may regulate replicative lifespan by altering mRNA translation [93]. This idea is supported and extended to other eukaryotes, including vertebrates, by several recent publications [3, 94–96]. However, many questions remain; foremost among these: by what mechanism does translation affect lifespan? Differential translation of specific target messages could be one possible explanation [90]. Curiously, reduced formation of 60S ribosomal subunits, but not blocking of translation *per se*, is sufficient to increase replicative lifespan. In mutants where formation of 60S ribosomal subunits is compromised, general translation is greatly reduced but translation of *GCN4* is relatively increased. GCN4 is a starvation-responsive transcription factor and, importantly, its deletion prevents full lifespan extension in these mutants [92]. The relevant transcription targets of GCN4 are not known, nor is it yet known if functional homologs of GCN4 similarly influence longevity in other organisms. Lastly, although translation appears to play an important role in replicative aging in yeast, other TORC1 outputs may additionally contribute.

1.4. Regulation of Metabolism

As a central controller of cell growth, it has been rather assumed that TORC1 signals influence cellular metabolism [16]. This is almost certain to be true but the published evidence supporting this assumption, especially in yeast, is not particularly robust. The best support for this assumption comes from genome-wide expression studies of rapamycin-treated cells [26].

1.4.1.   Indirect Regulation of Metabolism—Transcription. Expression profiling using microarrays has been an informative approach to elucidate pathways downstream of TORC1 [26]. In addition to diminishing *Ribosomal protein* and *RiRi* gene expression, TORC1 inhibition also reduces the expression of genes encoding glycolytic enzymes, but increases the expression of genes encoding citric acid cycle enzymes, suggesting that rapamycin causes a metabolic switch from fermentation to respiration. This prediction has now been confirmed [97]. However one might surmise that even though the abundance of these mRNAs is dramatically altered, this would not be reflected at the protein level given the continued general depression of translation in the presence of rapamycin. Presumably, changes in transcript levels would be more biologically significant during transient inhibition of TORC1 (and subsequent transient depression of translation) following environmentally relevant shifts in nutrient quality. Similar arguments can be made for the regulation of genes involved in "nitrogen discrimination" and "retrograde response" signaling.

1.4.1.1.   Nitrogen Discrimination. In the presence of preferred nitrogen sources (e.g., glutamine, asparagine, or ammonia), the GATA-family transcription factor GLN3 is bound by its cytoplasmic anchor URE2 [26]. In the absence of these high-quality nitrogen sources, or in the presence of rapamycin, GLN3 is released from URE2, translocates to the nucleus and induces expression of "nitrogen discrimination" genes encoding proteins required to import and catabolize poor nitrogen sources such as proline and allantoin. It is important to note here that the nitrogen quality signals to URE2·GLN3 are transmitted in part by TORC1 and in part by independent pathways. This regulation presumably involves direct TAP42·PPase action on URE2 and GLN3, both of which are phosphoproteins. However, mounting evidence demonstrates that this regulation is complex and additionally varies depending on the yeast strain employed ([98]; and references therein). Thus, caution should be employed when using the expression of GLN3 targets as a proxy for TORC1 activity [98, 99].

1.4.1.2.   Retrograde Response. The retrograde response is a mitochondria-to-nucleus signaling pathway that transmits changes in mitochondrial function to nuclear gene expression [100]. This pathway culminates with a pair of bZip/HLH transcription factors, RTG1 and RTG3, which activate genes encoding citric acid cycle and peroxisomal enzymes required for *de novo* amino acid biosynthesis in general and glutamine/glutamate homeostasis in particular. TORC1 antagonizes RTG1/3 activity presumably by promoting their interaction with a cytoplasmic anchor composed of MKS1 and the redundant 14-3-3 proteins, BMH1 and BMH2. TORC1 inactivation causes MKS1 to associate with an inhibitor, RTG2, and a subsequent release of RTG1/3. How TORC1 signals impinge upon MKS1 and/or RTG2 has not been completely elucidated but presumably involves TAP42·PPases [26, 49].

1.4.2. Direct Regulation of Metabolism. TORC1 can also influence metabolism through more direct routes that do not involve transcription.

1.4.2.1. Permease Sorting. Budding yeast is a very choosy eater—when provided a mixture of nutrients, preferred ones will be exclusively consumed before suboptimal ones are utilized. TORC1 signals influence this hierarchical consumption by regulating the sorting/activity of various nutrient permeases [26]. Under optimal growth conditions, many high-affinity, substrate-specific permeases are active at the plasma membrane. In contrast, in nutrient-poor conditions, a few low-affinity, broad-specificity permeases are active at the plasma membrane. Nutrient shifts (up or down) result in the ubiquitination and subsequent vacuolar degradation of plasma membrane permeases. TORC1 appears to regulate these sorting events via the TAP42·PPase target NPR1 [26]. NPR1 is a Ser/Thr kinase that becomes hyperactive upon TORC1 inhibition but how NPR1 regulates permease sorting is not understood. Interestingly, at least a dozen (sugar, amino acid, phosphate, sulfate, proton, zinc, etc.), permeases/transporters were identified as rapamycin-sensitive phosphoproteins [53]. If and how phosphorylation of these permeases affects their localization and/or activity remains to be determined.

1.4.2.2. Inhibition of AMP-Activated Protein Kinase AMPK/SNF1. SNF1 is the yeast ortholog of mammalian AMPK. SNF1 functions in a heterotrimeric complex and its kinase activity is induced by a number of cellular stresses. In general terms, active SNF1 stimulates ATP-producing catabolic reactions and inhibits ATP-consuming anabolic reactions both directly, through phosphorylation of metabolic enzymes, and indirectly, through changes in transcription [101]. Full activation of SNF1 requires phosphorylation on Thr210. This residue is phosphorylated by three kinases (SAK1, TOS3, and ELM1) and dephosphorylated by REG1-GLC7 protein phosphatase 1, but the mechanistic details of how these enzymes are themselves regulated is not known. Interestingly, Thr210 becomes hyperphosphorylated upon rapamycin treatment [102]. This result suggests that TORC1 additionally regulates metabolism via inhibition of SNF1. Interestingly, although AMPK is best characterized as an upstream regulator of metazoan TORC1 [103], a recent study has suggested that analogous to the situation in yeast, mTORC1 also inhibits AMPK [95].

1.4.2.3. Phosphorylation of Metabolic Enzymes. Lastly, the recent elucidation of the rapamycin-sensitive phosphoproteome identified a number of metabolic enzymes that appear to be differentially phosphorylated upon rapamycin treatment (Table 9.3; [53]). As can be seen from this list, a large number of metabolic pathways appear to be affected, suggesting that we have only a rudimentary understanding of how TORC1 influences metabolic processes.

2. TORC1 antagonizes Stress Responses

In addition to stimulating cell growth, signals from TORC1 play an equally important role in suppressing stress-response programs. Indeed, constitutive activation of any of several stress-responsive pathways is incompatible with rapid growth and toxic for yeast cells. Thus, it is essential that these pathways be repressed if cells are to realize their maximal growth potential. Analogous to the discussion above, much of the repressive signals emanating from TORC1 impinge upon transcription programs but additional, more direct targets, also have been described.

2.1. Stress-Activated Transcription

TORC1 regulates a number of stress-responsive transcription factors [26]. The transcription factors that control the expression of "nitrogen discrimination" and "retrograde response" genes have been described above. In addition, nutrient stresses can trigger different developmental responses in yeast. For example, the absence of glucose leads to induction of filamentous/invasive growth of haploid cells whereas the absence of both fermentable carbon and high-quality nitrogen sources triggers meiosis and sporulation in diploids [104]. The suppression of meiosis and sporulation by fermentable carbon and high-quality nitrogen sources appears to be mediated by TORC1 as TORC1 antagonizes the nuclear localization and stabilization of IME1, a transcription factor and master regulator of developmental process [105]. TORC1 may regulate filamentous growth by regulating the phosphorylation and presumably the activity of the Ser/Thr kinase KSP1 [53].

2.1.1. Environmental Stress Response. Transcription profiling has demonstrated that yeast cells induce a stereotypical transcription program in response to the presence of any of a number of noxious stresses [106, 107]. This so-called "environmental stress response" is mediated in large

TABLE 9.3

Metabolic Enzymes Differentially Phosphorylated upon
Rapamycin Treatment

Gene name	Function of encoded protein
AMD1	AMP deaminase
PFK1,2	Phosphofructokinase
FOL1	Folic acid biosynthesis
PRS5	5-Phospho-ribosyl-1a-pyrophosphate synthetase
HOM3	Aspartate kinase
LYS12	Lysine biosynthesis
FBA1	Fructose 1,6-bisphosphate aldolase
TGL5	Triacylglycerol lipase
LPX1	Lipase of peroxisomes
DGK1	Diacylglycerol kinase
MET12	Methionine biosynthesis

Bolded genes have obvious mammalian homologs; YOGY and http://
www.yeastgenome.org/.

part by the partially redundant Zn^{2+}-finger transcription factors MSN2 and
MSN4 which bind *stress-response elements* (STREs) in the promoters of
~200 genes. As one might imagine, the regulation of MSN2/4 is rather
complex [108]. TORC1 signals to MSN2/4 via both the TAP42·PPase and
SCH9 branches as well as an additional Ser/Thr kinase RIM15 [35, 109].
Inactivation of TORC1 results in both the activation and nuclear
localization of RIM15 which activates, by unknown mechanisms, MSN2/4
and thus stimulates expression of STRE-regulated genes. Although growth-
inhibitory, the expression of these genes (encoding chaperones, superoxide
dismutase, etc.) will enable the cell to better cope with environmental
stress.

2.1.1.1. Environmental Stress Response and Chronological Lifespan.
Interestingly, induction of the environmental stress response appears to be an
important determinant for survival of yeast cells during stationary phase
(chronological lifespan). Previous studies had shown that partial inhibition
of TORC1 or deletion of *SCH9* increases chronological lifespan in a RIM15-
and MSN2/4-dependent manner [35, 110, 111]. In a well designed study, the
Kennedy and Kaeberlein groups have recently identified acetic acid as a cell-
extrinsic mediator of dell death during chronological aging [112]. It now seems
clear that activation of the environmental stress response (e.g., by partial
TORC1 inhibition or deletion of *SCH9*) allows cells to better tolerate

exposure to acetic acid and thus remain viable longer when in a quiescent state. Suppression of the environmental stress response, however, is not the only mechanism by which TORC1 influences chronological lifespan— regulation of autophagy also appears to play an important role.

2.2. Autophagy

Literally, "self-eating," autophagy is a catabolic process where cellular components are degraded by the lysomal/vacuolar system to liberate and thus reallocate molecular building blocks. Many forms of autophagic processes have now been described, most are highly regulated, many by TORC1. Like induction of the environmental stress response, induction of autophagy also extends chronological aging in yeast; and blocking autophagy reduces the lifespan extension normally attained with dietary restriction or TORC1 inhibition [90]. Captivatingly, a very recent study demonstrated that spermidine, a naturally occurring essential polyamine, extends lifespan in yeast, worms, and flies, likely by inducing autophagy [113] and it will of course be exciting to see if spermidine is an efficacious longevity drug in man.

The fascinating subject of the regulation of autophagy by TORC1 is the subject of a separate chapter and thus is not described further here.

VIII. Conclusions

It is now clear that TORC1 acts as a messaging hub: it receives the signals generated from numerous environmental cues and transmits this information to coordinate cellular mass accumulation and stress responses. Fascinating biology aside, given the role that mTORC1 plays in numerous pathologies makes a complete understanding of TORC1 signaling imperative.

Relative to studies of most other signal transduction pathways, the study of TORC1 signaling has benefitted tremendously from having a potent and specific small molecule inhibitor. Indeed, treating cells with rapamycin has made it relatively straightforward to characterize the direct effectors and distal readouts downstream of TORC1. Elucidating the molecular details coupling TORC1 effectors to these readouts will be demanding, but genome-wide studies such as the recent characterization of the rapamycin-sensitive phosphoproteome will greatly facilitate these efforts. These genome-wide studies will undoubtedly also illuminate new functions of TORC1 as well.

Dissection of the pathways upstream of TORC1 has presented a much greater challenge to the field. In particular, the lack of an unambiguous reporter of TORC1 activity in yeast has made it very difficult to define when TORC1 is, or is not, active *in vivo*. This challenge was recently overcome with the characterization of the *bona fide* TORC1 substrate SCH9. Now yeast researchers are poised to unleash the full potential of yeast genetics upon the problem of what is upstream of TORC1. As has been demonstrated over and over again in the TOR field, yeast still has much to contribute.

ACKNOWLEDGMENTS

Work in the Loewith lab is supported by the Canton of Geneva, the Swiss National Science Foundation, the Fondation Leenaards, and the European Research Council. The author thanks D. Shore, C. De Virgilio, and members of the Loewith lab for helpful comments on this work and N. Roggli for help with the artwork.

REFERENCES

1. Sehgal, S.N., Baker, H., and Vezina, C. (1975). Rapamycin (AY-22, 989), a new antifungal antibiotic. II. Fermentation, isolation and characterization. *J Antibiot (Tokyo)* 28:727–732.
2. Vezina, C., Kudelski, A., and Sehgal, S.N. (1975). Rapamycin (AY-22, 989), a new antifungal antibiotic. I. Taxonomy of the producing streptomycete and isolation of the active principle. *J Antibiot (Tokyo)* 28:721–726.
3. Harrison, D.E., *et al.* (2009). Rapamycin fed late in life extends lifespan in genetically heterogeneous mice. *Nature* 460:392–395.
4. Houchens, D.P., Ovejera, A.A., Riblet, S.M., and Slagel, D.E. (1983). Human brain tumor xenografts in nude mice as a chemotherapy model. *Eur J Cancer Clin Oncol* 19:799–805.
5. Martel, R.R., Klicius, J., and Galet, S. (1977). Inhibition of the immune response by rapamycin, a new antifungal antibiotic. *Can J Physiol Pharmacol* 55:48–51.
6. Ramcharitar, S., Gaster, A.L., Daemen, J., and Serruys, P. (2007). Drug-eluting stents, restenosis and revascularization. *Herz* 32:287–295.
7. Thomson, A.W., Turnquist, H.R., and Raimondi, G. (2009). Immunoregulatory functions of mTOR inhibition. *Nat Rev Immunol* 9:324–337.
8. Huber, A., Rubio, A., and Loewith, R. (2009). Mammalian target of rapamycin signaling and cancer treatment. *Cancer Chemoth Rev* 4:210–223.
9. Heitman, J., Movva, N.R., and Hall, M.N. (1991). Targets for cell cycle arrest by the immunosuppressant rapamycin in yeast. *Science* 253:905–909.
10. Cafferkey, R., *et al.* (1993). Dominant missense mutations in a novel yeast protein related to mammalian phosphatidylinositol 3-kinase and VPS34 abrogate rapamycin cytotoxicity. *Mol Cell Biol* 13:6012–6023.
11. Helliwell, S.B., Wagner, P., Kunz, J., Deuter-Reinhard, M., Henriquez, R., and Hall, M.N. (1994). TOR1 and TOR2 are structurally and functionally similar but not identical phosphatidylinositol kinase homologues in yeast. *Mol Biol Cell* 5:105–118.
12. Kunz, J., Henriquez, R., Schneider, U., Deuter-Reinhard, M., Movva, N.R., and Hall, M.N. (1993). Target of rapamycin in yeast, TOR2, is an essential phosphatidylinositol kinase homolog required for G1 progression. *Cell* 73:585–596.

13. Stan, R., McLaughlin, M.M., Cafferkey, R., Johnson, R.K., Rosenberg, M., and Livi, G.P. (1994). Interaction between FKBP12-rapamycin and TOR involves a conserved serine residue. *J Biol Chem* 269:32027–32030.
14. Crespo, J.L., and Hall, M.N. (2002). Elucidating TOR signaling and rapamycin action: lessons from *Saccharomyces cerevisiae*. *Microbiol Mol Biol Rev* 66:579–591, Table of contents.
15. Keith, C.T., and Schreiber, S.L. (1995). PIK-related kinases: DNA repair, recombination, and cell cycle checkpoints. *Science* 270:50–51.
16. Wullschleger, S., Loewith, R., and Hall, M.N. (2006). TOR signaling in growth and metabolism. *Cell* 124:471–484.
17. Schmidt, A., Kunz, J., and Hall, M.N. (1996). TOR2 is required for organization of the actin cytoskeleton in yeast. *Proc Natl Acad Sci USA* 93:13780–13785.
18. Zheng, X.F., Florentino, D., Chen, J., Crabtree, G.R., and Schreiber, S.L. (1995). TOR kinase domains are required for two distinct functions, only one of which is inhibited by rapamycin. *Cell* 82:121–130.
19. Chen, E.J., and Kaiser, C.A. (2003). LST8 negatively regulates amino acid biosynthesis as a component of the TOR pathway. *J Cell Biol* 161:333–347.
20. Loewith, R., *et al.* (2002). Two TOR complexes, only one of which is rapamycin sensitive, have distinct roles in cell growth control. *Mol Cell* 10:457–468.
21. Reinke, A., *et al.* (2004). TOR complex 1 includes a novel component, Tco89p (YPL180w), and cooperates with Ssd1p to maintain cellular integrity in *Saccharomyces cerevisiae*. *J Biol Chem* 279:14752–14762.
22. Wedaman, K.P., Reinke, A., Anderson, S., Yates, J., 3rd, McCaffery, J.M., and Powers, T. (2003). Tor kinases are in distinct membrane-associated protein complexes in *Saccharomyces cerevisiae*. *Mol Biol Cell* 14:1204–1220.
23. Hara, K., *et al.* (2002). Raptor, a binding partner of target of rapamycin (TOR), mediates TOR action. *Cell* 110:177–189.
24. Jacinto, E., *et al.* (2004). Mammalian TOR complex 2 controls the actin cytoskeleton and is rapamycin insensitive. *Nat Cell Biol* 6:1122–1128.
25. Kim, D.H., *et al.* (2002). mTOR interacts with raptor to form a nutrient-sensitive complex that signals to the cell growth machinery. *Cell* 110:163–175.
26. De Virgilio, C., and Loewith, R. (2006). Cell growth control: little eukaryotes make big contributions. *Oncogene* 25:6392–6415.
27. Dunlop, E.A., and Tee, A.R. (2009). Mammalian target of rapamycin complex 1: signalling inputs, substrates and feedback mechanisms. *Cell Signal* 21:827–835.
28. Peterson, T.R., *et al.* (2009). DEPTOR is an mTOR inhibitor frequently overexpressed in multiple myeloma cells and required for their survival. *Cell* 137:873–886.
29. Berchtold, D., and Walther, T.C. (2009). TORC2 plasma membrane localization is essential for cell viability and restricted to a distinct domain. *Mol Biol Cell* 20:1565–1575.
30. Binda, M., *et al.* (2009). The Vam6 GEF controls TORC1 by activating the EGO complex. *Mol Cell* 35:563–573.
31. Sturgill, T.W., Cohen, A., Diefenbacher, M., Trautwein, M., Martin, D.E., and Hall, M.N. (2008). TOR1 and TOR2 have distinct locations in live cells. *Eukaryot Cell* 7:1819–1830.
32. Urban, J., *et al.* (2007). Sch9 is a major target of TORC1 in *Saccharomyces cerevisiae* [see comment]. *Mol Cell* 26:663–674.
33. Li, H., Tsang, C.K., Watkins, M., Bertram, P.G., and Zheng, X.F. (2006). Nutrient regulates Tor1 nuclear localization and association with rDNA promoter. *Nature* 442:1058–1061.
34. Kuranda, K., Leberre, V., Sokol, S., Palamarczyk, G., and Francois, J. (2006). Investigating the caffeine effects in the yeast *Saccharomyces cerevisiae* brings new insights into the

connection between TOR, PKC and Ras/cAMP signalling pathways. *Mol Microbiol* 61:1147–1166.

35. Wanke, V., *et al.* (2008). Caffeine extends yeast lifespan by targeting TORC1. *Mol Microbiol* 69:277–285.

36. Beugnet, A., Tee, A.R., Taylor, P.M., and Proud, C.G. (2003). Regulation of targets of mTOR (mammalian target of rapamycin) signalling by intracellular amino acid availability. *Biochem J* 372:555–566.

37. Lempiainen, H., *et al.* (2009). Sfp1 interaction with TORC1 and Mrs6 reveals feedback regulation on TOR signaling. *Mol Cell* 33:704–716.

38. Dubouloz, F., Deloche, O., Wanke, V., Cameroni, E., and De Virgilio, C. (2005). The TOR and EGO protein complexes orchestrate microautophagy in yeast. *Mol Cell* 19:15–26.

39. Gao, M., and Kaiser, C.A. (2006). A conserved GTPase-containing complex is required for intracellular sorting of the general amino-acid permease in yeast. *Nat Cell Biol* 8:657–667.

40. Zinzalla, V., and Hall, M.N. (2008). Signal transduction: linking nutrients to growth. *Nature* 454:287–288.

41. Sancak, Y., *et al.* (2008). The Rag GTPases bind raptor and mediate amino acid signaling to mTORC1. *Science* 320:1496–1501.

42. Aronova, S., Wedaman, K., Anderson, S., Yates, J., 3rd, and Powers, T. (2007). Probing the membrane environment of the TOR kinases reveals functional interactions between TORC1, actin, and membrane trafficking in *Saccharomyces cerevisiae*. *Mol Biol Cell* 18:2779–2794.

43. Zurita-Martinez, S.A., Puria, R., Pan, X., Boeke, J.D., and Cardenas, M.E. (2007). Efficient Tor signaling requires a functional class C Vps protein complex in *Saccharomyces cerevisiae*. *Genetics* 176:2139–2150.

44. Jorgensen, P., Rupes, I., Sharom, J.R., Schneper, L., Broach, J.R., and Tyers, M. (2004). A dynamic transcriptional network communicates growth potential to ribosome synthesis and critical cell size. *Genes Dev* 18:2491–2505.

45. Di Como, C.J., and Arndt, K.T. (1996). Nutrients, via the Tor proteins, stimulate the association of Tap42 with type 2A phosphatases. *Genes Dev* 10:1904–1916.

46. Zheng, Y., and Jiang, Y. (2005). The yeast phosphotyrosyl phosphatase activator is part of the Tap42-phosphatase complexes. *Mol Biol Cell* 16:2119–2127.

47. Jordens, J., *et al.* (2006). The protein phosphatase 2A phosphatase activator is a novel peptidyl-prolyl cis/trans-isomerase. *J Biol Chem* 281:6349–6357.

48. Jiang, Y., and Broach, J.R. (1999). Tor proteins and protein phosphatase 2A reciprocally regulate Tap42 in controlling cell growth in yeast. *EMBO J* 18:2782–2792.

49. Duvel, K., Santhanam, A., Garrett, S., Schneper, L., and Broach, J.R. (2003). Multiple roles of Tap42 in mediating rapamycin-induced transcriptional changes in yeast. *Mol Cell* 11:1467–1478.

50. Yan, G., Shen, X., and Jiang, Y. (2006). Rapamycin activates Tap42-associated phosphatases by abrogating their association with Tor complex 1. *EMBO J* 25:3546–3555.

51. Jacinto, E., Guo, B., Arndt, K.T., Schmelzle, T., and Hall, M.N. (2001). TIP41 interacts with TAP42 and negatively regulates the TOR signaling pathway. *Mol Cell* 8:1017–1026.

52. Kong, M., Ditsworth, D., Lindsten, T., and Thompson, C.B. (2009). alpha4 is an essential regulator of PP2A phosphatase activity. *Mol Cell* 36:51–60.

53. Huber, A., *et al.* (2009). Characterization of the rapamycin-sensitive phosphoproteome reveals that Sch9 is a central coordinator of protein synthesis. *Genes Dev* 23:1929–1943.

54. Shin, C.S., Kim, S.Y., and Huh, W.K. (2009). TORC1 controls degradation of the transcription factor Stp1, a key effector of the SPS amino-acid-sensing pathway in *Saccharomyces cerevisiae*. *J Cell Sci* 122:2089–2099.

55. Marion, R.M., *et al.* (2004). Sfp1 is a stress- and nutrient-sensitive regulator of ribosomal protein gene expression. *Proc Natl Acad Sci USA* 101:14315–14322.
56. Lempiainen, H., and Shore, D. (2009). Growth control and ribosome biogenesis. *Curr Opin Cell Biol.* 21:855–863.
57. Warner, J.R. (1999). The economics of ribosome biosynthesis in yeast. *Trends Biochem Sci* 24:437–440.
58. Powers, T., and Walter, P. (1999). Regulation of ribosome biogenesis by the rapamycin-sensitive TOR-signaling pathway in *Saccharomyces cerevisiae. Mol Biol Cell* 10:987 1000.
59. Zaragoza, D., Ghavidel, A., Heitman, J., and Schultz, M.C. (1998). Rapamycin induces the G₀ program of transcriptional repression in yeast by interfering with the TOR signaling pathway. *Mol Cell Biol* 18:4463–4470.
60. Claypool, J.A., *et al.* (2004). Tor pathway regulates Rrn3p-dependent recruitment of yeast RNA polymerase I to the promoter but does not participate in alteration of the number of active genes. *Mol Biol Cell* 15:946–956.
61. Freckleton, G., Lippman, S.I., Broach, J.R., and Tavazoie, S. (2009). Microarray profiling of phage-display selections for rapid mapping of transcription factor-DNA interactions. *PLoS Genet* 5:e1000449.
62. Liko, D., Slattery, M.G., and Heideman, W. (2007). Stb3 binds to ribosomal RNA processing element motifs that control transcriptional responses to growth in *Saccharomyces cerevisiae. J Biol Chem* 282:26623–26628.
63. Zhu, C., *et al.* (2009). High-resolution DNA-binding specificity analysis of yeast transcription factors. *Genome Res* 19:556–566.
64. Gavin, A.C., *et al.* (2002). Functional organization of the yeast proteome by systematic analysis of protein complexes. *Nature* 415:141–147.
65. Humphrey, E.L., Shamji, A.F., Bernstein, B.E., and Schreiber, S.L. (2004). Rpd3p relocation mediates a transcriptional response to rapamycin in yeast. *Chem Biol* 11:295–299.
66. Kasten, M.M., and Stillman, D.J. (1997). Identification of the *Saccharomyces cerevisiae* genes STB1-STB5 encoding Sin3p binding proteins. *Mol Gen Genet* 256:376–386.
67. Lippman, S.I., and Broach, J.R. (2009). Protein kinase A and TORC1 activate genes for ribosomal biogenesis by inactivating repressors encoded by Dot6 and its homolog Tod6. *Proc Natl Acad Sci USA.*
68. Cipollina, C., van den Brink, J., Daran-Lapujade, P., Pronk, J.T., Porro, D., and de Winde, J.H. (2008). *Saccharomyces cerevisiae* SFP1: at the crossroads of central metabolism and ribosome biogenesis. *Microbiology* 154:1686–1699.
69. Cipollina, C., van den Brink, J., Daran-Lapujade, P., Pronk, J.T., Vai, M., and de Winde, J.H. (2008). Revisiting the role of yeast Sfp1 in ribosome biogenesis and cell size control: a chemostat study. *Microbiology* 154:337–346.
70. Singh, J., and Tyers, M. (2009). A Rab escort protein integrates the secretion system with TOR signaling and ribosome biogenesis. *Genes Dev* 23:1944–1958.
71. Honma, Y., *et al.* (2006). TOR regulates late steps of ribosome maturation in the nucleoplasm via Nog1 in response to nutrients. *EMBO J* 25:3832–3842.
72. Zhao, Y., McIntosh, K.B., Rudra, D., Schawalder, S., Shore, D., and Warner, J.R. (2006). Fine-structure analysis of ribosomal protein gene transcription. *Mol Cell Biol* 26:4853–4862.
73. Martin, D.E., Soulard, A., and Hall, M.N. (2004). TOR regulates ribosomal protein gene expression via PKA and the Forkhead transcription factor FHL1. *Cell* 119:969–979.
74. Rudra, D., Mallick, J., Zhao, Y., and Warner, J.R. (2007). Potential interface between ribosomal protein production and pre-rRNA processing. *Mol Cell Biol* 27:4815–4824.
75. Reid, J.L., Iyer, V.R., Brown, P.O., and Struhl, K. (2000). Coordinate regulation of yeast ribosomal protein genes is associated with targeted recruitment of Esa1 histone acetylase. *Mol Cell* 6:1297–1307.

76. Rohde, J.R., and Cardenas, M.E. (2003). The tor pathway regulates gene expression by linking nutrient sensing to histone acetylation. *Mol Cell Biol* 23:629–635.
77. Willis, I.M., and Moir, R.D. (2007). Integration of nutritional and stress signaling pathways by Maf1. *Trends Biochem Sci* 32:51–53.
78. Cook, M., and Tyers, M. (2007). Size control goes global. *Curr Opin Biotechnol* 18:341–350.
79. Barbet, N.C., Schneider, U., Helliwell, S.B., Stansfield, I., Tuite, M.F., and Hall, M.N. (1996). TOR controls translation initiation and early G1 progression in yeast. *Mol Biol Cell* 7:25–42.
80. Costanzo, M., *et al.* (2004). CDK activity antagonizes Whi5, an inhibitor of G1/S transcription in yeast. *Cell* 117:899–913.
81. Ashe, M., de Bruin, R.A., Kalashnikova, T., McDonald, W.H., Yates, J.R., 3rd, and Wittenberg, C. (2008). The SBF- and MBF-associated protein Msa1 is required for proper timing of G1-specific transcription in *Saccharomyces cerevisiae. J Biol Chem* 283:6040–6049.
82. Nakashima, A., *et al.* (2008). The yeast Tor signaling pathway is involved in G2/M transition via polo-kinase. *PLoS ONE* 3:e2223.
83. Petersen, J., and Nurse, P. (2007). TOR signalling regulates mitotic commitment through the stress MAP kinase pathway and the Polo and Cdc2 kinases. *Nat Cell Biol* 9:1263–1272.
84. Albig, A.R., and Decker, C.J. (2001). The target of rapamycin signaling pathway regulates mRNA turnover in the yeast *Saccharomyces cerevisiae. Mol Biol Cell* 12:3428–3438.
85. Hinnebusch, A.G. (2005). Translational regulation of GCN4 and the general amino acid control of yeast. *Annu Rev Microbiol* 59:407–450.
86. Cherkasova, V.A., and Hinnebusch, A.G. (2003). Translational control by TOR and TAP42 through dephosphorylation of eIF2alpha kinase GCN2. *Genes Dev* 17:859–872.
87. Linder, P. (2003). Yeast RNA helicases of the DEAD-box family involved in translation initiation. *Biol Cell* 95:157–167.
88. Johnson, S.P., and Warner, J.R. (1987). Phosphorylation of the *Saccharomyces cerevisiae* equivalent of ribosomal protein S6 has no detectable effect on growth. *Mol Cell Biol* 7:1338–1345.
89. Proud, C.G. (2009). mTORC1 signalling and mRNA translation. *Biochem Soc Trans* 37:227–231.
90. Stanfel, M.N., Shamieh, L.S., Kaeberlein, M., and Kennedy, B.K. (2009). The TOR pathway comes of age. *Biochim Biophys Acta* 1790:1067–1074.
91. Kaeberlein, M., *et al.* (2005). Regulation of yeast replicative life span by TOR and Sch9 in response to nutrients. *Science* 310:1193–1196.
92. Steffen, K.K., *et al.* (2008). Yeast life span extension by depletion of 60s ribosomal subunits is mediated by Gcn4. *Cell* 133:292–302.
93. Steinkraus, K.A., Kaeberlein, M., and Kennedy, B.K. (2008). Replicative aging in yeast: the means to the end. *Annu Rev Cell Dev Biol* 24:29–54.
94. Colman, R.J., *et al.* (2009). Caloric restriction delays disease onset and mortality in rhesus monkeys. *Science* 325:201–204.
95. Selman, C., *et al.* (2009). Ribosomal protein S6 kinase 1 signaling regulates mammalian life span. *Science* 326:140–144.
96. Zid, B.M., *et al.* (2009). 4E-BP extends lifespan upon dietary restriction by enhancing mitochondrial activity in *Drosophila. Cell* 139:149–160.
97. Bonawitz, N.D., Chatenay-Lapointe, M., Pan, Y., and Shadel, G.S. (2007). Reduced TOR signaling extends chronological life span via increased respiration and upregulation of mitochondrial gene expression. *Cell Metab* 5:265–277.

98. Georis, I., Feller, A., Tate, J.J., Cooper, T.G., and Dubois, E. (2009). Nitrogen catabolite repression-sensitive transcription as a readout of Tor pathway regulation: the genetic background, reporter gene and GATA factor assayed determine the outcomes. *Genetics* 181:861–874.

99. Neklesa, T.K., and Davis, R.W. (2009). A genome-wide screen for regulators of TORC1 in response to amino acid starvation reveals a conserved Npr2/3 complex. *PLoS Genet* 5:e1000515.

100. Liu, Z., and Butow, R.A. (2006). Mitochondrial retrograde signaling. *Annu Rev Genet* 40:159–185.

101. Hedbacker, K., and Carlson, M. (2008). SNF1/AMPK pathways in yeast. *Front Biosci* 13:2408–2420.

102. Orlova, M., Kanter, E., Krakovich, D., and Kuchin, S. (2006). Nitrogen availability and TOR regulate the Snf1 protein kinase in *Saccharomyces cerevisiae*. *Eukaryot Cell* 5:1831–1837.

103. Shaw, R.J. (2009). LKB1 and AMP-activated protein kinase control of mTOR signalling and growth. *Acta Physiol (Oxf)* 196:65–80.

104. Zaman, S., Lippman, S.I., Zhao, X., and Broach, J.R. (2008). How Saccharomyces responds to nutrients. *Annu Rev Genet* 42:27–81.

105. Colomina, N., Liu, Y., Aldea, M., and Gari, E. (2003). TOR regulates the subcellular localization of Ime1, a transcriptional activator of meiotic development in budding yeast. *Mol Cell Biol* 23:7415–7424.

106. Causton, H.C., et al. (2001). Remodeling of yeast genome expression in response to environmental changes. *Mol Biol Cell* 12:323–337.

107. Gasch, A.P., and Werner-Washburne, M. (2002). The genomics of yeast responses to environmental stress and starvation. *Funct Integr Genomics* 2:181–192.

108. Swinnen, E., et al. (2006). Rim15 and the crossroads of nutrient signalling pathways in *Saccharomyces cerevisiae*. *Cell Div* 1:3.

109. Pedruzzi, I., et al. (2003). TOR and PKA signaling pathways converge on the protein kinase Rim15 to control entry into G_0. *Mol Cell* 12:1607–1613.

110. Powers, R.W., 3rd, Kaeberlein, M., Caldwell, S.D., Kennedy, B.K., and Fields, S. (2006). Extension of chronological life span in yeast by decreased TOR pathway signaling. *Genes Dev* 20:174–184.

111. Wei, M., et al. (2008). Life span extension by calorie restriction depends on Rim15 and transcription factors downstream of Ras/PKA, Tor, and Sch9. *PLoS Genet* 4:e13.

112. Burtner, C.R., Murakami, C.J., Kennedy, B.K., and Kaeberlein, M. (2009). A molecular mechanism of chronological aging in yeast. *Cell Cycle* 8:1256–1270.

113. Eisenberg, T., et al. (2009). Induction of autophagy by spermidine promotes longevity. *Nat Cell Biol.*

114. Berger, A.B., Decourty, L., Badis, G., Nehrbass, U., Jacquier, A., and Gadal, O. (2007). Hmo1 is required for TOR-dependent regulation of ribosomal protein gene transcription. *Mol Cell Biol* 27:8015–8026.

115. Hannan, K.M., et al. (2003). mTOR-dependent regulation of ribosomal gene transcription requires S6K1 and is mediated by phosphorylation of the carboxy-terminal activation domain of the nucleolar transcription factor UBF. *Mol Cell Biol* 23:8862–8877.

116. Kasahara, K., et al. (2007). Assembly of regulatory factors on rRNA and ribosomal protein genes in *Saccharomyces cerevisiae*. *Mol Cell Biol* 27:6686–6705.

10

TORC2 and Sphingolipid Biosynthesis and Signaling: Lessons from Budding Yeast

TED POWERS • SOFIA ARONOVA • BRAD NILES

Department of Molecular and Cellular Biology
College of Biological Sciences
University of California, Davis
California, USA

I. Abstract

Studies in budding yeast, *Saccharomyces cerevisiae*, have established an important role for the target of rapamycin (TOR) signaling network in regulating cell growth and cell architecture through the control of gene expression, protein biosynthesis, protein trafficking, as well as organization of the actin cytoskeleton. The central component of this network is the TOR kinase, which assembles into two distinct membrane associated protein complexes, termed TORC1 and TORC2, where TORC1 is uniquely inhibited by rapamycin. A novel role for this network, in particular involving TORC2, has emerged recently with respect to the regulation of the biosynthesis of sphingolipids. Complex sphingolipids are integral components of cell membranes and their biosynthesis involves the production of important bioactive intermediates, including sphingoid long-chain bases (LCBs) and ceramides, which play distinct signaling roles crucial for cell growth and survival. Here, we review recent results demonstrating a role for TORC2 in the balanced production of these intermediates and discuss

ISSN NO: 1874-6047
DOI: 10.1016/S1874-6047(10)27010-5

how this regulation appears to involve two distinct downstream effector branches of TORC2, namely, the plextrin homology (PH) domain containing proteins SLM1 and SLM2, as well as the AGC kinases YPK1 and YPK2. We describe how these components interact with other effectors of sphingolipid metabolism and impinge on actin cytoskeletal organization, stress response, and cell survival. Finally, we discuss the likely relevance of these findings for mammalian cells.

II. Introduction

Remarkable progress has been made in our understanding of the target of rapamycin (TOR) signaling network in recent years, from the identification of components that define its elaborate molecular architecture to an increased understanding of complex physiological functions under its control in distinct cell types and organisms. In this respect, model organisms have provided an important means for probing TOR function, in particular where forward genetic approaches are available, such as in *Saccharomyces cerevisiae*. Indeed, through the isolation of mutations that confer dominant rapamycin resistance, studies of budding yeast led to the initial discovery of TOR1 and TOR2, the two TOR homologs in yeast [1, 2]. The fact that yeast possesses two TOR kinases allowed for detailed genetic analyses that revealed that TOR2, but not TOR1, is involved in a essential function that is rapamycin-insensitive [3, 4]. These studies were crucial as they laid the groundwork for thinking about TOR as being involved in multiple functions and thus foreshadowed the subsequent discoveries of distinct TOR complexes 1 and 2 (TORC1 and TORC2) in yeast as well as in higher eukaryotes [2]. A remaining important goal is to attain a complete description of cellular processes that are regulated by these complexes, either independently or collaboratively. Here, we review recent work in yeast that extends the reach of the TORC2-specific arm of TOR signaling into the realm of lipid metabolism, in particular the biosynthesis of sphingolipids, the intermediates of which play important roles critical to cell growth, differentiation, and death in a wide spectrum of eukaryotic cells. To place these results in context, we first briefly review the relationship between TORC1 and TORC2 and their activities in yeast and how this led to a consideration of sphingolipid biology.

III. TORC1 Versus TORC2

In yeast, TORC1 contains TOR1 or TOR2 as well as three additional proteins, KOG1, LST8, and TCO89 [5–7]. This complex mediates a large number of rapamycin-sensitive activities related to cell growth, including control of translation, nutrient-related gene expression, autophagy, and protein trafficking and stability [2]. Many events downstream from TORC1 are mediated either by a TAP42/PP2A phosphatase regulatory module or involve functional interactions with the AGC kinases PKA and SCH9 [8–18]. TORC2 contains TOR2 as well as several additional proteins, LST8, AVO1–AVO3, and BIT61 [5–7]. This complex plays a role in polarized cell growth and cytoskeletal organization, in particular signaling to components required for the proper positioning of actin, in the form of cortical patches, at the site of bud emergence [3, 4, 19, 20]. These components include the RHO1 GTPase and its associated regulatory partners, which function in part by signaling to PKC1, an upstream activator of the mitogen-activated protein kinase (MAPK) SLT2/MPK1 [4, 20, 21]. More recently, additional downstream effectors of TORC2 have been identified, including the AGC kinase YPK2 and two novel proteins SLM1 and SLM2 [22–24]. The role of these effectors will be discussed in later sections of this review.

Parallel studies in mammalian cells and other higher eukaryotes have demonstrated that a single mTOR similarly assembles into two major complexes, mTORC1 and mTORC2, with both complexes containing identifiable orthologs of most TOR1/TOR2 binding partners [2]. Additional studies have demonstrated further that these complexes function in part by recognizing and phosphorylating distinct AGC kinases. Thus, mTORC1 recognizes S6K1 and mTORC2 recognizes AKT (also known as PKB) [2, 25–27]. The fact that the site(s) phosphorylated by TORC1/2 within these kinases include what are referred to as a hydrophobic motif (HM) and turn motif (TM) is important because this indicates that the TOR complexes represent the long sought-after "PDK2" or "Site 2" kinase that collaborates with PDK1 for activation of these AGC kinases [26, 27]. Remarkably, this has turned out to be a conserved feature in yeast as well, in that TORC1 phosphorylates SCH9 [17] and TORC2 phosphorylates YPK2 [22] at similar sites. Thus, TORC1/2 collaborates with the yeast orthologs of PDK1, termed PKH1 and PKH2, to activate distinct AGC kinase family members [26, 28].

The apparent specificity of TORC1 and TORC2 for distinct substrates is consistent with the prevailing view in the field that each complex regulates distinct growth-related processes, for example ribosome biogenesis and cytoskeletal organization, respectively, and thereby collaborate to control "temporal" versus "spatial" aspects of growth [2]. In retrospect, this view was initially based on genetic findings that mutations in TOR2 but not TOR1 result in depolarization of the actin cytoskeleton and was therefore considered to be a TOR2-unique, presumably rapamycin-insensitive function [3, 4, 29, 30]. Subsequently, mutations in TOR2 have also been shown to result in defects in receptor mediated endocytosis and cell integrity signaling, where the latter represents a mechanism whereby cell envelope and/or cell wall stability is monitored in response to external stress, and both are processes that rely on an intact actin cytoskeleton [4, 29, 31, 32]. A number of subsequent findings, however, suggest that this view of a strict division of labor by TORC1 and TORC2 is an oversimplification of their activities. For example, loss of nonessential subunits of TORC1, or mutations in the downstream TAP42/PP2A phosphatase complex, cause defects in cell integrity signaling [7, 33–36]. In addition, inhibition of TORC1 with rapamycin in fact results in actin depolarization [37]. In this regard, we have determined that rapamycin-induced actin depolarization occurs well before any observable signs of a cell cycle arrest, suggesting this may be a direct consequence of TORC1 inhibition [37]. Moreover, we have identified both genetic as well as physical interactions between components of TORC1 and proteins involved in actin cytoskeletal dynamics and endocytosis, suggesting there is significant overlap with respect to processes regulated by TORC1 and TORC2 [37]. This conclusion is consistent with studies in higher eukaryotes as well, where both mTORC1 and mTORC2 have now been implicated in the regulation of polarized cell growth, cell migration, and endocytosis [38–42].

Taken together, these results raise the important question as to whether there is a cellular process that is uniquely influenced by TORC2 in yeast. In this respect, it is significant that a number of studies point to a novel role for the TORC2 branch of the network as being important for the regulation of the biosynthesis of precursors to complex sphingolipids, many of which represent important bioactive lipids that regulate cell growth and survival in both yeast and mammals. To place these studies in proper context, a brief overview of the sphingolipid pathway is first provided.

IV.   Sphingolipid Biosynthesis: A Brief Primer

Complex sphingolipids are an essential component of eukaryotic membranes, constituting approximately 10% of lipids within the plasma membrane, where they play important roles in actin dynamics and endocytosis, as well as have been proposed to organize membrane domains into detergent resistant "rafts" important for signaling events [43, 44]. The early steps of sphingolipid biosynthesis take place within the endoplasmic reticulum (ER) and are relatively well conserved between yeast and mammals [43–45]. These include the first and rate-limiting step of the condensation of serine and a fatty acyl-CoA, most often C16-CoA, by the enzyme serine palmitoyltransferase (SPTase), to form 3-ketodihydro-sphingosine, as well as the subsequent reduction of this product to form dihydrophingosine (DHS) (Figure 10.1A). DHS and its derivatives are commonly referred to as sphingoid long-chain bases (LCBs). In yeast, DHS can be hydroxylated by SUR2 to form phytosphingosine (PHS) and these represent the two major LCBs in yeast (Figure 10.1A). Both DHS and PHS are substrates for the next step in the pathway, which involves their N-acylation with a second, generally much longer fatty acyl-CoA, primarily C26-CoA, to form DHS- and PHS-ceramides. This step is carried out by the enzyme ceramide synthase (CerS), which in yeast consists of two catalytic subunits LAC1 and LAG1 and a regulatory subunit LIP1 [46–48] (Figure 10.1A). The fatty acyl-CoAs utilized by CerS are produced through a complex elongation cycle that extends the length of shorter fatty acyl-CoA substrates (Figure 10.1B); a detailed molecular mechanism of this process has been described recently [49]. Following their synthesis, ceramides can undergo additional hydroxylation reactions within the fatty acyl moiety [50–52].

Synthesis of ceramides represents one of the most evolutionarily con-served steps within the pathway, where close homologs of the yeast subunits of CerS, LAG1 and LAC1, have been identified in mammalian cells [53–55]. At least six distinct isoforms have been identified, CerS1–6, where each isoform is believed to utilize acyl CoA substrates that have different chain lengths, ranging generally from 16 to 24 carbons [55, 56]. Structural models indicate that the conserved functional core of CerS consists of eight trans-membrane helices where the active site is likely to lie within helices 5 and 6 [57]. Interestingly, no ortholog for LIP1 has been identified in mammalian cells, and one report has demonstrated that yeast cells can survive without LIP1 if they overexpress human CerS5 [58]. This result suggests that mam-malian cells may not need LIP1, however, this has not been exhaustively examined. Interestingly, an analogous situation has existed for the compo-nents of SPTase, where mammalian orthologs of the catalytic subunits LCB1 and LCB2 are readily identifiable but no ortholog of the small regulatory subunit TSC3 can be found. Nevertheless, Dunn and coworkers have recently described the identification novel small regulators of mammalian SPTase that function similarly to TSC3, despite sharing no structural simi-larity [59]. Thus, it is possible that an analogous situation will turn out to exist for mammalian CerS.

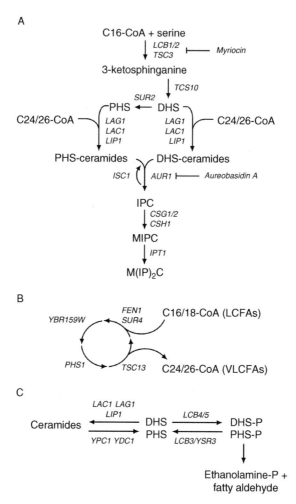

FIG. 10.1. Sphingolipid metabolism in *S. cerevisiae*. (A) Schematic of the sphingoid long-chain base (LCB) branch of sphingolipid biosynthesis. (B) Elongation cycle for conversion of long chain fatty acyl-CoAs (LCFAs) to very long chain fatty acyl-CoAs (VLCFAs) for utilization by ceramide synthase (i.e., LAG1/LAC1/LIP1 in (A)). (C) Schematic of the ceramide/LCBP rheostat in yeast, depicting interconversion of LCBs (DHS and PHS) into either LCBPs (DHS-P and PHS-P) or ceramides. In (A)–(C), italics refer to enzymes and associated regulatory proteins that catalyze each step. Not shown are additional hydroxylation steps that can occur prior to formation of IPC to generate distinct hydroxylated species of complex sphingolipids.

Once synthesized in the ER, ceramides are transported to the Golgi where they are further modified by incorporation of distinct polar head groups, to form what are termed complex sphingolipids, by the action of enzymes located within the lumen of this organelle [43, 44, 60, 61]. In yeast, these modifications consist first of the addition of inositol phosphate to form inositol phosphoceramide (IPC) by AUR1, also known as IPC synthase (Figure 10.1A). In a second step, a mannose sugar is transferred to the inositol ring to form mannose inositol phosphoceramide (MIPC). The enzyme that carries out this step consists of either one of two catalytic subunits, CSG1 or CSH1, together with a regulatory subunit CSG2 (Figure 10.1A). The final step involves the addition of a second inositol phosphate to form mannose-(inositol phosphate)$_2$ ceramide [M(IP)$_2$C] by the action of IPT1, where the completed complex sphingolipids exit the Golgi via vesicle trafficking to the plasma membrane [61].

The major differences between yeast and mammals with respect to the early steps of sphingolipid biosynthesis is that PHS is not formed in mammalian cells and DHS-ceramides are instead acted upon by a desaturase that introduces a double bond at the C4–C5 position with the sphingoid base to form the major species of ceramides found within mammalian cells [62–64]. Ceramides are then modified into a different set of complex sphingolipids within the Golgi, including sphingomyelin, glucosylceramides, and galactosylceramides [62–64]. Ceramides can also be catabolized by the action of ceramidases to yield the LCB sphingosine. Interestingly, sphingosine is considered to be the major LCB in mammalian cells, which means that prior formation of ceramides is an obligate step for its production, which is another important distinction from yeast.

In addition to being converted to ceramides, LCBs can also be phosphorylated to form LCBPs (e.g., DHS-1P or PHS-1P in yeast and SPH-1P in mammals). In yeast, both LCBs and LCBPs have been identified as important mediators of heat stress [43, 44, 65–67]. LCBs have also been shown to be involved in the direct activation of PKH1 and PKH2, as well as to their downstream signaling partners YPK1, YPK2, and SCH9 [68–71]. A sensitive balance is believed to exist between the levels of LCBPs versus ceramides that is important for proper cell growth and survival [45, 64, 72, 73]. The regulation of this balance has been termed the "ceramide/LCBP rheostat" [74, 75]. In yeast, in addition to CerS, this rheostat is formed by the action of several enzymes, including the LCB kinases LCB4/5 and the LCBP phosphatases LCB3/YSR3 (Figure 10.1C).

V. Regulation of Sphingolipid Metabolism: Connections to TOR

Yeast genetics has proven instrumental to deciphering the enigmatic or "sphinx-like" nature of sphingolipids [76], including the identification of most of the enzymes that comprise the sphingolipid biosynthetic pathway. As importantly, genetics has provided valuable insight into components that are likely to be involved in the regulation of different steps within the pathway. The results of several studies, summarized below, suggest that TOR signaling is intimately involved with sphingolipid metabolism.

A.   CSG2 AS AN EARLY INROAD TO SPHINGOLIPID REGULATION

A seminal study by Dunn and coworkers [77] utilized an elegant genetic screen based on the prior observation that deletion of either CSG1 or CSG2 renders cells hypersensitive to growth in the presence of high concentrations of calcium (100 mM), due to the accumulation of a specific IPC, termed IPC-C, that possesses a single fatty acyl hydroxylation [77–79]. It is not completely clear why these mutants are Ca^{2+} hypersensitive, although it has been speculated to be related to Ca^{2+}-induced expression of CSG2 required for the concomitant rapid conversion of IPCs to MIPCs in high Ca^{2+} [61]. Nevertheless, the observation that reducing levels of IPC-C ameliorates the Ca^{2+} hypersensitivity of a csg2 mutant led to the concept that isolation of suppressors of this phenotype could identify components required for the biosynthesis of sphingolipids at steps upstream of IPC formation [77, 78].

A very clever aspect of one of these screens was to identify suppressors that are at the same time temperature-sensitive, to enable the isolation of mutations within essential genes [77]. Using this approach, a number of temperature-sensitive suppressors of CSG2 (TSC) mutants were isolated (not to be confused with the TSC1/2 proteins that function upstream of TORC1 in mammalian cells [2], an unfortunate coincidence in nomenclature for the field). In addition to identifying novel sphingolipid biosynthetic enzymes, this screen also identified MSS4, an enzyme involved in the formation of the lipid second messenger phosphatidylinositol-4,5-bisphosphate ($PI4,5P_2$), as well as two components of TORC2, namely TOR2 and TSC11/AVO3 (referred to here simply as AVO3). In retrospect, these early findings suggested an intimate relationship between phosphoinositide signaling, TORC2, and sphingolipid biosynthesis/signaling. However, it has taken multiple groups and many years to begin to make sense of these connections.

B. SLM PROTEINS LINK PHOSPHOINOSITIDES, TORC2, AND SPHINGOLIPID BIOSYNTHESIS

MSS4 catalyzes the formation of PI4,5P$_2$ by phosphorylating the precursor lipid PI4P and is the sole PI4 5-kinase in yeast [80, 81]. Once formed, PI4,5P$_2$ plays diverse roles related to actin cytoskeletal organization, cell integrity signaling, as well as endocytosis, in part by associating with proteins via interaction with a motif termed a pleckstrin homology (PH) domain, which then helps these proteins associate with the plasma membrane [82]. To explore the function of PI4,5P$_2$ in yeast further, Emr and coworkers identified genes that when deleted give rise to a synthetically lethal phenotype in combination with a hypomorphic temperature-sensitive allele of MSS4 [23]. Several of these so-called synthetically lethal with MSS4 (SLM) mutants corresponded to known regulators of actin cytoskeleton organization and endocytosis, including the AVO2 subunit of TORC2. A number of previously uncharacterized genes were also identified, including SLM1, which encodes a novel PH-domain containing protein. While SLM1 mutants are viable, these investigators also identified a close homolog of SLM1, termed SLM2, where the combined deletion of both genes is lethal, an indication that these two proteins play an overlapping essential cellular function. Subsequent studies established that SLM1/2 are phosphorylated by TORC2 and that these phosphorylation events regulate their activity, in part for signaling to the RHO1/PKC1 pathway to maintain proper actin polarization and cell integrity [23] (Figure 10.2). An independent

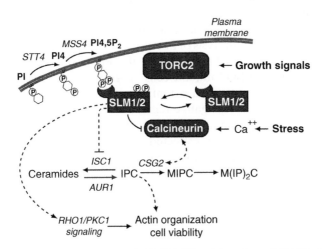

FIG. 10.2. Relationship between TORC2, SLM1/2, calcineurin, and PI4,5P$_2$ during sphingolipid biosynthesis and signaling. See text for details.

study by Kunz and coworkers also identified SLM1/2 as novel PH domain containing proteins and demonstrated their role in actin cytoskeletal organization, as well as their regulation by TORC2 [24]. Also consistent with these findings, overexpression of SLM1/2 can suppress temperature-sensitive mutations in AVO3 [83].

A connection between sphingolipids and $PI4,5P_2$ was strengthened when two components involved in fatty acyl-CoA elongation, SUR4 and FEN1, were identified in a different genetic screen by the Emr group looking for mutants that confer a synthetic lethal phenotype with a hypormorphic mutant form of STT4, the PI 4-kinase that generates PI4P at the plasma membrane for modification to $PI4,5P_2$ by MSS4 [84]. Given their prior observations regarding the role of TORC2/SLM1/2 in $PI4,5P_2$ mediated events, as well as earlier results of Dunn and coworkers, these investigators examined closely a role for TORC2 and SLM1/2 in sphingolipid biosynthesis. Consistent with a positive role in the formation of sphingolipids, these authors found that temperature-sensitive mutations in STT4, MSS4, SLM1/2, and TOR2 all blocked the heat-induced accumulation of complex sphingolipids, in particular IPC-C [84]. Interestingly, these authors found that SLM1/2 mutants cannot rescue the Ca^{2+} sensitivity of a CSG2 mutant but rather that these mutations display a synthetically lethal phenotype. In an independent study, Kunz and coworkers demonstrated that SLM1/2 mutants are both hypersensitive to myriocin as well as display synthetic negative interactions with several sphingolipid biosynthetic mutants [85].

Interestingly, Tabuchi *et al.* [84] observed that the lethality of a SLM1/2 CSG2 mutant could be rescued by concomitant deletion of ISC1, the yeast ortholog of mammalian neutral sphingomyelinase (SMase), which produces ceramide by hydrolysis of different inositol phosphosphingolipids. Deletion of ISC1 also partially rescued the low levels of IPC that are observed in SLM1/2 CSG2 mutant cells. Together these results led Emr and coworkers to propose that a potential role for SLM1/2 in the sphingolipid pathway is to act as a negative regulator of ISC1 (Figure 10.2). However, an increase in ceramides was not observed in SLM1/2 or TOR2 mutants, where ISC1 activity is expected to have become activated, according to this scenario. By contrast, ceramide accumulation was readily detected by mutations in AUR1, which catalyzes the formation of IPCs and is thus the reverse of the step carried out by ISC1 [84]. Together these findings suggest that the connection between TORC2/SLM1/2 and sphingolipid metabolism is more complex than simply to suppress ISC1 activity.

C. CALCINEURIN: ANOTHER LEVEL OF CONTROL FOR TORC2/SLM1/2 AND SPHINGOLIPIDS

Calcineurin is an evolutionarily conserved Ca^{2+}/calmodulin serine/threonine protein phosphatase that plays an important role in cellular responses to stress, including environmental stress in yeast [86, 87]. In yeast, this enzyme is composed to two redundant catalytic subunits, CNA1 and CNA2, and an essential positive regulatory subunit, CNB1 [87]. In response to stress, increased cytoplasmic Ca^{2+} levels activate calcineurin, which then dephosphorylates targets by interacting with a specific docking site that contains the consensus sequence PxIxIT or a related element [88]. One important target in yeast is the stress-induced transcription factor Crz1, which controls the expression of a wide array of genes that promote cell survival under stress conditions [87, 89, 90]. It is clear that calcineurin regulates other events independently of transcription in yeast, including stress-induced depolarization of the actin cytoskeleton, and thus is likely to interact with other targets as well [87]. Indeed, a number of studies have now demonstrated that calcineurin interacts with and dephosphorylates both SLM1 and SLM2, via interactions with a variant of the PxIxIT motif [91, 92]. Thus, a picture has emerged wherein SLM1/2 undergoes complex phosphorylation and dephosphorylation reactions to regulate their activity under stress conditions [84, 85, 91, 92] (Figure 10.2). A reciprocal situation has also been described wherein TORC2 activity negatively regulates calcineurin activity, potentially via SLM1/2, and is likely to contribute to the suppression of stress responses during optimal growth conditions [91].

With respect to sphingolipids and calcineurin, there appear to be multiple levels of interaction. First, calcineurin has been suggested to regulate levels of SLM1/2 phosphorylation in response to IPC levels. Thus, whereas heat shock increases SLM1 phosphorylation in WT cells, no increase is observed in csg2 mutant cells, due to IPC-dependent activation of calcineurin [84]. Second, in addition to inducing calcineurin, heat stress also causes a transient increase in levels of LCBs, in particular PHS, which stimulates phosphorylation of SLM1/2, likely via activation of PKH1/2 but independent of TORC2 [92]. This activation is important for mediating stress-induced changes in actin polarization as well as nutrient-regulated permease endocytosis and turnover [92]. The complex interplay between calcineurin, sphingolipids, and SLM1/2 is demonstrated by the fact that reducing SLM1/2 activity leads to myriocin hypersensitivity in a manner that is dependent upon the presence of an intact calcineurin interaction motif [85]. While these interactions are clearly important for cell survival under stress conditions, it remains unclear the extent to which calcineurin and/or SLM1/2 play essential roles in sphingolipid biosynthesis under normal growth conditions [92].

D. TORC2, YPK1/2, AND THE CERAMIDE/LCBP RHEOSTAT

Many of studies described above emphasize a role for TORC2/SLM1/2 signaling for the production of sphingolipids under conditions of heat stress. This is a necessary consequence of analyzing temperature-sensitive mutations that possess a nonpermissive temperature of 37 °C or higher. However, as described above, heat shock alone causes the transient increase in the levels of many sphingolipid intermediates, including LCBs, LCBPs, and ceramides. While the mechanism for increased levels of ceramides has been shown to be in part dependent upon increased ISC1 phospholipase activity, the means by which LCB and LCBP levels increase following heat shock remains largely unexplored [61]. Thus, the use of temperature-sensitive mutants to explore the regulation of sphingolipid biosynthesis possesses some important caveats.

In the course of our studies of TORC2, we isolated a unique temperature-sensitive allele of AVO3, a TORC2-specific component and ortholog of mammalian Rictor, that displays a significant growth defect at 30 °C, the standard growth temperature for yeast [93]. Thus, we reasoned we could utilize this allele to explore the role of TORC2 in sphingolipid biosynthesis under conditions that did not result in heat-induced changes in sphingolipid intermediates. Accordingly, we used liquid chromatography coupled with tandem mass spectrometry (LC–MS/MS) to profile sphingolipids in wild-type and TORC2-deficient cells, following a temperature shift from 25 to 30 °C for a period of time that preceded any observable growth defect for the mutant. Given the conservation of TOR signaling and the fact that the early steps of sphingolipid biosynthesis are among the most conserved between diverse eukaryotic species, we focused on steps occurring prior to IPC formation.

We observed that TORC2-deficient cells are severely compromised for the production of ceramides, including both PHS- and DHS-ceramide species, as well as their hydroxylated forms, within 3 h following a shift to 30 °C [93]. A number of observations suggest that this defect is specifically at the step of *de novo* ceramide formation by CerS (i.e., LAG1/LAC1/LIP1). First, we have assayed the incorporation of an unnatural LCB, SPH-C17, into ceramides following addition to cells *in vivo*; here we observed significantly fewer ceramides formed in TORC2-deficient cells compared to wild-type cells (our unpublished results). Second, when we assayed ceramide synthesis using microsomal membranes isolated from temperature shifted TORC2-deficient cells, the activity of CerS in these membranes was significantly reduced compared to wild-type membrane [93].

In addition to decreased levels of ceramides, we observed changes in the levels of other sphingolipid metabolites in TORC2-deficient cells, in particular decreased LCBs and increased LCBPs [93]. One possibility is that these represent secondary effects of DHS and PHS being shunted to DHS-P and PHS-P due to their build up in the mutant. Consistent with this conclusion is the fact that a strong increase in the levels of DHS and PHS is observed in TORC2-deficient cells upon deletion of the major LCB kinase, LCB4, which is severely compromised for LCBP formation [93]. Deleting LCB4 and/or providing exogenous PHS also rectifies other defects in TORC2-deficient cells, in particular actin polarization, suggesting alterations in LCB/LCBP levels contributes to phenotypes observed in this mutant. Interestingly, treating cells with Fumonisin B, a specific inhibitor of CerS, does not by itself lead to an increase in intracellular levels of LCBPs, which is in stark contrast to our findings with TORC2-deficient cells. These findings suggest that TORC2 influences more than ceramide synthesis and may play a more general role within the ceramide/LCBP rheostat. Whether this involves regulation of LCB kinases, LCBP phosphatases or other components of the rheostat (Figure 10.1C) is currently under investigation.

Our findings suggest further that TORC2 regulates ceramide levels at least in part by signaling through the kinases YPK1 and YPK2 ([93] and our unpublished results). This conclusion is based on the fact that a constitutively active allele of YPK2, which does not require TORC2 for its activity [22], restores ceramide levels within TORC2-deficient cells, as well as suppresses the growth and actin polarization defects in this mutant [93]. We observe similar results using a constitutively active allele of YPK1 as well (our unpublished results). We also find that YPK2 is hypophosphorylated in TORC2-deficient cells, in agreement with previous findings that TORC2 is an important positive regulator of YPK2 [93]. Given the ability of LCBs to positively regulate the activity of PKH1/2 and, to a lesser extent, YPK1/2, our findings suggest the existence of a "feed forward" regulatory loop wherein LCBs act both as a substrate for CerS as well as stimulates a signaling pathway involved in ceramide production (Figure 10.3).

This point of regulation we have described appears to be distinct from the findings described above for SLM1/2 where it is proposed these act as a negative regulator of ISC1 [84]. Thus, a prediction of the model proposed by Emr and coworkers is that loss of TORC2 activity should result in elevated levels of ceramides, whereas we observe a significant reduction of these species. Also, whereas deletion of ISC1 is able to rescue several

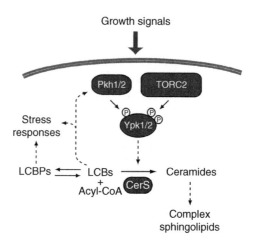

F<small>IG</small>. 10.3. Model depicting positive regulation of *de novo* ceramide synthesis by TORC2, via signaling to YPK1 as well as YPK2. See text for details.

phenotypes associated with loss of SLM1/2 activity, we observe no such rescue of any TORC2-deficient phenotypes by deleting ISC1 (our unpublished results). On the other hand, we observe that deletion of CNB1 is able to partially rescue growth and ceramide defects in TORC2-deficient cells [93], in agreement with suppression of several SLM1/2 mutant phenotypes by deletion of CNB1 [84]. We are presently exploring the relationship between TORC2/YPK1/2 and the TORC2/SLM1/2 branches of this network, in particular with respect to sphingolipid biosynthesis.

Interestingly, both LCBs and the PKH1/2/YPK1/2 kinases have been implicated in the regulated assembly and disassembly of eisosomes, which are novel structures that exist at the plasma membrane and have been proposed to control entry within cells during endocytosis [94–96]. Given the similarity between the localization of eisosomes and the punctate-like structures formed by TORC2 at the plasma membrane and at adjacent sites [6, 97, 98], it has been suggested that TORC2 may regulate endocytosis via eisosome association [98]. A more recent study suggests, however, that TORC2 localizes to a microdomain within the plasma membrane that is distinct from eisosomes [99]. These findings are consistent with our cell fractionation data where TORC2 components behave distinct from the major eisosome subunits PIL1 and LSP1 [37] (our unpublished results). Thus, the relationship between TORC2 and eisosomes remains to be clarified.

VI.   Implications for Mammalian Cells

An important question raised by the findings described above is whether mTORC2 signaling is involved in the regulation of sphingolipid biosynthesis, including ceramides, in mammalian cells. With respect to ceramides, this raises an apparent paradox given that TORC2 is generally considered to act in a "progrowth" manner whereas ceramides are conventionally viewed as being associated with inducing stress-induced growth inhibition and apoptosis [2, 62–64, 100]. Superimposed on this view are two complexities. First, an adequate level of ceramides must be maintained in normally growing cells to facilitate the synthesis of complex sphingolipids, which are essential for normal cell growth and development [62–64]. Thus, it is reasonable to assume that progrowth signals will turn out to be involved in regulating *de novo* ceramide biosynthesis. Second, there are six different CerS isoforms in mammalian cells, each of which produce ceramides containing different N-acyl chain lengths that are likely to have distinct biological properties [55, 56]. Indeed, it is becoming increasingly clear that different ceramides produced by these isoforms can have both pro- or antigrowth properties, which can depend further upon the exact cell type involved [101–103]. Consistent with this conclusion, we have found that mTORC2 differentially regulates distinct CerS isoforms in mammalian cells and, moreover, that this regulation is achieved in part through AKT activation (our unpublished results). Thus, we anticipate that there are likely to be parallels between yeast and mammals with respect to the regulation of ceramide biosynthesis by TORC2, a view that is consistent with the phylogenetic conservation of both TOR signaling as well as the early steps of the sphingolipid biosynthetic pathway.

VII.   Conclusions and Perspective

As summarized in this review, studies in budding yeast have both aided in the identification of sphingolipid biosynthetic enzymes and have revealed the broad outlines of a complex regulatory system that controls the production and turnover of sphingolipid intermediates. We have learned that this regulation includes key roles for TORC2 and its downstream effectors, the SLM1/2 proteins and the YPK1/2 kinases, and is integrated with both phosphoinositide signaling and calcineurin activity. This progress now gives rise to a number of crucial questions that are to be addressed within the immediate future, including both a mechanistic understanding of how the enzymes involved in this biosynthetic pathway are regulated by the TORC2 network, including CerS and other components of ceramide/LCBP rheostat. In addition, we wish to understand more about how sphingolipids regulate actin dynamics, protein trafficking, and endocytosis, beyond the known role for LCBs in modulating the activity of the kinases PKH1/2 and YPK1/2. Finally, we need to understand the extent to which these regulatory events are conserved in higher eukaryotes, where it is clear that sphingolipids and their intermediates play important roles relevant to human health and disease. Together with recent findings that mTOR plays an important role in fatty acid biosynthesis in mammalian cells [104, 105], studies summarized here suggest that the TOR network will continue to expand to include regulation of diverse aspects of lipid biosynthesis.

REFERENCES

1. Heitman, J., Movva, N.R., and Hall, M.N. (1991). Targets for cell cycle arrest by the immunosuppressant rapamycin in yeast. *Science* 253:905–909.
2. Wullschleger, S., Loewith, R., and Hall, M.N. (2006). TOR signaling in growth and metabolism. *Cell* 124:471–484.
3. Schmidt, A., Kunz, J., and Hall, M.N. (1996). TOR2 is required for organization of the actin cytoskeleton in yeast. *Proc Natl Acad Sci USA* 93:13780–13785.
4. Helliwell, S.B., Howald, I., Barbet, N., and Hall, M.N. (1998). TOR2 is part of two related signaling pathways coordinating cell growth in *Saccharomyces cerevisiae*. *Genetics* 148:99–112.
5. Loewith, R., *et al.* (2002). Two TOR complexes, only one of which is rapamycin sensitive, have distinct roles in cell growth control. *Mol Cell* 10:457–468.
6. Wedaman, K.P., Reinke, A., Anderson, S., Yates, J., 3rd, McCaffery, J.M., and Powers, T. (2003). Tor kinases are in distinct membrane-associated protein complexes in *Saccharomyces cerevisiae*. *Mol Biol Cell* 14:1204–1220.
7. Reinke, A., *et al.* (2004). TOR complex 1 includes a novel component, Tco89p (YPL180w), and cooperates with Ssd1p to maintain cellular integrity in *Saccharomyces cerevisiae*. *J Biol Chem* 279:14752–14762.
8. Di Como, C.J., and Arndt, K.T. (1996). Nutrients, via the Tor proteins, stimulate the association of Tap42 with type 2A phosphatases. *Genes Dev* 10:1904–1916.
9. Jiang, Y., and Broach, J.R. (1999). Tor proteins and protein phosphatase 2A reciprocally regulate Tap42 in controlling cell growth in yeast. *EMBO J* 18:2782–2792.
10. Jacinto, E., Guo, B., Arndt, K.T., Schmelzle, T., and Hall, M.N. (2001). TIP41 interacts with TAP42 and negatively regulates the TOR signaling pathway. *Mol Cell* 8:1017–1026.
11. Düvel, K., Santhanam, A., Garrett, S., Schneper, L., and Broach, J.R. (2003). Multiple roles of Tap42 in mediating rapamycin-induced transcriptional changes in yeast. *Mol Cell* 11:1467–1478.
12. Jorgensen, P., Rupes, I., Sharom, J.R., Schneper, L., Broach, J.R., and Tyers, M. (2004). A dynamic transcriptional network communicates growth potential to ribosome synthesis and critical cell size. *Genes Dev* 18:2491–2505.
13. Marion, R.M., *et al.* (2004). Sfp1 is a stress- and nutrient-sensitive regulator of ribosomal protein gene expression. *Proc Natl Acad Sci USA* 101:14315–14322.

14. Schmelzle, T., Beck, T., Martin, D.E., and Hall, M.N. (2004). Activation of the RAS/cyclic AMP pathway suppresses a TOR deficiency in yeast. *Mol Cell Biol* 24:338–351.
15. Zurita-Martinez, S.A., and Cardenas, M.E. (2005). Tor and cyclic AMP-protein kinase A: two parallel pathways regulating expression of genes required for cell growth. *Eukaryot Cell* 4:63–71.
16. Chen, J.C., and Powers, T. (2006). Coordinate regulation of multiple and distinct biosynthetic pathways by TOR and PKA kinases in *S. cerevisiae. Curr Genet*1–13.
17. Urban, J., *et al.* (2007). Sch9 is a major target of TORC1 in *Saccharomyces cerevisiae. Mol Cell* 26:663–674.
18. Huber, A., *et al.* (2009). Characterization of the rapamycin-sensitive phosphoproteome reveals that Sch9 is a central coordinator of protein synthesis. *Genes Dev* 23:1929–1943.
19. Schmidt, A., Bickle, M., Beck, T., and Hall, M.N. (1997). The yeast phosphatidylinositol kinase homolog TOR2 activates RHO1 and RHO2 via the exchange factor ROM2. *Cell* 88:531–542.
20. Crespo, J.L., and Hall, M.N. (2002). Elucidating TOR signaling and rapamycin action: lessons from *Saccharomyces cerevisiae. Microbiol Mol Biol Rev* 66:579–591, Table of contents.
21. Pruyne, D., and Bretscher, A. (2000). Polarization of cell growth in yeast. *J Cell Sci* 113:571–585.
22. Kamada, Y., *et al.* (2005). Tor2 directly phosphorylates the AGC kinase Ypk2 to regulate actin polarization. *Mol Cell Biol* 25:7239–7248.
23. Audhya, A., *et al.* (2004). Genome-wide lethality screen identifies new PI4,5P$_2$ effectors that regulate the actin cytoskeleton. *EMBO J* 23:3747–3757.
24. Fadri, M., Daquinag, A., Wang, S., Xue, T., and Kunz, J. (2005). The pleckstrin homology domain proteins Slm1 and Slm2 are required for actin cytoskeleton organization in yeast and bind phosphatidylinositol-4, 5-bisphosphate and TORC2. *Mol Biol Cell* 16:1883–1900.
25. Dann, S.G., Selvaraj, A., and Thomas, G. (2007). mTOR complex1-S6K1 signaling: at the crossroads of obesity, diabetes and cancer. *Trends Mol Med* 13:252–259.
26. Jacinto, E., and Lorberg, A. (2008). TOR regulation of AGC kinases in yeast and mammals. *Biochem J* 410:19–37.
27. Sarbassov, D.D., Guertin, D.A., Ali, S.M., and Sabatini, D.M. (2005). Phosphorylation and regulation of Akt/PKB by the rictor-mTOR complex. *Science* 307:1098–1101.
28. Powers, T. (2007). TOR signaling and S6 kinase 1: yeast catches up. *Cell Metab* 6:1–2.
29. Schmidt, A., and Hall, M.N. (1998). Signaling to the actin cytoskeleton. *Annu Rev Cell Dev Biol* 14:305–338.
30. Helliwell, S.B., Schmidt, A., Ohya, Y., and Hall, M.N. (1998). The Rho1 effector Pkc1, but not Bni1, mediates signalling from Tor2 to the actin cytoskeleton. *Curr Biol* 8:1211–1214.
31. Bickle, M., Delley, P.A., Schmidt, A., and Hall, M.N. (1998). Cell wall integrity modulates RHO1 activity via the exchange factor ROM2. *EMBO J* 17:2235–2245.
32. deHart, A.K.A., Schnell, J.D., Allen, D.A., Tsai, J.-Y., and Hicke, L. (2003). Receptor internalization in yeast requires the Tor2-Rho1 signaling pathway. *Mol Biol Cell* 14:4676–4684.
33. Torres, J., Di Como, C.J., Herrero, E., and Angeles de la Torre-Ruiz, M. (2002). Regulation of the cell integrity pathway by rapamycin-sensitive TOR function in budding yeast. *J Biol Chem* 277:43495–43502.
34. Angeles de la Torre-Ruiz, M., Torres, J., Arino, J., and Herrero, E. (2002). Sit4 is required for proper modulation of the biological functions mediated by Pkc1 and the cell integrity pathway in *Saccharomyces cerevisiae. J Biol Chem* 277:33468–33476.

35. Reinke, A., Chen, J.C., Aronova, S., and Powers, T. (2006). Caffeine targets TOR complex I and provides evidence for a regulatory link between the FRB and kinase domains of Tor1p. *J Biol Chem* 281:31616–31626.

36. Wang, H., and Jiang, Y. (2003). The Tap42-protein phosphatase type 2A catalytic subunit complex is required for cell cycle-dependent distribution of actin in yeast. *Mol Cell Biol* 23:3116–3125.

37. Aronova, S., Wedaman, K., Anderson, S., Yates, J., 3rd, and Powers, T. (2007). Probing the membrane environment of the TOR kinases reveals functional interactions between TORC1, actin, and membrane trafficking in *Saccharomyces cerevisiae*. *Mol Biol Cell* 18:2779–2794.

38. Jacinto, E., *et al.* (2004). Mammalian TOR complex 2 controls the actin cytoskeleton and is rapamycin insensitive. *Nat Cell Biol* 6:1122–1128.

39. Sarbassov, D.D., *et al.* (2004). Rictor, a novel binding partner of mTOR, defines a rapamycin-insensitive and raptor-independent pathway that regulates the cytoskeleton. *Curr Biol* 14:1296–1302.

40. Berven, L.A., Willard, F.S., and Crouch, M.F. (2004). Role of the p70(S6K) pathway in regulating the actin cytoskeleton and cell migration. *Exp Cell Res* 296:183–195.

41. Hennig, K.M., Colombani, J., and Neufeld, T.P. (2006). TOR coordinates bulk and targeted endocytosis in the Drosophila melanogaster fat body to regulate cell growth. *J Cell Biol* 173:963–974.

42. Neufeld, T.P. (2007). TOR regulation: sorting out the answers. *Cell Metab* 5:3–5.

43. Cowart, L.A., and Obeid, L.M. (2007). Yeast sphingolipids: recent developments in understanding biosynthesis, regulation, and function. *Biochim Biophys Acta* 1771:421–431.

44. Dickson, R.C., Sumanasekera, C., and Lester, R.L. (2006). Functions and metabolism of sphingolipids in *Saccharomyces cerevisiae*. *Prog Lipid Res* 45:447–465.

45. Sims, K.J., Spassieva, S.D., Voit, E.O., and Obeid, L.M. (2004). Yeast sphingolipid metabolism: clues and connections. *Biochem Cell Biol* 82:45–61.

46. Schorling, S., Vallee, B., Barz, W.P., Riezman, H., and Oesterhelt, D. (2001). Lag1p and Lac1p are essential for the Acyl-CoA-dependent ceramide synthase reaction in *Saccharomyces cerevisae*. *Mol Biol Cell* 12:3417–3427.

47. Guillas, I., *et al.* (2001). C26-CoA-dependent ceramide synthesis of *Saccharomyces cerevisiae* is operated by Lag1p and Lac1p. *EMBO J* 20:2655–2665.

48. Vallee, B., and Riezman, H. (2005). Lip1p: a novel subunit of acyl-CoA ceramide synthase. *EMBO J* 24:730–741.

49. Denic, V., and Weissman, J.S. (2007). A molecular caliper mechanism for determining very long-chain fatty acid length. *Cell* 130:663–677.

50. Lester, R.L., and Dickson, R.C. (1993). Sphingolipids with inositolphosphate-containing head groups. *Adv Lipid Res* 26:253–274.

51. Dunn, T.M., Haak, D., Monaghan, E., and Beeler, T.J. (1998). Synthesis of monohydroxylated inositolphosphorylceramide (IPC-C) in *Saccharomyces cerevisiae* requires Scs7p, a protein with both a cytochrome b5-like domain and a hydroxylase/desaturase domain. *Yeast* 14:311–321.

52. Haak, D., Gable, K., Beeler, T., and Dunn, T. (1997). Hydroxylation of *Saccharomyces cerevisiae* ceramides requires Sur2p and Scs7p. *J Biol Chem* 272:29704–29710.

53. Guillas, I., *et al.* (2003). Human homologues of LAG1 reconstitute Acyl-CoA-dependent ceramide synthesis in yeast. *J Biol Chem* 278:37083–37091.

54. Obeid, L.M., and Hannun, Y.A. (2003). Ceramide, stress, and a "LAG" in aging. *Sci Aging Knowledge Environ* 2003:PE27.

55. Pewzner-Jung, Y., Ben-Dor, S., and Futerman, A.H. (2006). When do Lasses (longevity assurance genes) become CerS (ceramide synthases)?: insights into the regulation of ceramide synthesis *J Biol Chem* 281:25001–25005.
56. Lahiri, S., *et al.* (2007). Kinetic characterization of mammalian ceramide synthases: determination of K(m) values towards sphinganine. *FEBS Lett* 581:5289–5294.
57. Kageyama-Yahara, N., and Riezman, H. (2006). Transmembrane topology of ceramide synthase in yeast. *Biochem J* 398:585–593.
58. Lahiri, S., and Futerman, A.H. (2005). LASS5 is a bona fide dihydroceramide synthase that selectively utilizes palmitoyl-CoA as acyl donor. *J Biol Chem* 280:33735–33738.
59. Han, G., *et al.* (2009). Identification of small subunits of mammalian serine palmitoyl-transferase that confer distinct acyl-CoA substrate specificities. *Proc Natl Acad Sci USA* 106:8186–8191.
60. Funato, K., and Riezman, H. (2001). Vesicular and nonvesicular transport of ceramide from ER to the Golgi apparatus in yeast. *J Cell Biol* 155:949–959.
61. Dickson, R.C. (2008). Thematic review series: sphingolipids. New insights into sphingolipid metabolism and function in budding yeast. *J Lipid Res* 49:909–921.
62. Lahiri, S., and Futerman, A.H. (2007). The metabolism and function of sphingolipids and glycosphingolipids. *Cell Mol Life Sci* 64:2270–2284.
63. Bartke, N., and Hannun, Y.A. (2009). Bioactive sphingolipids: metabolism and function. *J Lipid Res* 50(Suppl.):S91–S96.
64. Zheng, W., *et al.* (2006). Ceramides and other bioactive sphingolipid backbones in health and disease: lipidomic analysis, metabolism and roles in membrane structure, dynamics, signaling and autophagy. *Biochim Biophys Acta* 1758:1864–1884.
65. Jenkins, G.M., Richards, A., Wahl, T., Mao, C., Obeid, L., and Hannun, Y. (1997). Involvement of yeast sphingolipids in the heat stress response of *Saccharomyces cerevisiae*. *J Biol Chem* 272:32566–32572.
66. Skrzypek, M.S., Nagiec, M.M., Lester, R.L., and Dickson, R.C. (1999). Analysis of phosphorylated sphingolipid long-chain bases reveals potential roles in heat stress and growth control in Saccharomyces. *J Bacteriol* 181:1134–1140.
67. Zhang, X., Skrzypek, M.S., Lester, R.L., and Dickson, R.C. (2001). Elevation of endogenous sphingolipid long-chain base phosphates kills *Saccharomyces cerevisiae* cells. *Curr Genet* 40:221–233.
68. Friant, S., Lombardi, R., Schmelzle, T., Hall, M.N., and Riezman, H. (2001). Sphingoid base signaling via Pkh kinases is required for endocytosis in yeast. *EMBO J* 20:6783–6792.
69. Liu, K., Zhang, X., Lester, R.L., and Dickson, R.C. (2005). The sphingoid long chain base phytosphingosine activates AGC-type protein kinases in *Saccharomyces cerevisiae* including Ypk1, Ypk2, and Sch9. *J Biol Chem* 280:22679–22687.
70. Sun, Y., *et al.* (2000). Sli2 (Ypk1), a homologue of mammalian protein kinase SGK, is a downstream kinase in the sphingolipid-mediated signaling pathway of yeast. *Mol Cell Biol* 20:4411–4419.
71. Liu, K., Zhang, X., Sumanasekera, C., Lester, R.L., and Dickson, R.C. (2005). Signalling functions for sphingolipid long-chain bases in *Saccharomyces cerevisiae*. *Biochem Soc Trans* 33:1170–1173.
72. Hannun, Y.A. (1996). Functions of ceramide in coordinating cellular responses to stress. *Science* 274:1855–1859.
73. Taha, T.A., Hannun, Y.A., and Obeid, L.M. (2006). Sphingosine kinase: biochemical and cellular regulation and role in disease. *J Biochem Mol Biol* 39:113–131.
74. Kobayashi, S.D., and Nagiec, M.M. (2003). Ceramide/long-chain base phosphate rheostat in *Saccharomyces cerevisiae*: regulation of ceramide synthesis by Elo3p and Cka2p. *Eukaryot Cell* 2:284–294.

75. Spiegel, S., and Milstien, S. (2000). Sphingosine-1-phosphate: signaling inside and out. *FEBS Lett* 476:55–57.
76. Futerman, A.H., and Hannun, Y.A. (2004). The complex life of simple sphingolipids. *EMBO Rep* 5:777–782.
77. Beeler, T., *et al.* (1998). The *Saccharomyces cerevisiae* TSC10/YBR265w gene encoding 3-ketosphinganine reductase is identified in a screen for temperature-sensitive suppressors of the Ca2+-sensitive csg2Delta mutant. *J Biol Chem* 273:30688–30694.
78. Zhao, C., Beeler, T., and Dunn, T. (1994). Suppressors of the Ca(2+)-sensitive yeast mutant (csg2) identify genes involved in sphingolipid biosynthesis. Cloning and characterization of SCS1, a gene required for serine palmitoyltransferase activity. *J Biol Chem* 269:21480–21488.
79. Beeler, T.J., Fu, D., Rivera, J., Monaghan, E., Gable, K., and Dunn, T.M. (1997). SUR1 (CSG1/BCL21), a gene necessary for growth of *Saccharomyces cerevisiae* in the presence of high Ca2+ concentrations at 37 degrees C, is required for mannosylation of inositolphosphorylceramide. *Mol Gen Genet* 255:570–579.
80. Yoshida, S., Ohya, Y., Nakano, A., and Anraku, Y. (1994). Genetic interactions among genes involved in the STT4-PKC1 pathway of *Saccharomyces cerevisiae*. *Mol Gen Genet* 242:631–640.
81. Audhya, A., and Emr, S.D. (2003). Regulation of PI4,5P$_2$ synthesis by nuclear-cytoplasmic shuttling of the Mss4 lipid kinase. *EMBO J* 22:4223–4236.
82. Lemmon, M.A., and Ferguson, K.M. (2000). Signal-dependent membrane targeting by pleckstrin homology (PH) domains. *Biochem J* 350(Pt 1):1–18.
83. Ho, H.L., Shiau, Y.S., and Chen, M.Y. (2005). *Saccharomyces cerevisiae*TSC11/AVO3 participates in regulating cell integrity and functionally interacts with components of the Tor2 complex. *Curr Genet* 47:273–288.
84. Tabuchi, M., Audhya, A., Parsons, A.B., Boone, C., and Emr, S.D. (2006). The phosphatidylinositol 4, 5-biphosphate and TORC2 binding proteins Slm1 and Slm2 function in sphingolipid regulation. *Mol Cell Biol* 26:5861–5875.
85. Daquinag, A., Fadri, M., Jung, S.Y., Qin, J., and Kunz, J. (2007). The yeast PH domain proteins Slm1 and Slm2 are targets of sphingolipid signaling during the response to heat stress. *Mol Cell Biol* 27:633–650.
86. Aramburu, J., Rao, A., and Klee, C.B. (2000). Calcineurin: from structure to function. *Curr Top Cell Regul* 36:237–295.
87. Cyert, M.S. (2001). Genetic analysis of calmodulin and its targets in *Saccharomyces cerevisiae*. *Annu Rev Genet* 35:647–672.
88. Boustany, L.M., and Cyert, M.S. (2002). Calcineurin-dependent regulation of Crz1p nuclear export requires Msn5p and a conserved calcineurin docking site. *Genes Dev* 16:608–619.
89. Stathopoulos, A.M., and Cyert, M.S. (1997). Calcineurin acts through the CRZ1/TCN1-encoded transcription factor to regulate gene expression in yeast. *Genes Dev* 11:3432–3444.
90. Matheos, D.P., Kingsbury, T.J., Ahsan, U.S., and Cunningham, K.W. (1997). Tcn1p/Crz1p, a calcineurin-dependent transcription factor that differentially regulates gene expression in *Saccharomyces cerevisiae*. *Genes Dev* 11:3445–3458.
91. Mulet, J.M., Martin, D.E., Loewith, R., and Hall, M.N. (2006). Mutual antagonism of target of rapamycin and calcineurin signaling. *J Biol Chem* 281:33000–33007.
92. Bultynck, G., Heath, V.L., Majeed, A.P., Galan, J.M., Haguenauer-Tsapis, R., and Cyert, M.S. (2006). Slm1 and slm2 are novel substrates of the calcineurin phosphatase required for heat stress-induced endocytosis of the yeast uracil permease. *Mol Cell Biol* 26:4729–4745.

93. Aronova, S., *et al.* (2008). Regulation of ceramide biosynthesis by TOR complex 2. *Cell Metab* 7:148–158.
94. Walther, T.C., Brickner, J.H., Aguilar, P.S., Bernales, S., Pantoja, C., and Walter, P. (2006). Eisosomes mark static sites of endocytosis. *Nature* 439:998–1003.
95. Walther, T.C., *et al.* (2007). Pkh-kinases control eisosome assembly and organization. *EMBO J* 26:4946–4955.
96. Luo, G., Gruhler, A., Liu, Y., Jensen, O.N., and Dickson, R.C. (2008). The sphingolipid long-chain base-Pkh1/2-Ypk1/2 signaling pathway regulates eisosome assembly and turn-over. *J Biol Chem* 283:10433–10444.
97. Kunz, J., Schneider, U., Howald, I., Schmidt, A., and Hall, M.N. (2000). HEAT repeats mediate plasma membrane localization of Tor2p in yeast. *J Biol Chem* 275:37011–37020.
98. Sturgill, T.W., Cohen, A., Diefenbacher, M., Trautwein, M., Martin, D.E., and Hall, M.N. (2008). TOR1 and TOR2 have distinct locations in live cells. *Eukaryot Cell* 7:1819–1830.
99. Berchtold, D., and Walther, T.C. (2009). TORC2 plasma membrane localization is essential for cell viability and restricted to a distinct domain. *Mol Biol Cell* 20:1565–1575.
100. Sabatini, D.M. (2006). mTOR and cancer: insights into a complex relationship. *Nat Rev Cancer* 6:729–734.
101. Senkal, C.E., Ponnusamy, S., Bielawski, J., Hannun, Y.A., and Ogretmen, B. (2010). Antiapoptotic roles of ceramide-synthase-6-generated C16-ceramide via selective regula-tion of the ATF6/CHOP arm of ER-stress-response pathways. *FASEB J* 24:296–308.
102. Senkal, C.E., *et al.* (2007). Role of human longevity assurance gene 1 and C18-ceramide in chemotherapy-induced cell death in human head and neck squamous cell carcinomas. *Mol Cancer Ther* 6:712–722.
103. Menuz, V., *et al.* (2009). Protection of *C. elegans* from anoxia by HYL-2 ceramide synthase. *Science* 324:381–384.
104. Porstmann, T., *et al.* (2008). SREBP activity is regulated by mTORC1 and contributes to Akt-dependent cell growth. *Cell Metab* 8:224–236.
105. Porstmann, T., Santos, C.R., Lewis, C., Griffiths, B., and Schulze, A. (2009). A new player in the orchestra of cell growth: SREBP activity is regulated by mTORC1 and contributes to the regulation of cell and organ size. *Biochem Soc Trans* 37:278–283.

11

TORC1 Signaling in the Budding Yeast Endomembrane System and Control of Cell–Cell Adhesion in Pathogenic Fungi

ROBERT J. BASTIDAS • MARIA E. CARDENAS

Department of Molecular Genetics and Microbiology
Duke University Medical Center, Durham
North Carolina, USA

I. Abstract

The rapamycin-sensitive TORC1 protein kinase is the central component of a conserved signal transduction cascade controlling cell growth in
response to nutrients and growth factors. Groundbreaking studies are
uncovering novel roles for the endomembrane vesicular trafficking system
as a platform for TORC1 signaling. TORC1 components, regulators, and
major effectors have been localized to late endosomes and the vacuolar
periphery in yeast. Mutations in the class C Vps, Ego/Gse, and PAS protein
complexes, involved in vesicular trafficking and protein sorting, in combination with mutation of the nonessential Tor1 kinase severely compromise
or abolish cell growth. Class C Vps complex function sustains amino acid
homeostasis for efficient TORC1 signaling. The Ego/Gse complex (EGOC)
GTPase activity responds to amino acids to regulate permease sorting and
activate TORC1 signaling. The emerging view is that amino acids are

THE ENZYMES, Vol. XXVII   199   ISSN NO: 1874-6047
DOI: 10.1016/S1874-6047(10)27011-7

sensed in intimate association with the endomembrane system, which facilitates molecular interactions to enable TORC1 activation and signaling. Novel roles for TORC1 are also surfacing throughout the fungal kingdom. TORC1 regulates filamentous growth in fungi. In the human pathogen *Candida albicans*, TORC1 controls transcriptional programs including those involved in ribosome biogenesis and nutritional control. Remarkably, in *C. albicans* TORC1 governs expression of adhesins, which promote tissue adherence, biofilm formation, and virulence. These studies reveal TORC1 pathway wiring plasticity as pathogenic yeasts adapted to their host niche environments. A preeminent role for TORC1 signaling in fungal virulence has attracted interest in the use of rapamycin and its analogs in antifungal therapy.

II. TORC1 Signaling from the Budding Yeast Endomembrane System

As with real estate three key factors with respect to signal transduction are: location, location, location. Much of what we know about signal transduction is focused on events that transpire at the plasma membrane or in the nucleus and yet we know that cells have the ability to sense and respond to cues generated in the cytoplasm and on the endomembrane network that lies within. For this purpose, the cell must adapt the architectural organization of signaling machineries to ensure their asymmetric distribution and prompt but spatially confined signaling. Spatial restriction of signaling is emerging as a central theme via which fundamental cell processes are accomplished, as illustrated by the following examples. From studies of programed cell death we now appreciate that the surface of the mitochondria serves as a staging area for cellular life and death decisions (reviewed in Ref. [1]). Moreover, diverse cues sensed at the plasma membrane are transduced into the cytoplasm via endocytic vesicles in which downstream effectors are activated to process signals and discharge local outputs (reviewed in Ref. [2]). Novel roles for the endomembrane vesicular system in spatially restricted signaling that occurs in response to mating pheromone, alkaline pH stress, and during feast or famine in budding yeasts and other fungi have recently been characterized [3–6]. Similarly, an increasing number of compelling observations, discussed below, lead to the view that nutrient sensing via TORC1 signaling occurs in intimate association with the cellular endomembrane system, which (in this case) includes endosomal, prevacuolar, and vacuolar membranes.

III. TORC1 Components and Its Major Downstream Effectors Localize to Endomembranes

Most components of the TORC1 signaling cascade have been characterized as novel residents of the endomembrane network. The first evidence was provided by early immunofluorescence and cellular fractionation approaches that showed Tor2 localizes to the vacuolar surface [7]. Subsequent studies have extended these findings to the remaining components of TORC1, including Tor1, Lst8, Kog1, and Tco89, all of which have been localized to endosomal, prevacuolar, and vacuolar compartments [8–13]. In contrast, the rapamycin-insensitive TORC2 is restricted to distinct foci on the plasma membrane via the Avo1 pleckstrin homology-like lipid binding domain [13]. In addition, it has been reported that under nutrient-replete conditions a significant fraction of Tor1 localizes to the nucleus where it controls rRNA expression and condensation [14–16]. Localization of TORC2 in the plasma membrane and Tor1 in the nucleus is dynamic and in the later case is controlled by nutrient availability [13, 14].

The striking connection between TORC1 components and the endomembrane vacuolar system is not only physical but also functional and, as discussed below, appears to enable TORC1 activation and signaling. Aside from TORC1 components, two major characterized TORC1 downstream effectors, the Tap42–Sit4 protein phosphatase and the AGC kinase Sch9, have also been found associated with the vacuolar membranes [17, 18]. Tap42–Sit4, which is involved in controlling expression of nutrient and stress responsive genes and Gcn2-regulated translation, is associated with light endomembranes. During growth under optimal nitrogen conditions, Tap42–Sit4 interacts with TORC1 and this interaction may mediate Tap42–Sit4 vacuolar membrane association [17]. Interestingly, TORC1 and Tap42–Sit4 association to endomembranes is resistant to nonanionic detergent extraction and, as characterized for TORC1, is confined to discrete microdomains [17, 19]. Sch9 is a direct substrate of TORC1 involved in the regulation of stress, ribosome biogenesis (Ribi) and ribosomal protein (RP) gene expression, and in controlling entry into G_0, chronological life span, and translational initiation [18, 20–22]. Sch9 is known to reside at the vacuolar surface, and indirect evidence has suggested that this localization is a prerequisite for phosphorylation by TORC1 [18, 20]. Taken together, these observations underscore a central role for the vacuolar membrane in facilitating and spatially restricting molecular interactions between TORC1 and its downstream effectors.

IV. Genetic and Functional Interactions Between Tor1 and Protein Trafficking Regulators Provide Insights into TORC1 Activation by Amino Acids

Spatial restriction of TORC1 in the vacuolar periphery also has important implications with regard to its activation by amino acids, as illustrated in several studies. In yeast the homologous, Tor1 and Tor2 proteins can substitute for each other as members of the TORC1 [8, 23]. However, Tor1 populates TORC1 to a greater extent than Tor2 [8, 12]. Moreover, while Tor2 is essential for viability, Tor1 is not [24]. Capitalizing upon these observations, a genome-wide screen was performed aimed at identifying genes that when mutated in combination with a *tor1* mutation result in synthetic lethal or decreased fitness defects. This screen yielded over 200 genes representing several functional categories involved in actin organization, transcription, and organellar (ribosomal, mitochondrial, ER, Golgi, and vacuolar) functions (Figure 11.1A) [25]. Those genes that compromised vacuolar-related functions were remarkable in that mutation of 96% of these genes increased the sensitivity of cells to rapamycin, underscoring their link to TORC1 function (Figure 11.1B) [25–27]. Moreover, several of these genes encode proteins that belong to well-characterized multiprotein complexes, including the class C Vps (vacuolar protein secretion), Ego/Gse (exit from G_0/Gap amino acid permease sorting in endosomes), and PAS (preautophagosomal) complexes. Previously, a unique Tor1, but not Tor2, connection to components of the class C Vps complex was reported based on genetic interactions and rapamycin hypersensitivity conferred by class C Vps mutations [7, 27]. These results challenge the model that Tor2 is capable of providing all of the rapamycin-sensitive functions associated with TORC1 signaling and indicated that there are Tor1/2-shared functions as well as both Tor1- and Tor2-unique functions.

The class C Vps complex is required for docking and membrane fusion of Golgi-derived vesicles at the endosome, and between endosomes and vacuoles, and thereby actively regulates protein and membrane traffic between these cell compartments [28, 29]. Synthetic lethal interaction can arise between two genes whose products act in parallel or compensatory pathways. However, mutation of Tor1 or rapamycin exposure did not have detectable effects in forward and retrograde Golgi to endosome protein transport or on any of the protein sorting routes to the vacuole assisted by the class C Vps complex, including the endosomal, Cvt, or the nonendosomal pathways [25]. An alternative study has reported a modest delay in a late step of fluid phase endocytosis upon TORC1 inhibition by rapamycin; however, the physiological significance of this effect is unknown [19].

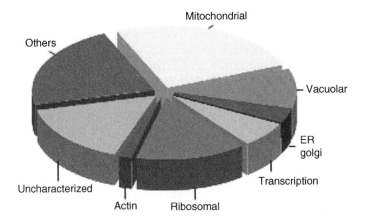

A

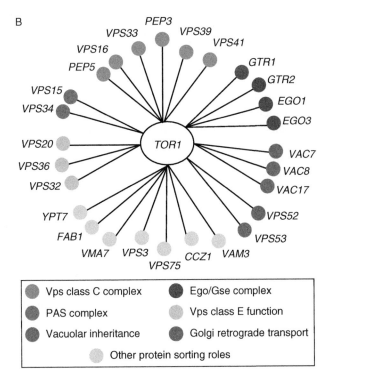

B

FIG. 11.1. Schematic representation of the genetic synthetic Tor1 interaction network. (A) Functional categories represented by the genes that when mutated in combination with *tor1* mutation confer synthetic lethal or decrease fitness defects. (B) The genetic synthetic interaction Tor1 network involved in vacuolar, vesicular trafficking, and protein sorting functions includes distinctive protein complexes. Mutation of 23 out of 27 of these genes results in rapamycin hypersensitivity [25–27]. Modified from Figure 2 [79]. (See color plate section in the back of the book.)

These results excluded models that invoke a role for TORC1 in regulating the class C Vps complex, or acting in a parallel compensatory pathway to provide a known class C Vps complex function.

Mutations in class C Vps genes result in vacuolar fragmentation, low amino acid levels, and in defects in endosome and vacuolar acidification, poor nitrogen source utilization, and nitrogen starvation survival [5, 25, 30]. It has been postulated that TORC1 is activated by amino acids in both mammals and yeasts. It has long been known that the vacuole is the major reservoir of amino acids in yeast [31]. These observations prompted studies to consider the impact of the class C Vps complex functions in TORC1 activation. Remarkably, the growth defect of class C *vps* and the double *tor1 pep3* mutants was alleviated by increasing the concentration of particular amino acids in the culture media. The most effective amino acids were glutamic acid and glutamine, which are central to ammonium assimilation and distribution, and earlier studies demonstrated to activate signaling pathways, including the Tor pathway ([32] and reviewed in Ref. [33]). These studies support a role for the class C Vps complex function in promoting amino acid homeostasis for efficient TORC1 signaling and further suggest that Tor1 is more efficient than Tor2 in supporting growth under limiting amino acid conditions. Whether this is attributable to the superior ability of Tor1, as compared to Tor2, to interact with TORC1 components remains to be established. Collectively, these studies also support a model in which association of TORC1 with the vacuolar membrane is essential to sense amino acid signals and to relay these signals to its several effectors.

Despite demonstration that amino acids have a critical role in activating Tor signaling, and the fact that in mammals hormones and other growth factors have little impact on mTor activity in the absence of amino acids [34], until recently little was known about the mechanism involved in this activation. In mammals, the Rheb GTPase and its GAP regulators Tsc1/Tsc2 integrate hormonal and energy signals transmitted by the PI3K/PKB and LKB1/AMPK signaling modules to activate mTORC1 (reviewed in Ref. [35]). Conserved Rheb, Tsc1, and Tsc2 orthologs also modulate TORC1 activity in *Schizosaccharomyces pombe* (reviewed in Ref. [36]). In contrast, in *Saccharomyces cerevisiae* the Rheb ortholog does not appear to be linked to TORC1 function and Tsc1 and Tsc2 are not conserved (Ref. [36], and our own unpublished results).

The EGOC responds to amino acids to regulate the sorting of the general amino acid permease Gap1 [37], although this function may operate only in certain genetic strain backgrounds [42]. Together with TORC1, EGOC is required to resume growth following a rapamycin-induced cell cycle arrest and similar to TORC1, is also restricted to prevacuolar and vacuolar compartments [37, 38]. The EGOC contains a lipid anchored protein (Ego3/Gse2), and four peripheral proteins including two small GTPases (Gtr1 and Gtr2) and two other components (Ego1/Gse1 and Ltv1) [37, 38]. From these early studies in yeast, the EGOC appeared uniquely poised to convey vacuole-derived amino acid signals to TORC1. The mammalian and drosophila orthologs of Gtr1 and Gtr2 are a family of four proteins known as the Rag GTPases. Recent reports show that the Rag GTPases are necessary for mTORC1 activation by amino acids [39, 40]. Regulation of Gap1 sorting in response to amino acids by the Ego complex involves direct physical interaction [37]. Similarly, the Rag GTPases convey amino acid signals to mTORC1 through physical association with the mTORC1 complex subunit Raptor [39]. A recurring theme in the roles of GTPases and their GEFs and GAPs is regulation of protein localization (reviewed in Ref. [41]). Cascades of small GTPases driving protein localization and regulation have been characterized, and Rag, Gtr1 and Gtr2 belong to this family of related small G-proteins. Remarkably, it was demonstrated that upon amino acid addition the RagA,B GTPase dimer is loaded with GTP thereby recruiting mTORC1 to vesicular compartments, where it colocalizes with Rheb1 [39]. Although the identity of these vesicles is not yet fully established, they contain Rab7, a bona fide marker of late endosomes and lysosomes, which are the mammalian counterpart of yeast prevacuolar compartments and vacuoles. Whether the Rag GTPase also localizes to the membrane of Rheb1-containing vesicles, perhaps tethered to membranes by interaction with a transmembrane protein as is the case for the Gtr1,2 GTPases in the context of the EGOC, is an important question that will require further study. In a recent study, the EGOC was also shown to mediate amino acid sensing and TORC1 activation in yeast [42]. Similar to mTORC1 activation by Rag GTPases, this effect involves physical interaction of EGOC with TORC1. However, in contrast with Rag A–B^{GTP} alleles, which fully activate mTORC1 in amino acid free medium, the EGOC GTPase subunit Gtr1^{GTP} allele only partially restored TORC1 activity in yeast. In addition, unlike Rag A–B, which recruits mTORC1 to Rheb-containing lysosomes, amino acid signaling via the EGOC does not alter TORC1 vacuolar localization. This study also identified the class C Vps complex component Vac6/Vps39, previously shown to be required for TORC1 activation [25], as a GEF for Gtr1 [42]. In summary, these results are compelling and reveal that in both yeasts and mammals the vesicular endomembrane network plays center stage in facilitating molecular interactions for TORC1 activation by amino acids.

V.   Interactions Between Vesicular System Components and TORC1-Controlled Transcriptional Regulators are Required for Balanced Cell Growth

1. GLN3 nuclear translocation elicited by poor nitrogen source requires functional vesicular trafficking and rapamycin bypasses this requirement.

Further evidence for endomembrane vesicular involvement in TORC1-regulated signaling incidentally transpired from studies examining the nutrient utilization defects that underlie the vesicular trafficking mutants discussed in Section IV. It was observed that class C *vps* mutants exhibit severe defects in activating expression of NCR genes and, consistent with this finding, are unable to grow in poor nitrogen sources [5, 25]. It has been proposed that under optimal growth conditions TORC1 activity represses the expression of the NCR genes, whose products are involved in the transport and utilization of suboptimal nitrogen sources such as proline [43–46]. Under nutrient-replete growth conditions the NCR transactivator Gln3 and its repressor Ure2 are hyperphosphorylated and form a complex that is restricted to the cytosol. Rapamycin treatment leads to Gln3–Ure2 dephosphorylation, complex dissociation, and Gln3 nuclear translocation [43, 46]. The TORC1 effector in NCR regulation is the Tap42–Sit4 protein phosphatase. Either *sit4* mutation or inactivation of Tap42 prevents Gln3 nuclear translocation in response to rapamycin [43]. However, two recent pieces of evidence indicate this model of regulation does not operate for nitrogen source quality: first, Gln3 phosphorylation status does not correlate with nuclear localization and second, in certain strain backgrounds Gln3 is controlled through an unknown Sit4-independent mechanism [47, 48].

The inability of class C *vps* mutants to utilize poor nitrogen sources results from their marked defect in Gln3 nuclear translocation and NCR gene expression [5]. Similar phenotypes were also observed in class D *vps* mutants, which are impaired in Golgi to endosome transport, but not in mutants affected in other protein trafficking steps. Intriguingly, Gln3 nuclear translocation elicited by rapamycin was independent of class C and D Vps function, indicating differential mechanisms for NCR gene induction in response to rapamycin versus poor nitrogen sources. Detailed examination of Gln3 cytoplasmic distribution in wild-type cells found a significant fraction of Gln3 and Ure2 peripherally associated with low-density membranes and Gln3 colocalized with Golgi and endosomal markers both by sucrose gradient fractionation as well as by immunofluorescent microscopy [5]. In the Class C and D *vps* mutants the insoluble Gln3 fraction increased and comigrated with lighter fractions of the gradient, consistent with entrapment of Gln3 in the small vesicles that are known to accumulate in these mutants [5, 49]. The nature of the Gln3 association with membranes remains to be determined, however, association via interaction with Ure2 has been excluded. An interaction of Gln3 with Tor1 has been previously reported; however, deletion of the Tor1 binding domain in Gln3 did not perturb its membrane association [5]. In summary, these studies indicate that Golgi to endosome trafficking is a step in the route that Gln3 takes to the nucleus in response to nitrogen source discrimination and that rapamycin bypasses this step. These results also suggest that TORC1 signaling to control Gln3 regulation is staged on the endomembrane system. However, given recent evidence that the effects of rapamycin and Sit4 on Gln3 could extend beyond nuclear localization [50], these studies do not exclude the possibility that TORC1 regulation could be executed at multiple levels and compartments such as the nucleus.

Although elucidation of the mechanisms involved will require further experimentation, an independent study provides some clues for postulating a working model. It has been shown that Tap42–Sit4 is tethered to light membranes via interaction with TORC1 during growth under optimal nitrogen conditions [17]. Complete nitrogen starvation or rapamycin exposure elicits release of Tap42–Sit4 into the cytosol and Gln3 dephosphorylation, which in turn leads to Ure2–Gln3 complex dissociation and Gln3 nuclear translocation [17, 51]. Interestingly, unlike rapamycin or nitrogen starvation, shifting cells from rich to poor nitrogen medium (proline) does not cause release of Tap42–Sit4 from membranes [51]. These results support a model in which, under poor nitrogen source conditions, Gln3–Ure2 associated with Golgi-derived vesicles is transported to endosomes or a later compartment where activated Tap42–Sit4 complexes reside in association with TORC1 and it is there that Gln3 is dephosphorylated [52]. It remains to be determined if Tap42–Sit4 complexes associated with TORC1 are active. In contrast to rapamycin, which elicits pronounced Gln3 dephosphorylation, in response to poor nitrogen source Gln3 is only modestly dephosphorylated [53].

2. The Rab escort Mrs6 and Sfp1 integrate vesicular fluxes with TORC1-controlled ribosome biogenesis.

A compelling demonstration of the vesicular trafficking system impact on TORC1 signaling was uncovered in recent studies showing the Rab escort Mrs6 interacts with the Ribi and RP genes transactivator Sfp1. During active growth in rich nutrient conditions Sfp1 is localized in the nucleus and nutrient depletion or rapamycin treatment result in Sfp1 export to the cytoplasm where it is restrained by interaction with Mrs6 [18, 54–56].

TORC1 directly phosphorylates Sfp1 and thereby regulates its nuclear localization and binding to RP promoters [56]. Mrs6 also interacts with the family of Rab GTPases that mediate vesicle fluxes in the secretory pathway (reviewed in Ref. [57]). Thus, it has been proposed that Mrs6 and TORC1 integrate information from vesicle fluxes and nutritional status to control ribosome biogenesis via Sfp1 [55]. Taken together these observations further elaborate the model that balanced growth is coordinated by interaction of multiple system sensors, among which the vesicular system and the TORC1 pathway play fundamental roles.

VI.   Tor Signaling in Fungal Pathogens

The Tor signaling cascade has been largely conserved throughout the fungal kingdom. However, whereas in *S. cerevisiae* Tor responds to nutrient-derived signals to coordinate cell growth and proliferation, in *S. pombe* Tor signaling modulates sexual differentiation in response to nutritional cues [58]. These differences provide a framework to consider Tor roles in other fungal organisms, particularly those that are pathogens of humans.

Recent findings demonstrate that Tor signaling promotes vegetative and hyphal growth, and controls cell–cell adherence in the human fungal pathogen *C. albicans* [59]. Although the Tor kinases are broadly conserved, these studies underscore contrasting strategies fungal organisms employ in utilizing Tor signaling. In the second part of this chapter, we discuss the implications and novel findings for Tor signaling among divergent fungal species, in particular those that are pathogenic to humans. Note that because rapamycin-sensitive effects have been attributed to TORC1 activity in other organisms, hereafter we will refer to all rapamycin-sensitive *C. albicans* Tor effects as TORC1 effects.

VII.   Control of Filamentous Differentiation by TORC1 Signaling in Divergent Fungi

Many fungi differentiate into a filamentous form during their life cycle, and nutrient starvation is a strong stimulator of this dimorphic transition. The human pathogen *Cryptococcus neoformans* mates in response to nitrogen limitation and forms filaments and infectious basidiospores during its sexual cycle [60]. *C. albicans* and the emerging opportunistic pathogen *Candida lusitaniae* undergo filamentous growth when cultured on nitrogen limited (SLAD) medium [61–63]. Giving the prominent role of TORC1 signaling in controlling cellular responses to nutrients, it is not surprising that sublethal concentrations of rapamycin prevent filamentous differentiation of *S. cerevisiae*, *C. neoformans*, *C. lusitaniae*, and *C. albicans*. These results illustrate that TORC1 signaling plays a conserved role in regulating filamentous differentiation in response to nutritional cues in diverse fungi [59, 64, 65].

This paradigm was originally established in *S. cerevisiae*, in which the Tor cascade drives yeast pseudohyphal differentiation in response to nitrogen availability via the protein phosphatase Sit4 and its associated regulatory protein Tap42 as discussed in Section III [64]. The Tap42–Sit4 role in regulating the NCR transcriptional response likely underlies TORC1's ability to promote pseudohyphal differentiation [43, 46, 66, 67]. Paradoxically, constitutive expression as well as repression of NCR-regulated genes (generated by *ure2* or *gln3* mutations, respectively) both block pseudohyphal growth [64, 68, 69]. Based on these lines of evidence a proposed model suggests that the ability to turn TORC1, and thereby the NCR genes, ON and OFF is required for filamentous differentiation [64].

Pseudohyphal differentiation is also regulated by the MAP kinase and the cAMP-PKA signaling cascades and through a transcription factor cascade comprised of Sok2, Phd1, Ash1, and Swi5 [69–73]. Based on epistasis analysis, the TORC1 cascade most likely acts in parallel with these pathways [64]. Other proteins that do not function in these pathways also regulate filamentous growth, including the Mep2 ammonium permease [68].

As is the case with *S. cerevisiae*, rapamycin also elicits a response similar to the NCR transcriptional induction in *C. albicans* (Figure 11.2; [59]). Furthermore, loss of function mutations in *MEP2*, *GLN3*, and *GAT1* (another NCR transactivator) homologs, completely block the ability of *C. albicans* to undergo the yeast to hyphae transition during nitrogen limiting conditions, strongly indicating conservation of TORC1 signaling effectors in regulating nitrogen-induced dimorphic transitions [74–76]. In contrast to *S. cerevisiae*, induction of *C. albicans* pseudohyphal and hyphal growth can be readily observed in both agar- and liquid-based growth mediums. Interestingly, rapamycin inhibits *C. albicans* filamentation on a variety of hyphal inducing agar-based mediums, whereas in liquid medium rapamycin has negligible effects on hyphal induction [59]. Thus, these studies implicate the TORC1 cascade as a regulator of hyphal differentiation on agar surfaces. Remarkably, this pathogen evolved a repertoire of responses triggered by contact with surfaces, such as invasive hyphal growth, which occurs during contact with semisolid surfaces or during infection of the human host [77, 78]. Another contact dependent-response occurs during *C. albicans* biofilm formation on solid surfaces. In *C. albicans*, it is thought that growth during contact with a surface results in mechanical perturbations of the cell wall or plasma membrane that is detected by cell wall integrity pathways. Indeed, the cell integrity kinase Mkc1 is

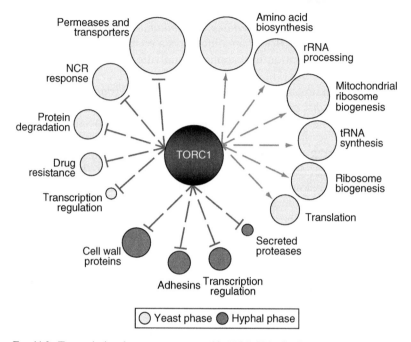

FIG. 11.2. Transcriptional programs governed by TORC1 in *C. albicans*. TORC1 stimulates transcriptional programs devoted to protein synthesis, similar to the Ribi regulon governed by TORC1 in *S. cerevisiae*. TORC1 also inhibits the NCR transcriptional response and expression of genes coding for permeases and transporters. In contrast to *S. cerevisiae*, TORC1 in *C. albicans* modulates the expression of hyphae-specific genes coding for cell wall constituents, including several adhesins and their transcriptional regulators, as well as key virulence factors such as secreted proteases. Adapted from Ref. [59], Supplementary Tables S1, S2, and S4. (See color plate section in the back of the book.)

activated when *C. albicans* cells are grown on surfaces, and *mkc1/mkc1* null mutants are defective in contact-dependent responses [78]. In *S. cerevisiae*, Tor regulates the cell integrity pathway (reviewed in Ref. [79]), and it would be reasonable to hypothesize a similar role for Tor in *C. albicans*. The cell integrity pathway controlled by Mkc1 could, therefore, be a potential avenue by which Tor controls hyphal differentiation on semisolid surfaces.

The observation that rapamycin inhibits hyphal differentiation in divergent fungal pathogens implicates TORC1 as a key element in governing critical virulence traits. The human fungal pathogens, *Blastomyces dermatitidis*, *Histoplasma capsulatum*, *Paraccocidioides brasiliensis*, *Coccidioides immitis*, and *Sporothrix schenckii* are thermally dimorphic, primarily growing as saprobic molds in the environment. Upon inhalation of asexual conidia by the host, these fungi adopt a parasitic yeast-like form induced by a temperature dependent mycelium-to-yeast morphological transition. Strains or mutants locked as hyphae are avirulent [80]. It will be interesting to assess whether TORC1 plays a role in thermal dimorphism as well. In *C. neoformans*, dimorphic transitions also occur during its life cycle. Cellular fusion triggered by pheromones results in a dikaryon that undergoes a dimorphic transition into hyphae decorated with terminal basidia. Nuclear fusion and meiosis occur in the basidium, and budding from the basidial surface generates chains of infectious spores that infect humans (reviewed in Ref. [81]). Whereas in dimorphic fungi yeasts play a prominent role in infection, in *C. albicans* hyphae is associated with pathogenesis. During *C. albicans* evolution into a commensal in the mammalian gastrointestinal tract, hyphae formation was likely necessary for niche colonization and to compete with bacteria. It also enables formation of drug resistant biofilms, survival, and escape from the host immune system. Exploring the Tor-based circuitry controlling dimorphism, therefore, promises to be important for understanding fungal pathogenesis.

VIII.　The TORC1 Cascade and Cellular Adhesion

Adherence of pathogenic microorganisms to host tissue is regarded as an important requisite for host niche colonization and for their ability to invade and proliferate within internal organs. For example, the ability of *Candida* species to gain access to tissues in humans results from cellular adhesion to plastic devices such as catheters, prosthetic heart valves, and dentures [82]. It is known that proliferation of *Candida* on inert surfaces involves multicellular colony growth known as biofilm formation; a cooperative behavior involving collective adherence among cells, thus allowing for sharing of scarce resources and for fending off antimicrobial attacks. *Candida* biofilms also form on biotic surfaces such as the oral cavity and gastrointestinal tract, allowing *Candida* to compete with the local microbiota and serving as a driving force for its success as a mammalian commensal.

Global transcriptional profiling of *C. albicans* cells exposed to rapamycin has revealed evolutionary conserved roles for TORC1 in regulating the expression of genes involved in nitrogen starvation response and ribosome biogenesis (Figure 11.2; [59]). Surprisingly, a novel role for TORC1 in controlling the expression of several cell wall and hyphal-specific genes (including surface adhesins mediating cellular adherence) was uncovered embedded within the Tor controlled gene expression programs (Table 11.1). This cohort of genes comprises the agglutinin like sequences (ALS) *ALS1* and *ALS3* (named for their resemblance to sexual agglutinins

TABLE 11.1

RAPAMYCIN INDUCTION OF HYPHAE-SPECIFIC ADHESINS AND CELL WALL TRANSCRIPTS

orf19 Id	Locus name	S. cerevisiae best hit	Fold change	Description
Hyphal induced				
orf19.5741	*ALS1*	*FLO9*	29.7	Adhesin
orf19.1327	*RBT1*	*MUC1*	7.2	Predicted cell wall protein
orf19.2355	*ALS3*	*FLO1*	7.0	Adhesin
orf19.3374	*ECE1*	–	6.1	Cell wall protein of unknown function
orf19.3642	*SUN41*	*SIM1*	4.1	Putative cell wall protein
Transcription regulation				
orf19.2823	*RFG1*	*ROX1*	6.2	Transcriptional regulator of filamentous growth
orf19.3127	*CZF1*	*UME6*	5.9	Transcriptional regulator of filamentous growth
orf19.5908	*TEC1*	*TEC1*	4.2	Transcriptional regulator of filamentous growth
orf19.454	*SFL1*	*SFL1*	5.9	Putative transcriptional repressor
Cell wall proteins				
orf19.1097	*ALS2*	*FLO1*	9.2	Adhesin
orf19.7586	*CHT3*	*CTS1*	7.5	Chitinase
orf19.348	–	*KRE6*	5.4	Predicted beta glucan synthase
orf19.2706	*CRH11*	*CRH1*	5.4	Putative chitin transglycosidase
orf19.3893	*SCW11*	*SCW11*	5.3	Protein similar to glucanase
orf19.1690	*TOS1*	*TOS1*	4.7	Similar to alpha agglutinin anchor subunit
orf19.242	*SAP8*	*YPS1*	4.5	Secreted aspartyl proteinase
orf19.2990	*XOG1*	*EXG1*	4.0	Exo-1,3-β-glucanase

Adapted from Ref. [59], Table 1.

of *S. cerevisiae*), the adhesins *HWP1* and *ECE1*, and their transcriptional repressors *NRG1* and *TUP1* [59].

. Both *ALS1* and *ALS3* are constituents of the *ALS* gene family that is comprised of at least eight family members, *ALS1–ALS7* and *ALS9* [83–88]. The structure of Als proteins consists of an N-terminal domain that mediates the adhesive activities of Als proteins, a repeat-rich central domain thought to be heavily glycosylated, a C-terminal domain containing multiple serine and threonine residues, and a glycosylphosphatidylinositol (GPI) anchor site, involved in cross-linking of adhesins to β-1,6-glucans in the cell wall [89]. The structure of *C. albicans* adhesins is very similar to that of the *S. cerevisiae* Flo family of flocculins and to *Candida glabrata* Epa family of adhesins; however, in contrast to *S. cerevisiae* and *C. glabrata* adhesin genes, *ALS* family members are not subtelomeric, and their products bind peptides instead of polysaccharides [90]. Hwp1, a surface protein with little similarity to the Als proteins is a well-known substrate for host transglutaminases and, in conjunction with Als1 and Als3, promotes adherence of *C. albicans* to mammalian epithelial and endothelial cells (reviewed in Ref. [91]).

Another role for the Als and Hwp1 proteins occurs in the context of biofilm formation. The typical architecture of a biofilm consists of basal layers of yeast cells adhered to a surface followed by layers of intertwined hyphal cells stabilized by adhesive interactions among neighboring cells. Recent evidence indicates that Hwp1 promotes interactions of the cell surface of *C. albicans* with other cell surfaces and that Als1 and Als3 facilitate this interaction [92]. The physical nature of these interactions is currently unknown; however, this study has led to a model in which these adhesins promote cell–cell interactions in the context of biofilm formation.

In accord with this transcriptional profile, rapamycin treatment resulted in increased expression of surface localized Als3 (Figure 11.3) and promoted extensive cellular aggregation of *C. albicans* cells in an adhesin-dependent fashion (Figure 11.4; [59]). Moreover, adhesin gene induction and cellular aggregation of rapamycin-treated cells were strongly dependent on the transactivators Efg1, a homolog of *S. cerevisiae* Sok2/Phd1 that is also involved in mediating cellular aggregation [73]. These findings support models in which TORC1 negatively controls cellular adhesion by governing the expression of cell surface adhesins (Figure 11.5; [59]). Rapamycin also modulates aggregation of *Candida guilliermondii* cells, and expression of adhesin molecules in mammalian endothelial cells. Thus

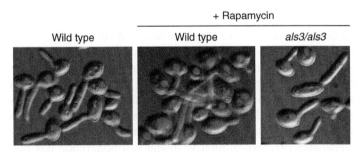

Fɪɢ. 11.3. Rapamycin stimulates surface expression of the Als3 adhesin in *C. albicans* hyphal cells. Exposure of wild-type hyphal cells to rapamycin results in increased expression of the Als3 adhesin at the surface of *C. albicans* hyphae, as assayed by indirect immunofluorescence (red) with an Als3 antiserum. Als3 resembles mammalian cadherins and mediates adherence of *C. albicans* hyphal cells to N-cadherins in endothelial cells and E-cadherins in oral epithelial cells [115]. Als3 is also required for cell–cell adhesion during biofilm development [92]. Adapted from Ref. [59], Figure 2C. (See color plate section in the back of the book.)

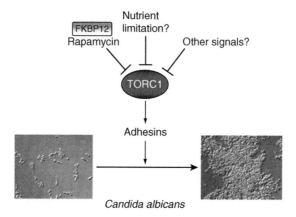

Candida albicans

Fɪɢ. 11.4. Cellular aggregation is stimulated in *C. albicans* hyphal cells exposed to rapamycin. Rapamycin inhibition of TORC1 results in robust aggregation of *C. albicans* hyphal cells. Interestingly, rapamycin-induced cell aggregation is primarily observed under particular growth conditions that promote hyphal differentiation (Spider medium, 37 °C). The triggers in this medium that prompt such a response when cells are treated with rapamycin remain unknown. Adapted from Ref. [59], Figure 1C.

TORC1 regulation of cell–cell adhesion might be conserved among organisms, including metazoans [59, 93].

These findings have potential global implications for a role of the TORC1 pathway in shaping the emergence of multicellularity in metazoa and fungi. Multicellularity appeared early and repeatedly in life evolution, and the transition from unicellularity to multicellularity is typically associated with an increase in the number of genes involved in cell differentiation, cell–cell communication, and cell–cell adhesion (reviewed in Refs. [94, 95]). Most multicellular lineages are simple in architecture (i.e., filamentous protists), while in others the transition to multicellularity has burgeoned into macroscopic body plans (plants, fungi, and metazoans). In animals, the evolution of cadherin and integrin gene families enabled new roles for these gene products in construction and patterning of animal bodies [96]. The proteome of the choanoflagellate suggests that cell–cell adhesion in animals was well developed in the unicellular ancestor of metazoans and choanoflagellates [97, 98]. Similarly, fungal cell adherence has been well documented and likely arose independently in this kingdom. Findings that rapamycin modulates cell aggregation in *C. albicans* and *C. guilliermondii*, and promotes expression of adhesion molecules in mammalian endothelial cells, opens the possibility that TORC1's regulation of cell–cell adherence might be broadly conserved among organisms possibly including metazoans and may have contributed to shaping multicellular transitions in eukaryotic organisms.

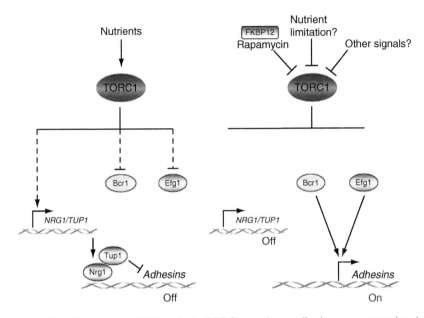

Fig. 11.5. Proposed model by which TORC1 regulates adhesin gene expression in *C. albicans*. (A) TORC1 promotes expression of the genes coding for the transcriptional repressors Nrg1 and Tup1 and inactivation of the transactivators Bcr1 and Efg1 by uncharacterized mechanisms. These events result in repression of adhesin gene expression. (B) Rapamycin treatment, nutrient limitation, and other unknown signals result in inactivation of TORC1 leading to induction of adhesin expression and cellular aggregation. Induction of adhesin expression results from a twofold mechanism: by downregulation of *NRG1* and *TUP1* expression and by activation of Bcr1 and Efg1. Adapted from Ref. [59], Figure 5.

IX.   Targeting the Tor Pathway: A Novel Therapeutic Antifungal Approach

Advances in genome sequencing and annotation technologies have become an invaluable tool in aiding our understanding of organismal biology. Capitalizing on this genomic revolution, the Fungal Genome Initiative has produced and analyzed the sequence of over 25 fungal organisms that are important to medicine, agriculture, and industry. These include fungi that are pathogens of humans (i.e., *C. albicans*, *C. neoformans*, *Aspergillus fumigatus*) and plants (i.e., *Magnaporthe grisea* and *Ustilago maydis*). Comparative genomics between closely related organisms has emerged as an important tool for understanding phenotypic differences, such as pathogenicity, and has facilitated the identification of conserved molecular pathways that can serve as targets for the development of broad-spectrum antimicrobial drugs.

Genome comparative analysis has now demonstrated a remarkable conservation of the Tor molecular cascade throughout the fungal kingdom. The Tor kinase, TORC1 and TORC2 constituents, and their regulators and effectors have been identified in the genomes of representative species of medical relevance (*C. albicans*, *C. neoformans*), in particular in basal lineages such as in the zygomycetes *Rhizopus oryzae* and *Mucor circinelloides* (Table 11.2, C. Shertz *et al.*, unpublished results). Both *R. oryzae* and *M. circinelloides* are common etiological agents of mucormycosis, an aggressive and invasive human fungal disease.

Remarkably, our own analysis and recent findings reveal a lack of a Tor homolog and all known Tor signaling components in the microsporidian pathogen *Encephalitozoon cuniculi*, representing the first eukaryote examined to date in which the entire Tor signaling cascade has been lost (C. Shertz *et al.*, unpublished results; [99]). Phylogenetic classification of these species has been controversial and ambiguous due their sparse and small genomes and rapidly evolving genes. While at first thought to be an ancient eukaryotic lineage closely related to fungi, recent studies provide evidence that they are true fungi that descended from a zygomycete ancestor and therefore represent a new and distinct basal fungal lineage [100]. Given that Tor controls essential processes in the cell, including protein synthesis, ribosomal biogenesis, autophagy, and cytoskeletal organization, it is unprecedented that a eukaryotic organism could survive in the absence of this essential signal transduction cascade. Strikingly, many other protein kinases and pathways involved in sensing nutrients and generating energy are absent from the *E. cuniculi* genome, and this is a reflection of the rampant gene loss that sculpted its 2.9 Mb genome, the smallest known for any eukaryote [99, 101]. The striking loss of this suite of kinases presumably arose during *E. cuniculi*'s streamlined and specialized adaptation as an obligate parasite, since the Tor cascade is also present in the intracellular pathogen *Trypanosoma cruzi*, one of the most ancient and evolutionarily divergent eukaryotes [102]. Within its parasitophorous vacuole, *E. cuniculi* relies on the host cell for acquisition of energy, nutrients, and for an osmotically stabilized environment that must be homeostatic relative to the changing environments of free-living fungi. Whole genome sequences for the microsporidian species *Enterocytozoon bieneusi* and *Antonospora locustae* will soon be available and it will be interesting to query whether these species have lost the Tor pathway as well.

Conservation of the Tor signaling signature network among pathogenic basal fungal lineages and its presence in trypanosomes suggests that this pathway arose early on in eukarya, in accord with its conservations in plants and metazoans (C. Shertz *et al.*, unpublished results; [102]). This evolutionary conservation serves as a platform for the design of novel antifungal

TABLE 11.2

Tor Cascade Signature Components and Putative Homologs in Pathogenic Fungi

Tor signaling components	Ascomycota		Basidiomycota	Zygomycota	Microsporidian
	Saccharomyces cerevisiae	Candida albicans	Cryptococcus neoformans	Mucor circinelloides	Encephalitozoon cuniculi
	Saprobe	Human fungal pathogens			
TORC1					
Tor1	TOR1	orf19.2290 (TOR1)	CNF03740 (TOR1)	PROT_ID: 72331	–
Tor2	TOR2	–	–	–	–
Kog1	KOG1	orf19.418 (KOG1)	CNDO1940	PROT_ID: 74940	–
Tco89	TCO89	orf19.761 (TCO89)	–	–	–
Lst8	LST8	or19.3862	CNM01130	PROT_ID: 47481	–
TORC2					
Tor2	TOR2	–	–	–	–
Lst8	LST8	or19.3862	CNM01130	PROT_ID: 47481	–
Avo1	AVO1	orf19.5221	CNJ01240	PROT_ID: 86151	–
Avo2	AVO2	orf19.215	–	PROT_ID: 84157	–
Avo3	AVO3	orf19.728 (TSC11)	CNM01660 (STE16)	PROT_ID: 19891	–
Bit61	BIT61	–	–	–	–

(Continued)

TABLE 11.2 (*CONTINUED*)

Tor signaling components	Ascomycota		Basidiomycota	Zygomycota	Microsporidian
	Saccharomyces cerevisiae	*Candida albicans*	*Cryptococcus neoformans*	*Mucor circinelloides*	*Encephalitozoon cuniculi*
	Saprobe	Human fungal pathogens			
Major regulators					
Tsc1	–	orf19.4110	–	PROT_ID: 76086	–
Tsc2	–	orf19.1798	CNC05620	PROT_ID: 39989	–
Rheb	*RHB1*[a]	orf19.5994 (*RHB1*)	CNB03660 (*RHEB*)	PROT_ID: 50694	–
Main effectors					
Sit4	*SIT4*	orf19.5200 (*SIT4*)	CND05500	–	–
Tap42	*TAP42*	orf19.4626	CNA03170	PROT_ID: 91310	–
Sch9	*SCH9*	orf19.829	CNN00360 (*SCH9*)	PROT_ID: 27458	–
Ypk2	*YPK2*	orf19.399	CNJ01100	PROT_ID: 87923	–

[a]Uncoupled from *S. cerevisiae* TORC1 signaling.
Tor pathway signaling homologs in pathogenic fungi identified through reciprocal best-hit BLASTp searches against characterized *S. cerevisiae* and *S. pombe* components.

therapies, which can also be applied to basal fungal pathogens. Over the last decade, the incidence and types of life-threatening fungal infections have raised due to the increasing number of immunocompromised individuals (resulting from HIV infection, neutropenia induced by chemotherapy, organ transplantation, and from the use of broad spectrum antibiotics and glucocorticosteroids), who are at risk for acquiring fungal infections. The present drug portfolio employed for treating systemic fungal infections consists of the polyene amphotericin B and its liposomal variants, as well as the azoles, allylamines, thiocarbamates, and fluorocytosine [103]. The need for new and broad spectrum antifungal agents with novel modes of action continues due to severe toxic side effects, fungistatic modes of action, and emergence of resistance to the current drug armamentarium.

The Tor kinase has received wide attention as an antifungal target due to its inhibition by the natural product rapamycin. Indeed, rapamycin was first identified for its potent antimicrobial activity against *C. albicans* [104, 105]. In comparison with amphotericin B, the mainstay antifungal used for combating fungal disease, rapamycin remains one of the most potent anti *Candida* drugs ever identified [106]. Subsequently, rapamycin was shown to have robust antifungal activity against several human fungal pathogens, including *Candida stelloidea, C. neoformans, A. fumigatus, Fusarium oxysporum*, and several pathogenic *Penicillium* species [107, 108]. However, the antifungal potential of rapamycin has been overshadowed by its potent immunosuppressive activity, which makes this compound less attractive as a therapeutic agent for treatment of fungal infections. Nevertheless, less immunosuppressive rapamycin analogs have been synthesized that retain antifungal activity against pathogenic *Candida* species as well as *C. neoformans* [109, 110].

The problem of systemic fungal infections will continue to grow as the number of individuals requiring immunosuppressive therapy increases. Less immunosuppressive rapamycin analogs offer new options in antifungal therapy. Topical applications and targeted delivery of these analogs are novel treatments that can also be explored for therapeutic use and can circumvent the immunosuppressive effect of rapamycin. Moreover, the use of rapamycin as an antifungal agent in an *in vivo* setting was reported to improve survival of mice with invasive aspergillosis [111]. Recent reports show that rapamycin encapsulated in lipid micelles retains high levels of potency *in vitro* [112]. In combination with solubilized amphotericin B and 5-flucytosine (5-FC), rapamycin synergistically increased the *in vitro* drug susceptibility of *C. albicans* isolates [112]. The synergistic activity of rapamycin in conjunction with amphotericin B and 5-FC combinations is encouraging as micelle encapsulation reduces the poor solubility of rapamycin in most drug vehicles and increases its compatibility with antifungal drugs. Furthermore, these *in vitro* results have promising therapeutic value since combinatorial therapy resulting in inhibition of multiple pathways simultaneously enhances efficacy of individual drugs by limiting exposure to toxic side effects and decreasing emergence of drug resistance. The challenge remains to exploit such combinatorial therapy by avoiding the immunosuppressive effects of rapamycin. The potential use of rapamycin and its analogs as antifungals appears promising and further development of new analogs is warranted.

X.   Remarks and Future Directions

The striking localization of TORC1 components, upstream activators, and downstream effectors to prevacuolar and vacuolar components anticipates a more general, central role for the endomembrane system as a staging platform for amino acid sensing and TORC1 signaling. Current studies on amino acid-derived signaling via Gtr1,2 and RagA–D GTPases to govern TORC1 in yeast and mammals, respectively lead to the view that amino acids are sensed internally. It remains to be determined if lysosomes to which mTORC1 is recruited also serve as amino acid reservoirs similar to the case of the yeast vacuole. Numerous challenges lie ahead, including: (1) further characterization of the amino acid sensing mechanism elucidating how the amino acid signal is transmitted to the Gtr/Rag GTPases, whether this mechanism involves GEF and GAP regulators, and the molecular effects triggered by Gtr interaction with TORC1; (2) does this mechanism involve amino acid receptors or transporters localized at the vacuolar membrane? These questions promise to capture our attention for the years ahead. The finding that Gln3 and Sfp1 nuclear import is dictated by their interaction with elements of the vesicular trafficking system were unexpected and establish a new paradigm that transient association of regulatory factors with endomembranes facilitates molecular interactions to enable TORC1 signaling outputs. Whether this is a general mode of regulation for other TORC1-governed targets awaits further studies. The model that rapamycin induces Gln3 localization via a different mechanism than those in response to nitrogen source emerged from several lines of investigation in different labs [5, 47, 50, 113] and sheds insight into the mechanism of rapamycin action. This finding raises the question of whether other rapamycin-activated effects are also processed via mechanisms that differ from those triggered by the natural inducers.

The Tor kinase has been conserved among eukaryotes and many of its physiological roles have been maintained in species as divergent as Baker's yeast and man. However, studies in budding and fission yeasts, and pathogenic fungi reveal divergent roles for the Tor signaling cascade. While the Tor kinases respond to nutrient-derived signals to promote cell growth in *S. cerevisiae* (reviewed in Ref. [79]), they control mating and mitotic entry in *S. pombe* [58, 114]. In addition to controlling nutrient-dependent cell growth, Tor regulates hyphal differentiation and cell aggregation in *C. albicans* [59]. Given the wealth of genomic information being generated at exponential rates, whole genome comparative analysis of the Tor pathway among basal eukaryotic organisms, especially fungi, are now feasible providing an unprecedented opportunity to understand the evolutionary trajectory of this signaling cascade and its impact on the evolution of fungal species and metazoans.

Finally, the natural product rapamycin has emerged as an important tool for characterizing the biological role of the Tor signaling cascade in fungal metabolism and virulence. Rapamycin's intrinsic potent antifungal activity provides a robust platform from which to develop novel antimicrobial therapies, which may include less immunosuppressive rapamycin analogs.

ACKNOWLEDGMENTS

We thank Joseph Heitman for critical reading of the manuscript. This work was supported by R01 CA114107 from the National Cancer Institute (to Maria E. Cardenas).

REFERENCES

1. Brunelle, J.K., and Letai, A. (2009). Control of mitochondrial apoptosis by the Bcl-2 family. *J Cell Sci* 122:437–441.
2. Sadowski, L., Pilecka, I., and Miaczynska, M. (2009). Signaling from endosomes: location makes a difference. *Exp Cell Res* 315:1601–1609.
3. Slessareva, J.E., Routt, S.M., Temple, B., Bankaitis, V.A., and Dohlman, H.G. (2006). Activation of the phosphatidylinositol 3-kinase Vps34 by a G protein alpha subunit at the endosome. *Cell* 126:191–203.
4. Boysen, J.H., and Mitchell, A.P. (2006). Control of Bro1-domain protein Rim20 localization by external pH, ESCRT machinery, and the *Saccharomyces cerevisiae* Rim101 pathway. *Mol Biol Cell* 17:1344–1353.
5. Puria, R., Zurita-Martinez, S.A., and Cardenas, M.E. (2008). Nuclear translocation of Gln3 in response to nutrient signals requires Golgi-to-endosome trafficking in *Saccharomyces cerevisiae. Proc Natl Acad Sci USA* 105:7194–7199.
6. Mitchell, A.P. (2008). A VAST staging area for regulatory proteins. *Proc Natl Acad Sci USA* 105:7111–7112.
7. Cardenas, M.E., and Heitman, J. (1995). FKBP12-rapamycin target TOR2 is a vacuolar protein with an associated phosphatidylinositol-4 kinase activity. *EMBO J* 14:5892–5907.
8. Wedaman, K.P., Reinke, A., Anderson, S., Yates, J., 3rd, McCaffery, J.M., and Powers, T. (2003). Tor kinases are in distinct membrane-associated protein complexes in *Saccharomyces cerevisiae. Mol Biol Cell* 14:1204–1220.

9. Kunz, J., Schneider, U., Howald, I., Schmidt, A., and Hall, M.N. (2000). HEAT repeats mediate plasma membrane localization of Tor2p in yeast. *J Biol Chem* 275:37011–37020.
10. Chen, E.J., and Kaiser, C.A. (2003). Lst8 negatively regulates amino acid biosynthesis as a component of the TOR pathway. *J Cell Biol* 161:333–347.
11. Araki, T., Uesono, Y., Oguchi, T., and Toh, E.A. (2005). Las24/Kog1, a component of the TOR complex 1 (TORC1), is needed for resistance to local anesthetic tetracaine and normal distribution of actin cytoskeleton in yeast. *Genes Genet Syst* 80:325–343.
12. Reinke, A., Anderson, S., McCaffery, J.M., Yates, J., 3rd, Aronova, S., Chu, S., Fairclough, S., Iverson, C., Wedaman, K.P., and Powers, T. (2004). TOR complex 1 includes a novel component, Tco89p (YPL180w), and cooperates with Ssd1p to maintain cellular integrity in *Saccharomyces cerevisiae*. *J Biol Chem* 279:14752–14762.
13. Berchtold, D., and Walther, T.C. (2009). TORC2 plasma membrane localization is essential for cell viability and restricted to a distinct domain. *Mol Biol Cell* 20:1565–1575.
14. Li, H., Tsang, C.K., Watkins, M., Bertram, P.G., and Zheng, X.F. (2006). Nutrient regulates Tor1 nuclear localization and association with rDNA promoter. *Nature* 442:1058–1061.
15. Wei, Y., Tsang, C.K., and Zheng, X.F. (2009). Mechanisms of regulation of RNA polymerase III-dependent transcription by TORC1. *EMBO J* 28:2220–2230.
16. Tsang, C.K., Li, H., and Zheng, X.S. (2007). Nutrient starvation promotes condensin loading to maintain rDNA stability. *EMBO J* 26:448–458.
17. Yan, G., Shen, X., and Jiang, Y. (2006). Rapamycin activates Tap42-associated phosphatases by abrogating their association with Tor complex 1. *EMBO J* 25:3546–3555.
18. Jorgensen, P., Rupes, I., Sharom, J.R., Schneper, L., Broach, J.R., and Tyers, M. (2004). A dynamic transcriptional network communicates growth potential to ribosome synthesis and critical cell size. *Genes Dev* 18:2491–2505.
19. Aronova, S., Wedaman, K., Anderson, S., Yates, J., 3rd, and Powers, T. (2007). Probing the membrane environment of the TOR kinases reveals functional interactions between TORC1, actin, and membrane trafficking in *Saccharomyces cerevisiae*. *Mol Biol Cell* 18:2779–2794.
20. Urban, J., Soulard, A., Huber, A., Lippman, S., Mukhopadhyay, D., Deloche, O., Wanke, V., Anrather, D., Ammerer, G., Riezman, H., Broach, J.R., De Virgilio, C., et al. (2007). Sch9 is a major target of TORC1 in *Saccharomyces cerevisiae*. *Mol Cell* 26:663–674.
21. Kaeberlein, M., Powers, R.W., 3rd, Steffen, K.K., Westman, E.A., Hu, D., Dang, N., Kerr, E.O., Kirkland, K.T., Fields, S., and Kennedy, B.K. (2005). Regulation of yeast replicative life span by TOR and Sch9 in response to nutrients. *Science* 310:1193–1196.
22. Huber, A., Bodenmiller, B., Uotila, A., Stahl, M., Wanka, S., Gerrits, B., Aebersold, R., and Loewith, R. (2009). Characterization of the rapamycin-sensitive phosphoproteome reveals that Sch9 is a central coordinator of protein synthesis. *Genes Dev* 23:1929–1943.
23. Loewith, R., Jacinto, E., Wullschleger, S., Lorberg, A., Crespo, J.L., Bonenfant, D., Oppliger, W., Jenoe, P., and Hall, M.N. (2002). Two TOR complexes, only one of which is rapamycin sensitive, have distinct roles in cell growth control. *Mol Cell* 10:457–468.
24. Heitman, J., Movva, N.R., and Hall, M.N. (1991). Targets for cell cycle arrest by the immunosuppressant rapamycin in yeast. *Science* 253:905–909.
25. Zurita-Martinez, S.A., Puria, R., Pan, X., Boeke, J.D., and Cardenas, M.E. (2007). Efficient Tor signaling requires a functional class C Vps protein complex in *Saccharomyces cerevisiae*. *Genetics* 176:2139–2150.

26. Chan, T.F., Carvalho, J., Riles, L., and Zheng, X.F. (2000). A chemical genomics approach toward understanding the global functions of the target of rapamycin protein (TOR). *Proc Natl Acad Sci USA* 97:13227–13232.
27. Xie, M.W., Jin, F., Hwang, H., Hwang, S., Anand, V., Duncan, M.C., and Huang, J. (2005). Insights into TOR function and rapamycin response: chemical genomic profiling by using a high-density cell array method. *Proc Natl Acad Sci USA* 102:7215–7220.
28. Srivastava, A., Woolford, C.A., and Jones, E.W. (2000). Pep3p/Pep5p complex: a putative docking factor at multiple steps of vesicular transport to the vacuole of *Saccharomyces cerevisiae*. *Genetics* 156:105–122.
29. Peterson, M.R., and Emr, S.D. (2001). The class C Vps complex functions at multiple stages of the vacuolar transport pathway. *Traffic* 2:476–486.
30. Raymond, C.K., Howald-Stevenson, I., Vater, C.A., and Stevens, T.H. (1992). Morphological classification of the yeast vacuolar protein sorting mutants: evidence for a prevacuolar compartment in class E vps mutants. *Mol Biol Cell* 3:1389–1402.
31. Kitamoto, K., Yoshizawa, K., Ohsumi, Y., and Anraku, Y. (1988). Mutants of *Saccharomyces cerevisiae* with defective vacuolar function. *J Bacteriol* 170:2687–2691.
32. Crespo, J.L., Powers, T., Fowler, B., and Hall, M.N. (2002). The TOR-controlled transcription activators Gln3, Rtg1, and Rtg3 are regulated in response to intracellular levels of glutamine. *Proc Natl Acad Sci USA* 99:6784–6789.
33. Liu, Z., and Butow, R.A. (2006). Mitochondrial retrograde signaling. *Annu Rev Genet* 40:159–185.
34. Hara, K., Yonezawa, K., Weng, Q.P., Kozlowski, M.T., Belham, C., and Avruch, J. (1998). Amino acid sufficiency and mTOR regulate p70 S6 kinase and eIF-4E BP1 through a common effector mechanism. *J Biol Chem* 273:14484–14494.
35. Avruch, J., Long, X., Ortiz-Vega, S., Rapley, J., Papageorgiou, A., and Dai, N. (2008). Amino acid regulation of TOR complex 1. *Am J Physiol Endocrinol Metab* 296:592–602.
36. Aspuria, P.J., and Tamanoi, F. (2004). The Rheb family of GTP-binding proteins. *Cell Signal* 16:1105–1112.
37. Gao, M., and Kaiser, C.A. (2006). A conserved GTPase-containing complex is required for intracellular sorting of the general amino-acid permease in yeast. *Nat Cell Biol* 8:657–667.
38. Dubouloz, F., Deloche, O., Wanke, V., Cameroni, E., and De Virgilio, C. (2005). The TOR and EGO protein complexes orchestrate microautophagy in yeast. *Mol Cell* 19:15–26.
39. Sancak, Y., Peterson, T.R., Shaul, Y.D., Lindquist, R.A., Thoreen, C.C., Bar-Peled, L., and Sabatini, D.M. (2008). The Rag GTPases bind raptor and mediate amino acid signaling to mTORC1. *Science* 320:1496–1501.
40. Kim, E., Goraksha-Hicks, P., Li, L., Neufeld, T.P., and Guan, K.L. (2008). Regulation of TORC1 by Rag GTPases in nutrient response. *Nat Cell Biol* 10:935–945.
41. Campbell, S.G., and Ashe, M.P. (2006). Localization of the translational guanine nucleotide exchange factor eIF2B: a common theme for GEFs? *Cell Cycle* 5:678–680.
42. Binda, M., Peli-Gulli, M.P., Bonfils, G., Panchaud, N., Urban, J., Sturgill, T.W., Loewith, R., and De Virgilio, C. (2009). The Vam6 GEF controls TORC1 by activating the EGO complex. *Mol Cell* 35:563–573.
43. Beck, T., and Hall, M.N. (1999). The TOR signalling pathway controls nuclear localization of nutrient-regulated transcription factors. *Nature* 402:689–692.
44. Cardenas, M.E., Cutler, N.S., Lorenz, M.C., Di Como, C.J., and Heitman, J. (1999). The TOR signaling cascade regulates gene expression in response to nutrients. *Genes Dev* 13:3271–3279.

45. Hardwick, J.S., Kuruvilla, F.G., Tong, J.K., Shamji, A.F., and Schreiber, S.L. (1999). Rapamycin-modulated transcription defines the subset of nutrient-sensitive signaling pathways directly controlled by the Tor proteins. *Proc Natl Acad Sci USA* 96:14866–14870.

46. Bertram, P.G., Choi, J.H., Carvalho, J., Ai, W., Zeng, C., Chan, T.F., and Zheng, X.F. (2000). Tripartite regulation of Gln3p by Tor, Ure2p, and phosphatases. *J Biol Chem* 275:35727–35733.

47. Cox, K.H., Kulkarni, A., Tate, J.J., and Cooper, T.G. (2004). Gln3 phosphorylation and intracellular localization in nutrient limitation and starvation differ from those generated by rapamycin inhibition of Tor1/2 in *Saccharomyces cerevisiae. J Biol Chem* 279:10270–10278.

48. Tate, J.J., Feller, A., Dubois, E., and Cooper, T.G. (2006). *Saccharomyces cerevisiae* Sit4 phosphatase is active irrespective of the nitrogen source provided, and Gln3 phosphorylation levels become nitrogen source-responsive in a sit4-deleted strain. *J Biol Chem* 281:37980–37992.

49. Gerrard, S.R., Bryant, N.J., and Stevens, T.H. (2000). Vps21 controls entry of endocytosed and biosynthetic proteins into the yeast prevacuolar compartment. *Mol Biol Cell* 11:613–626.

50. Georis, I., Tate, J.J., Cooper, T.G., and Dubois, E. (2008). Tor pathway control of the nitrogen-responsive *DAL5* gene bifurcates at the level of Gln3 and Gat1 regulation in *Saccharomyces cerevisiae. J Biol Chem* 283:8919–8929.

51. Di Como, C.J., and Jiang, Y. (2006). The association of Tap42 phosphatase complexes with TORC1: another level of regulation in Tor signaling. *Cell Cycle* 5:2729–2732.

52. Puria, R., and Cardenas, M.E. (2008). Rapamycin bypasses vesicle-mediated signaling events to activate Gln3 in *Saccharomyces cerevisiae. Commun Integr Biol* 1:23–25.

53. Tate, J.J., Georis, I., Feller, A., Dubois, E., and Cooper, T.G. (2009). Rapamycin-induced Gln3 dephosphorylation is insufficient for nuclear localization: Sit4 and PP2A phosphatases are regulated and function differently. *J Biol Chem* 284:2522–2534.

54. Marion, R.M., Regev, A., Segal, E., Barash, Y., Koller, D., Friedman, N., and O'Shea, E.K. (2004). Sfp1 is a stress- and nutrient-sensitive regulator of ribosomal protein gene expression. *Proc Natl Acad Sci USA* 101:14315–14322.

55. Singh, J., and Tyers, M. (2009). A Rab escort protein integrates the secretion system with TOR signaling and ribosome biogenesis. *Genes Dev* 23:1944–1958.

56. Lempiainen, H., Uotila, A., Urban, J., Dohnal, I., Ammerer, G., Loewith, R., and Shore, D. (2009). Sfp1 interaction with TORC1 and Mrs6 reveals feedback regulation on TOR signaling. *Mol Cell* 33:704–716.

57. Grosshans, B.L., Ortiz, D., and Novick, P. (2006). Rabs and their effectors: achieving specificity in membrane traffic. *Proc Natl Acad Sci USA* 103:11821–11827.

58. Weisman, R., and Choder, M. (2001). The fission yeast TOR homolog, Tor1+, is required for the response to starvation and other stresses via a conserved serine. *J Biol Chem* 276:7027–7032.

59. Bastidas, R.J., Heitman, J., and Cardenas, M.E. (2009). The protein kinase Tor1 regulates adhesin gene expression in *Candida albicans. PLoS Pathog* 5:e1000294.

60. Kwon-Chung, K.J. (1975). A new genus, filobasidiella, the perfect state of *Cryptococcus neoformans. Mycologia* 67:1197–1200.

61. Young, L.Y., Lorenz, M.C., and Heitman, J. (2000). A Ste12 homolog is required for mating but dispensable for filamentation in *Candida lusitaniae. Genetics* 155:17–29.

62. Csank, C., Schroppel, K., Leberer, E., Harcus, D., Mohamed, O., Meloche, S., Thomas, D.Y., and Whiteway, M. (1998). Roles of the *Candida albicans* mitogen-activated protein kinase

homolog, Cek1p, in hyphal development and systemic candidiasis. *Infect Immun* 66:2713–2721.

63. Tripathi, G., Wiltshire, C., Macaskill, S., Tournu, H., Budge, S., and Brown, A.J. (2002). Gcn4 co-ordinates morphogenetic and metabolic responses to amino acid starvation in *Candida albicans. EMBO J* 21:5448–5456.

64. Cutler, N.S., Pan, X., Heitman, J., and Cardenas, M.E. (2001). The TOR signal transduction cascade controls cellular differentiation in response to nutrients. *Mol Biol Cell* 12:4103 4113.

65. Bastidas, R.J., Reedy, J.L., Morales-Johansson, H., Heitman, J., and Cardenas, M.E. (2008). Signaling cascades as drug targets in model and pathogenic fungi. *Curr Opin Investig Drugs* 9:856–864.

66. Barbet, N.C., Schneider, U., Helliwell, S.B., Stansfield, I., Tuite, M.F., and Hall, M.N. (1996). Tor controls translation initiation and early G1 progression in yeast. *Mol Biol Cell* 7:25–42.

67. Di Como, C.J., and Arndt, K.T. (1996). Nutrients, via the Tor proteins, stimulate the association of Tap42 with type 2A phosphatases. *Genes Dev* 10:1904–1916.

68. Lorenz, M.C., and Heitman, J. (1998). The Mep2 ammonium permease regulates pseudohyphal differentiation in *Saccharomyces cerevisiae. EMBO J* 17:1236–1247.

69. Lorenz, M.C., and Heitman, J. (1997). Yeast pseudohyphal growth is regulated by GPA2, a G protein alpha homolog. *EMBO J* 16:7008–7018.

70. Liu, H., Styles, C.A., and Fink, G.R. (1993). Elements of the yeast pheromone response pathway required for filamentous growth of diploids. *Science* 262:1741–1744.

71. Robertson, L.S., and Fink, G.R. (1998). The three yeast A kinases have specific signaling functions in pseudohyphal growth. *Proc Natl Acad Sci USA* 95:13783–13787.

72. Pan, X., and Heitman, J. (1999). Cyclic AMP-dependent protein kinase regulates pseudohyphal differentiation in *Saccharomyces cerevisiae. Mol Cell Biol* 19:4874–4887.

73. Pan, X., and Heitman, J. (2000). Sok2 regulates yeast pseudohyphal differentiation via a transcription factor cascade that regulates cell-cell adhesion. *Mol Cell Biol* 20:8364–8372.

74. Biswas, K., and Morschhauser, J. (2005). The Mep2p ammonium permease controls nitrogen starvation-induced filamentous growth in *Candida albicans. Mol Microbiol* 56:649–669.

75. Dabas, N., and Morschhauser, J. (2007). Control of ammonium permease expression and filamentous growth by the GATA transcription factors Gln3 and Gat1 in *Candida albicans. Eukaryot Cell* 6:875–888.

76. Liao, W.L., Ramon, A.M., and Fonzi, W.A. (2008). Gln3 encodes a global regulator of nitrogen metabolism and virulence of *C. albicans. Fungal Genet Biol* 45:514–526.

77. Brown, D.H., Jr., Giusani, A.D., Chen, X., and Kumamoto, C.A. (1999). Filamentous growth of *Candida albicans* in response to physical environmental cues and its regulation by the unique *CZF1* gene. *Mol Microbiol* 34:651–662.

78. Kumamoto, C.A. (2005). A contact-activated kinase signals *Candida albicans* invasive growth and biofilm development. *Proc Natl Acad Sci USA* 102:5576–5581.

79. Rohde, J.R., Bastidas, R., Puria, R., and Cardenas, M.E. (2008). Nutritional control via Tor signaling in *Saccharomyces cerevisiae. Curr Opin Microbiol* 11:153–160.

80. Nemecek, J.C., Wuthrich, M., and Klein, B.S. (2006). Global control of dimorphism and virulence in fungi. *Science* 312:583–588.

81. Heitman, J. (2006). Sexual reproduction and the evolution of microbial pathogens. *Curr Biol* 16:R711–R725.

82. Kojic, E.M., and Darouiche, R.O. (2004). *Candida* infections of medical devices. *Clin Microbiol Rev* 17:255–267.

83. Zhao, X., Oh, S.H., Cheng, G., Green, C.B., Nuessen, J.A., Yeater, K., Leng, R.P., Brown, A.J., and Hoyer, L.L. (2004). *ALS3* and *ALS8* represent a single locus that encodes a *Candida albicans* adhesin; functional comparisons between Als3p and Als1p. *Microbiology* 150:2415–2428.
84. Zhao, X., Pujol, C., Soll, D.R., and Hoyer, L.L. (2003). Allelic variation in the contiguous loci encoding *Candida albicans* ALS5, ALS1 and ALS9. *Microbiology* 149:2947–2960.
85. Hoyer, L.L., and Hecht, J.E. (2001). The *ALS5* gene of *Candida albicans* and analysis of the Als5p N-terminal domain. *Yeast* 18:49–60.
86. Hoyer, L.L., and Hecht, J.E. (2000). The *ALS6* and *ALS7* genes of *Candida albicans*. *Yeast* 16:847–855.
87. Hoyer, L.L., Payne, T.L., and Hecht, J.E. (1998). Identification of *Candida albicans* Als2 and Als4 and localization of als proteins to the fungal cell surface. *J Bacteriol* 180:5334–5343.
88. Gaur, N.K., and Klotz, S.A. (1997). Expression, cloning, and characterization of a *Candida albicans* gene, *ALS1*, that confers adherence properties upon *Saccharomyces cerevisiae* for extracellular matrix proteins. *Infect Immun* 65:5289–5294.
89. Loza, L., Fu, Y., Ibrahim, A.S., Sheppard, D.C., Filler, S.G., and Edwards, J.E., Jr. (2004). Functional analysis of the *Candida albicans ALS1* gene product. *Yeast* 21:473–482.
90. Dranginis, A.M., Rauceo, J.M., Coronado, J.E., and Lipke, P.N. (2007). A biochemical guide to yeast adhesins: glycoproteins for social and antisocial occasions. *Microbiol Mol Biol Rev* 71:282–294.
91. Sundstrom, P. (2002). Adhesion in *Candida* spp. *Cell Microbiol* 4:461–469.
92. Nobile, C.J., Schneider, H.A., Nett, J.E., Sheppard, D.C., Filler, S.G., Andes, D.R., and Mitchell, A.P. (2008). Complementary adhesin function in *C. albicans* biofilm formation. *Curr Biol* 18:1017–1024.
93. Wood, S.C., Bushar, G., and Tesfamariam, B. (2006). Inhibition of mammalian target of rapamycin modulates expression of adhesion molecules in endothelial cells. *Toxicol Lett* 165:242–249.
94. Ruiz-Trillo, I., Burger, G., Holland, P.W., King, N., Lang, B.F., Roger, A.J., and Gray, M.W. (2007). The origins of multicellularity: a multi-taxon genome initiative. *Trends Genet* 23:113–118.
95. Rokas, A. (2008). The origins of multicellularity and the early history of the genetic toolkit for animal development. *Annu Rev Genet* 42:235–251.
96. Nichols, S.A., Dirks, W., Pearse, J.S., and King, N. (2006). Early evolution of animal cell signaling and adhesion genes. *Proc Natl Acad Sci USA* 103:12451–12456.
97. King, N., Westbrook, M.J., Young, S.L., Kuo, A., Abedin, M., Chapman, J., Fairclough, S., Hellsten, U., Isogai, Y., Letunic, I., Marr, M., Pincus, D., *et al.* (2008). The genome of the choanoflagellate *Monosiga brevicollis* and the origin of metazoans. *Nature* 451:783–788.
98. King, N., Hittinger, C.T., and Carroll, S.B. (2003). Evolution of key cell signaling and adhesion protein families predates animal origins. *Science* 301:361–363.
99. Miranda-Saavedra, D., Stark, M.J., Packer, J.C., Vivares, C.P., Doerig, C., and Barton, G.J. (2007). The complement of protein kinases of the microsporidium *Encephalitozoon cuniculi* in relation to those of *Saccharomyces cerevisiae* and *Schizosaccharomyces pombe*. *BMC Genomics* 8:309.
100. Lee, S.C., Corradi, N., Byrnes, E.J., 3rd, Torres-Martinez, S., Dietrich, F.S., Keeling, P.J., and Heitman, J. (2008). Microsporidia evolved from ancestral sexual fungi. *Curr Biol* 18:1675–1679.
101. Katinka, M.D., Duprat, S., Cornillot, E., Metenier, G., Thomarat, F., Prensier, G., Barbe, V., Peyretaillade, E., Brottier, P., Wincker, P., Delbac, F., El, Alaoui H., *et al.*

(2001). Genome sequence and gene compaction of the eukaryote parasite *Encephalitozoon cuniculi. Nature* 414:450–453.

102. Barquilla, A., Crespo, J.L., and Navarro, M. (2008). Rapamycin inhibits trypanosome cell growth by preventing TOR complex 2 formation. *Proc Natl Acad Sci USA* 105:14579–14584.

103. White, T.C., Marr, K.A., and Bowden, R.A. (1998). Clinical, cellular, and molecular factors that contribute to antifungal drug resistance. *Clin Microbiol Rev* 11:382–402.

104 Vezina, C., Kudelski, A., and Sehgal, S.N. (1975). Rapamycin (AY-22, 989), a new antifungal antibiotic. I. Taxonomy of the producing streptomycete and isolation of the active principle. *J Antibiot (Tokyo)* 28:721–726.

105. Baker, H., Sidorowicz, A., Sehgal, S.N., and Vezina, C. (1978). Rapamycin (AY-22, 989), a new antifungal antibiotic. III. In vitro and in vivo evaluation. *J Antibiot (Tokyo)* 31:539–545.

106. Ellis, D. (2002). Amphotericin B: spectrum and resistance. *J Antimicrob Chemother* 49 (Suppl 1):7–10.

107. Cruz, M.C., Cavallo, L.M., Gorlach, J.M., Cox, G., Perfect, J.R., Cardenas, M.E., and Heitman, J. (1999). Rapamycin antifungal action is mediated via conserved complexes with FKBP12 and Tor kinase homologs in *Cryptococcus neoformans. Mol Cell Biol* 19:4101–4112.

108. Wong, G.K., Griffith, S., Kojima, I., and Demain, A.L. (1998). Antifungal activities of rapamycin and its derivatives, prolylrapamycin, 32-desmethylrapamycin, and 32-desmethoxyrapamycin. *J Antibiot (Tokyo)* 51:487–491.

109. Dickman, D.A., Ding, H., Li, Q., Nilius, A.M., Balli, D.J., Ballaron, S.J., Trevillyan, J.M., Smith, M.L., Seif, L.S., Kim, K., Sarthy, A., Goldman, R.C., et al. (2000). Antifungal rapamycin analogues with reduced immunosuppressive activity. *Bioorg Med Chem Lett* 10:1405–1408.

110. Cruz, M.C., Goldstein, A.L., Blankenship, J., Del Poeta, M., Perfect, J.R., McCusker, J.H., Bennani, Y.L., Cardenas, M.E., and Heitman, J. (2001). Rapamycin and less immunosuppressive analogs are toxic to *Candida albicans* and *Cryptococcus neoformans* via FKBP12-dependent inhibition of TOR. *Antimicrob Agents Chemother* 45:3162–3170.

111. High, K.P., and Washburn, R.G. (1997). Invasive aspergillosis in mice immunosuppressed with cyclosporin A, tacrolimus (FK506), or sirolimus (rapamycin). *J Infect Dis* 175:222–225.

112. Vakil, R., Knilans, K., Andes, D., and Kwon, G.S. (2008). Combination antifungal therapy involving amphotericin B, rapamycin and 5-fluorocytosine using PEG-phospholipid micelles. *Pharm Res* 25:2056–2064.

113. Cox, K.H., Tate, J.J., and Cooper, T.G. (2004). Actin cytoskeleton is required for nuclear accumulation of Gln3 in response to nitrogen limitation but not rapamycin treatment in *Saccharomyces cerevisiae. J Biol Chem* 279:19294–19301.

114. Petersen, J., and Nurse, P. (2007). TOR signalling regulates mitotic commitment through the stress MAP kinase pathway and the Polo and Cdc2 kinases. *Nat Cell Biol* 9:1263–1272.

115. Phan, Q.T., Myers, C.L., Fu, Y., Sheppard, D.C., Yeaman, M.R., Welch, W.H., Ibrahim, A.S., Edwards, J.E., Jr., and Filler, S.G. (2007). Als3 is a *Candida albicans* invasin that binds to cadherins and induces endocytosis by host cells. *PLoS Biol* 5:e64.

12

TOR and Sexual Development in Fission Yeast

YOKO OTSUBO • MASAYUKI YAMAMOTO

Department of Biophysics and Biochemistry
Graduate School of Science
University of Tokyo, Hongo
Tokyo, Japan

I. Abstract

Fission yeast has two TOR (target of rapamycin) kinases, namely Tor1 and Tor2. Like TOR kinases in other eukaryotes, Tor1 and Tor2 form two distinct TOR complexes, TORC1 and TORC2—Tor1 composing TORC2 and Tor2 composing TORC1. Fission yeast cells initiate sexual development, that is, mating, meiosis, and sporulation, when starved for nutrients. However, cells defective in *tor1*, though viable, cannot arrest at G1 phase and enter sexual development under nitrogen starvation. In contrast, cells defective in *tor2* are lethal. Temperature-sensitive *tor2* mutant cells arrest at G1 phase and initiate sexual development at the restrictive temperature, even under nutritional conditions. Thus, with regard to the promotion of G1 arrest and sexual development, fission yeast TORC1 and TORC2 appear to execute opposite functions: TORC1 is suppressive whereas TORC2 is stimulatory. The entry to sexual development is regulated by several signaling pathways in fission yeast, including the TORC1 and TORC2 pathways. This chapter outlines the current knowledge of these pathways, with special focus on the TORC pathways. TORC1 is apparently involved in the signaling of nitrogen availability. Furthermore, recent findings suggest a direct connection between TORC1 and Mei2, the master regulator of meiosis in fission yeast.

THE ENZYMES, Vol. XXVII
229
ISSN NO: 1874-6047
DOI: 10.1016/S1874-6047(10)27012-9

II.   Introduction

Fission yeast *Schizosaccharomyces pombe* is a unicellular microorganism, which proliferates normally in the haploid state. Haploid cells bear one of the two mating types denoted as h^+ (*P*) or h^- (*M*). Under nutrient conditions, they go through the mitotic cell cycle. When they are starved of nutrients, especially nitrogen, they arrest at G1 phase and enter the stationary phase. If cells of the two mating types are in the neighborhood, h^+ and h^- cells make pairs and conjugate. Resulting diploid zygotes then initiate meiosis and generate asci containing four haploid spores. Zygotes can also grow as diploids if they are transferred to rich medium prior to the initiation of meiosis. These diploid cells undergo meiosis and sporulation without conjugation when deprived of nutrients. A more precise description of fission yeast sexual development has been published [1].

How cells recognize and adapt themselves to global nutritional conditions has been a longtime unsolved problem of molecular and cell biology. Recent studies have indicated that TOR (target of rapamycin) protein kinase plays an important role in nutritional signaling in fission yeast.

Fission yeast has two TOR kinases, namely Tor1 and Tor2. Tor1 is not absolutely necessary for vegetative growth, but is essential for growth under stressed conditions. It is also required for proper G1 arrest and onset of sexual development under nitrogen starvation. In contrast, Tor2 is indispensable for growth. Interestingly, temperature-sensitive *tor2* mutants arrest at G1 phase and initiate sexual development on nutrient medium at the restrictive temperature. Like other eukaryotes, fission yeast TOR proteins form two types of TOR complexes, namely TORC1 and TORC2. Fission yeast TORC1, containing Tor2, and TORC2, containing Tor1, apparently have opposite functions with respect to the control of G1 arrest and onset of sexual development, as pointed out previously [2].

In budding and fission yeast, it has been shown that sexual development is regulated by several signaling pathways, which are activated in response to external conditions such as nutrition, stresses, and mating pheromones. This chapter summarizes the current knowledge regarding these pathways in fission yeast, highlighting recent findings how TOR pathways control the entry to sexual development. We will then look at the central regulator of meiosis in fission yeast, Mei2 [3], and inspect the intimate relation between Mei2 and TORC1.

III.   Cell Cycle Regulation for Sexual Development

Because of its genetic and cytological tractability, fission yeast has been one of the most favored organisms for the study of the cell cycle. The mitotic cell cycle of fission yeast consists of G1, S, G2, and M phases like other eukaryotes, although the relative length of each phase is not quite the same as that in higher eukaryotes. G1 phase is the point of exit from the mitotic cycle to cell differentiation, namely sexual development in this unicellular organism [4].

In yeast, "Start" is a point in G1 at which the cell becomes committed to the initiation of the next cell cycle [5], which corresponds to the restriction point in mammalian cells [6]. Sufficient nutrients allow cells to begin a new mitotic cell cycle at Start, and once this commitment is taken, cells are obliged to complete one round of the cell cycle until they reach pre-Start G1 phase again [7].

Upon nitrogen starvation, haploid cells arrest at pre-Start G1 and enter either stationary phase or sexual development, as mentioned above. Down-regulation of the activity of CDK (cyclin-dependent kinase)–cyclin complexes, namely Cdc2–Cig2 and Cdc2–Cdc13, is a prerequisite for the G1 arrest. This is accomplished mainly by the accumulation of Rum1, a CDK inhibitor, and Ste9/Srw1, an activator of the E3 ubiquitin ligase APC/C (anaphase promoting complex/cyclosome) [8–12]. If *tor2-ts* mutant cells are shifted to the restriction temperature, they arrest at G1 and initiate sexual differentiation irrespective of nutritional conditions, and expression of *rum1* and *ste9* mRNA is increased in these cells [13]. By contrast, in *tor1*-deletion or *gad8*-deletion mutant cells, which cannot arrest properly at G1 phase, expression of *rum1* mRNA does not increase after nitrogen starvation [14].

IV.   Nutritional Signaling

A.   Ste11, a Key Transcription Factor for the Sexual Development

Nutritional starvation leads to the induction of expression of *ste11*, which encodes a transcription factor of the HMG (high mobility group) family. Ste11 is responsible for transcriptional activation of many genes required for the onset of sexual development, including genes for the mating pheromones and the master regulator of meiosis, Mei2 [15, 16]. These genes carry one to five copies of the common motif denoted as TR-box (TTCTTTGTTY) in their 5'-upstream region. Ste11 binds to the TR-box and activates transcription. Artificial expression of *ste11* causes ectopic expression of meiotic genes under nutritional conditions, and disruption of *ste11* renders the cell completely sterile, that is, defective in mating and meiosis. Thus, Ste11 is pivotal for the initiation of sexual development [16].

B.   The cAMP–PKA Pathway

Cyclic AMP (cAMP) serves as an important second messenger in the regulation of adaptation to the external conditions in many cell systems. In eukaryotic cells, the major target of cAMP is cAMP-dependent protein kinase (PKA). In fission yeast, active PKA promotes cell growth, suppressing initiation of sexual development, and the level of cAMP should be lowered when a cell is to enter sexual development [16–18].

When supply of glucose is sufficient, Gpa2, the alpha subunit of a heterotrimeric G-protein, becomes activated via the Git3 receptor [19–21]. Gpa2 in turn activates adenylate cyclase (Cyr1/Git2), which catalyzes production of the second messenger cAMP from ATP [22]. As in other eukaryotes, PKA is composed of the catalytic subunit (Pka1/Git6) and the regulatory subunit (Cgs1) in fission yeast. When cAMP is abundant, it associates with Cgs1. This allows the catalytic subunit Pka1 to be released from inhibitory Cgs1 and to exert its kinase activity [23]. Active Pka1 represses expression of *ste11* through inactivation of a zinc-finger transcription factor Rst2, which is indispensable for the transcription of *ste11* [24, 25]. While glucose is the major factor to affect the level of intracellular cAMP, this level also gradually decreases in cells starved for nitrogen over hours, and expression of *ste11* is eventually attained [18]. A summary of this pathway is schematically shown in Figure 12.1.

In metabolism, fission yeast PKA regulates gluconeogenesis negatively by repressing the transcription of *fbp1*, which encodes a key enzyme fructose-1,6-bis-phosphatse [26, 27]. Thus, the presence of ample glucose suppresses gluconeogenesis via the cAMP–PKA system. In addition to the mutants affected in sexual development, mutants named *git1* through *git10*, which were isolated as ones that could express *fbp1* constitutively, have facilitated analysis of the cAMP–PKA system in fission yeast [28].

C. TOR2 AND NITROGEN SENSING

Fission yeast Tor2 is essential for vegetative growth [29, 30]. Analyses of temperature-sensitive *tor2* (*tor2-ts*) mutants have revealed that they arrest at G1 phase even on nutrient medium, if shifted to the restrictive temperature. Furthermore, if the mutant strain is homothallic, that is, can produce both h^+ and h^- cells in clonal progeny, the cells initiate ectopic sexual development under nutritionally rich conditions at the restrictive temperature [13, 31–33] (Figure 12.2A). Conversely, constitutively active *tor2* mutants show delay in the response to nitrogen starvation and are defective in sexual development [34]. These observations indicate that Tor2 suppresses arrest at G1 and entry to sexual development when cells are to grow vegetatively on rich medium.

Fission yeast Tor2 forms a complex with Mip1 (Raptor homolog), Wat1/Pop3 (Lst8 homolog), Tco89 (budding yeast Tco89 homolog), and Toc1 (SPBP18G5.03), composing TORC1 (Figure 12.2B) [13, 33, 35]. Mip1 is essential for growth, and homothallic temperature-sensitive *mip1* mutants

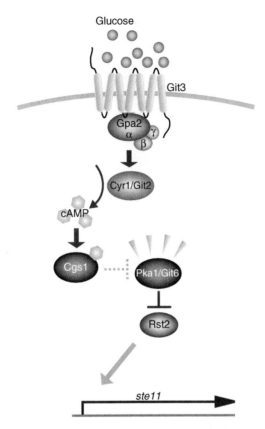

FIG. 12.1. The cAMP–PKA pathway. Abundant glucose stimulates production of cAMP, which in turn activates PKA (Pka1). Pka1 represses expression of *ste11*, which encodes a key transcription factor for the sexual development.

initiate sexual development on rich medium at the restrictive temperature, like *tor2-ts* mutants (our unpublished results). Wat1 is a common component of TORC1 and TORC2. Whereas budding yeast *LST8*, the *wat1* homolog, is essential for growth [36], deletion of *wat1* is not lethal in fission yeast. The reason for this is unclear. The *wat1*-deletion mutant is sterile and tends to diploidize [37, 38]. Tco89 and Toc1 have been identified by mass spectroscopic analysis. No homolog of Toc1 has been found in other species except for *Schizosaccharomyces japonicus* [35].

In *tor2-ts* cells, expression of such genes that encode cell-cycle inhibitors (e.g., *rum1*, *ste9*) or are responsive to nitrogen starvation (e.g., *ste11*, *isp6*, *fnx1*) is induced after the shift to the restrictive temperature. However,

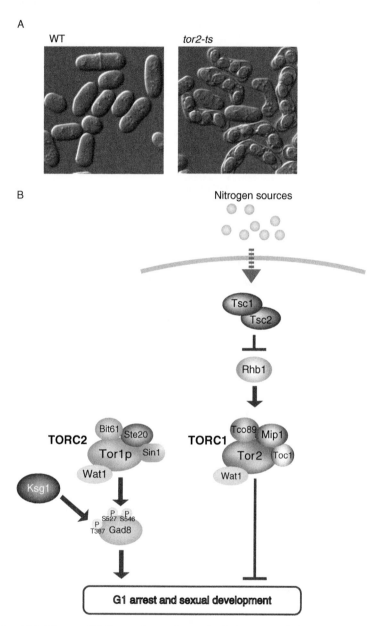

Fɪɢ. 12.2. The two TOR complex pathways in fission yeast. (A) Ectopic mating and sporulation exhibited by the *tor2-ts* mutant on nutrient medium. Wild-type cells (left) and *tor2-ts* cells (right) were grown on nutrient medium (YE) and shifted to the restrictive temperature 30 °C. (B) The TORC1 and TORC2 pathways. Like higher eukaryotes, fission

fbp1, a typical glucose starvation-responsive gene, shows no significant transcription in these mutants [13, 33]. Furthermore, microarray analysis has revealed that most, if not all, nitrogen starvation-responsive genes are induced in *tor2-ts* cells at the restrictive temperature, including *ste11* and *mei2* crucial for sexual development [33]. These results suggest that Tor2 mediates a signal from the nitrogen source that eventually regulates G1 arrest and sexual development.

Homologs of mammalian TSC1, TSC2, and Rheb have been identified in fission yeast. Fission yeast TSC1 (Tsc1) and TSC2 (Tsc2) form a complex [39] and function to downregulate Rheb (Rhb1), showing close similarity to the mammalian TSC–Rheb–TOR system [31–33, 40, 41]. Like Tor2, cells defective in Rhb1 arrest at G1 and express nitrogen starvation-responsive genes [42, 43]. Disruption of *tsc1* or *tsc2* causes inefficient mating [39]. Thus, it is presumable that the TSC–Rheb–TOR pathway responds to the supply of nitrogen and regulates sexual development in fission yeast (Figure 12.2B).

Unlike other eukaryotes, rapamycin does not affect vegetative growth of fission yeast. Rather, rapamycin inhibits sexual development at an early stage prior to mating, though it does not appear to interfere with the entry to stationary phase, under nitrogen starvation [44]. The reason for this unusual situation has not been fully understood, but recent studies suggest that fission yeast TOR can potentially interact with rapamycin [29, 30, 45, 46]. It has been reported that homothallic wild-type cells grown in glutamate medium initiate sexual development if rapamycin is added to the medium [47]. It has been also shown that a hypomorphic mutation in *tor2* makes the strain hypersensitive to rapamycin [35]. Thus, it is presumable that part of the Tor2 (TORC1) function may be downregulated by rapamycin under certain conditions.

D. HOMOLOGS OF SCH9 KINASE

Recent analysis has revealed that budding yeast Sch9 has functional similarities to mammalian S6K, which is one of the major targets of mTOR and a member of the AGC kinase family [48]. Originally, the

---

yeast has the TSC–Rheb–TORC1 pathway. Fission yeast TORC1 contains Mip1, Wat1, Tco89, Toc1, and Tor2. In the presence of ample nitrogen, TORC1 regulates G1 arrest and sexual development negatively, promoting vegetative growth. TORC2 contains Ste20, Wat1/Pop3, Sin1, Bit61, and Tor1, and it regulates G1 arrest and sexual development positively, in sharp contrast to TORC1. An AGC family kinase Gad8 is a principal target of TORC2. TORC2, in collaboration with a PDK1-like kinase Ksg1, phosphorylates Gad8 at Thr387, Ser527, and Ser546, which is essential for the function of Gad8.

SCH9 gene was identified as a high-copy suppressor of the complete loss-of-function of the three *TPK* genes, that is, *TPK1*, *TPK2*, and *TPK3*, which encode the catalytic subunit of PKA in budding yeast [49, 50].

There are at least three homologs of *SCH9* in the fission yeast genome, which are denoted respectively as *sck1*, *sck2*, and *psk1* [51–53]. The *sck1* (suppressor of loss of cAMP-dependent protein kinase) gene was isolated as a high-copy suppressor of the *git3* mutant, in which transcription of *fbp1* became constitutive, and was found to suppress also deficiency in *pka1*, which encodes the single ortholog of TPK1, TPK2, and TPK3 in fission yeast [51]. Unlike disruption of *SCH9*, however, disruption of none of *sck1*, *sck2*, and *psk1* causes obvious growth retardation or any remarkable phenotypes. It has been observed previously that the *sck1*-deletion mutant mates ectopically at a slightly higher frequency than the wild type on nutrient medium and that the *sck1 sck2* double mutant mates at a little higher frequency than *sck1* [52]. Our recent analysis has revealed that the *sck1 sck2 psk1* triple mutant mates at an even higher frequency on nutrient medium, although the majority of the cells continue mitotic growth as robustly as the wild type (our unpublished results). Meanwhile, whereas overexpression of either *sck1* or *sck2* can suppress defects of the *pka1*-deletion mutant [52], overexpression of *psk1* cannot do so. Thus, Psk1 seems to perform similar but not exactly the same functions as Sck1 and Sck2.

We have observed that the phosphorylation state of Psk1 is dependent on the activity of Tor2. Furthermore, its phosphorylation state appears also to depend on Ksg1 (kinase responsible for sporulation and growth), which is a PDK1 (phosphoinositide-dependent kinase-1)-like kinase (our unpublished results). It is known that TOR complexes phosphorylate certain AGC kinases in cooperation with PDK1. For example, mammalian TORC1 and TORC2 phosphorylate S6K1 and PKB/Akt, respectively, and fission yeast TORC2, which contains Tor1, phosphorylates an AGC kinase Gad8p (see below) [54]. Thus, it is presumable that Psk1 may constitute a pathway with TORC1 as Sch9 does. Upon nutritional starvation, the level of Psk1 phosphorylation decreases quickly, as is the case with Sch9 (our unpublished results).

E.   STRESS-ACTIVATED MAP KINASE CASCADE

The fission yeast Sty1/Spc1 mitogen-activated protein kinase (MAPK) is a member of the eukaryotic stress-activated MAPK (SAPK) family. The Sty1/Spc1 MAPK pathway can be activated by a variety of stresses, including heat shock, high-osmolarity, oxidative stress, and nutritional starvation. When cells are exposed to a stress, the MAPKK Wis1 phosphorylates Sty1 on neighboring threonine and tyrosine residues to activate it [55–58].

Once activated by phosphorylation, Sty1 accumulates in the nucleus [59] and phosphorylates several targets there, including a bZIP transcription factor Atf1/Gad7 [60, 61]. Atf1 binds to Pcr1, another bZIP protein, to form a heterodimer [62, 63]. Wis1, Sty1, and Atf1 are essential to prosecute proper G1 arrest and enter either sexual development or G0-like stationary phase upon nitrogen depletion. Induction of *ste11* expression in response to nitrogen starvation becomes inoperative by the disruption of either *wis1*, or *sty1*, or *atf1* [60, 64, 65]. These results indicate that the stress-responsive MAP kinase pathway stimulates G1 arrest and induction of *ste11* expression under nitrogen starvation (Figure 12.3).

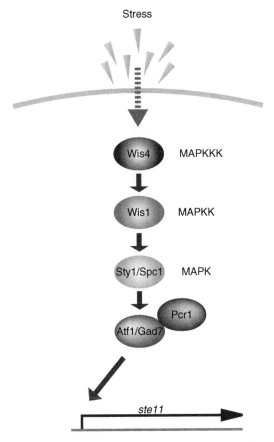

Fig. 12.3. The stress-activated MAP kinase cascade. This cascade, composed of two MAPKKKs Wis4/Wak1/Wik1 and Win1, MAPKK Wis1, and MAPK Sty1/Spc1, is activated in response to various stresses. Under nitrogen starvation, it fulfills function necessary for G1 arrest and induction of *ste11* expression.

F. The Tor1 Pathway

In contrast to Tor2, fission yeast Tor1 is not absolutely essential for growth. However, the *tor1*-deletion mutant is sterile and exhibits deficiency in the entry to G1 arrest and following sexual development in response to nitrogen starvation. Tor1 is also required for growth under stressed conditions, such as high osmotic pressure or high temperature [29, 30]. Originally, the *tor1* gene was identified by a database homology search and its function was analyzed subsequently by creating deletion mutants. It was also experimentally identified as the responsible gene for the *gad2-1* mutation, one of the *gad* (G1 arrest defective) mutations that brought a defect in G1 arrest and a defect in mating simultaneously [66]. The *gad8* gene was isolated as a high-copy suppressor of the sterility of the *gad2-1* mutant, and its product, Gad8, turned out to be a Ser/Thr kinase of the AGC family, which includes Ypk1 and Ypk2 of budding yeast and PKB/Akt, PKC, and S6K1 of mammal [54]. Like the *tor1*-deletion mutant, the *gad8*-deletion mutant fails to arrest at G1 under nitrogen starvation, is completely deficient in mating, and shows defects in growth under stressed conditions [66]. Members of the AGC kinase family are known to be phosphorylated at several conserved sites, which is critical for their activity. Gad8 also carries phosphorylation sites in the conserved regions called the activation loop (Thr387), the turn motif (Ser527), and the hydrophobic motif (Ser546). Site-directed mutagenesis of these phosphorylaion sites into alanine has shown that Tor1 is responsible for the phosphorylation of Ser527 and Ser546. PDK1 activates most AGC members through phosphorylation of the activation loop. Thr387 on the activation loop of Gad8 is phosphorylated by Ksg1, the PDK1-like kinase in fission yeast. The three phosphorylation sites are crucial for the activity and function of Gad8. As expression of a phosphorylation-mimic *gad8* allele can suppress stress sensitivity and mating deficiency of the *tor1*-deletion mutant, it has been proposed that Gad8 is the main target of TORC2 [66]. Substrates of Gad8, however, are still unknown. From above observations, conservation of the TOR (Tor1)–PDK1 (Ksg1)–AGC (Gad8) kinase signaling module among eukaryotes has been proposed, and indeed, this module has been found in mammalian and budding yeast cells. It is now considered that the TOR–PDK1–AGC trikinase system is a basic signaling module in eukaryotic cells [54] (Figure 12.2B).

Tor1 forms a complex with Ste20 (Rictor/Avo3 homolog), Wat1/Pop3, Sin1 (budding yeast Avo1 homolog), and Bit61 (budding yeast Bit61 homolog), composing TORC2 (Figure 12.2B) [13, 33, 35]. In *tor1*-deletion, *ste20*-deletion, or *wat1*-deletion cells, transcription of *ste11* is not inducible by nitrogen starvation (unpublished data). Like *tor1*-deletion, deletion of either *ste20*, or *wat1*, or *sin1* renders the cell sterile [38, 67, 68].

Sin1 was originally isolated as a protein interacting with the stress-activated MAPK Sty1/Spc1 and its naming regards this (SAPK interacting protein 1) [67]. However, neither *sin1*-deletion nor *tor1*-deletion cells show a defect in the stress-responsive gene expression regulated by the Sty1–Atf1 pathway [14]. Therefore, although TORC2 is required for the survival of cells under stresses, it appears that the TORC2 pathway contributes to overcome stresses independently of the Sty1 Atf1 pathway.

It has been recently reported that Tor1 participates in the regulation of the mitotic onset. Tor1 does this regulation by controlling the level of Pyp2, a phosphatase that regulates the Sty1 MAP kinase activity negatively [47, 69, 70]. Taken together, although Tor1 is not required for the Sty1-regulated stress-responsive gene expression, Tor1 seems to modulate the Sty1 MAP kinase activity at least in cell cycle regulation. How these observations can be reconciled remains to be seen.

V.   Mating Pheromone Signaling

A.   MATING-TYPE GENES

The mating type of a fission yeast haploid cell is determined by the *mat1* locus. If this locus carries the *P* sequence (*mat1-P*), the cell is h^+ (or *P*), and if it carries the *M* sequence (*mat1-M*), the cell is h^- (or *M*). Both *mat1-P* and *mat1-M* harbor two transcription units, named *mat1-Pc* and *mat1-Pi* (*mat1-Pm*), and *mat1-Mc* and *mat-Mi* (*mat1-Mm*), respectively [71]. In response to nitrogen starvation, an h^+ cell and an h^- cell mate to form a diploid zygote. Nitrogen starvation leads to the establishment of the pheromone signaling systems, through which h^+ and h^- cells communicate with each other and induce G1 arrest in the partner. Upon nitrogen starvation, Ste11 enhances expression of *mat1-Pc* in h^+ cells and *mat1-Mc* in h^- cells, both of which encode transcription factors [15, 16]. These transcription factors in turn stimulate expression of respective mating pheromones and their receptors. In h^- cells, Mat1-Mc and Ste11 physically interact and bind to the promoter of M-specific genes through the TR-box motif [72]. In h^+ cells, it is supposed that Mat1-Pc cooperates with Map1, a transcription factor of the MADS-box family [73, 74].

B.   PHEROMONE-RESPONSIVE MAP KINASE CASCADE AND THE RAS1 PATHWAY

The two mating pheromones of fission yeast, namely P-factor and M-factor, are small peptides secreted from h^+ cells and h^- cells, respectively. P-factor is a simple peptide of 23 amino acid residues, and M-factor is a farnesylated peptide of 9 amino acid residues [75, 76]. These pheromones bind to and activate seven-transmembrane receptors on the surface of opposite mating-type cells (Mam2 on h^- cells and Map3 on h^+ cells) [77, 78]. The receptors are coupled with an identical G protein, and once activated, its alpha subunit Gpa1 stimulates a MAP kinase cascade, composed of Byr2 (MAPKKK), Byr1 (MAPKK), and Spk1 (MAPK) [79–82] (Figure 12.4). The mating pheromone signal conveyed by the Byr2–Byr1–Spk1 MAP kinase cascade is essential for not only conjugation but also meiosis in fission yeast [83] (see below). Byr2 is also subject to the regulation by Ras1, the fission

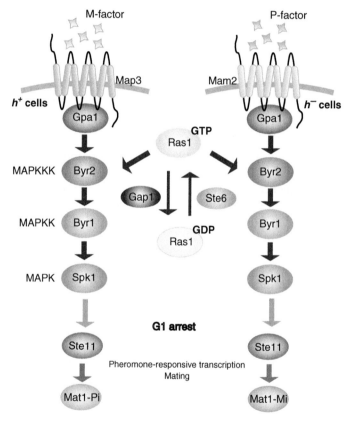

FIG. 12.4. The phremone-signaling pathway involving a MAP kinase cascade and the Ras1 protein. Binding of either P- or M-factor to the respective receptor results in the activation of the pheromone-responsive MAP kinase cascade Byr2–Byr1–Spk1. Ras1, the single homolog of the oncogenic protein Ras, modulates phremone-signaling in fission yeast. Ras1 is active in the GTP-bound state. Transcription of *mat1-Pi* and *mat1-Mi* is stimulated by the signal conveyed via the Byr2–Byr1–Spk1 cascade.

yeast homolog of oncogenic Ras protein [84, 85]. The activity of Ras1 is controlled by the GDP/GTP exchange factor (GEF) Ste6 and the GTPase-activating protein (GAP) Gap1, probably in response to nutritional deprivation [86, 87] (Figure 12.4). However, how nutrients regulate the Ras1 activity remains unknown.

VI.   Initiation of Meiosis

A.   Mei2, the Master Regulator of Meiosis

As mentioned above, nutritional starvation induces expression of *ste11*. Ste11 activates many genes required for mating and meiosis, including *mei2*, which encodes a crucial factor for the initiation of meiosis [16]. Mei2 is an RRM (RNA-recognition motif)-type RNA-binding protein of 750 amino acid residues, and is inactivated by Pat1(Ran1) kinase during the mitotic cell cycle [88, 89]. Inactivation of Pat1 causes the cell to initiate ectopic meiosis on nutrient medium, even in the haploid state. Pat1 phosphorylates Mei2 on residues Ser438 and Thr527. Artificial expression of a mutant form of Mei2 in which these two residues are substituted by alanine (mei2-SATA) also causes ectopic meiosis just like inactivation of Pat1. This indicates that the unphosphorylated form of Mei2 has the ability to switch the cell cycle and elicit meiotic differentiation. Thus, phosphorylation of Mei2 by Pat1 is a central control to prevent vegetative cells from entering ectopic meiosis (Figure 12.5). It has been shown that Pat1 also phosphorylates Ste11 [90]. This phosphorylation, however, appears to bring only an augmenting effect, because unphosphorylated Ste11, unlike unphosphorylated Mei2, does not induce ectopic meiosis by itself.

Phosphorylation of Mei2 by Pat1 results in acceleration of its degradation via ubiquitin-mediated proteolysis involving E2 Ubc2 and E3 Ubr1. Thus, Mei2 is unstable in growing wild-type cells and is stabilized in *pat1*-defective cells or cells undergoing meiosis [91]. Furthermore, it has been shown that Rad24, the major 14-3-3 protein in fission yeast, binds phosphorylated Mei2 preferentially, and thereby inhibits the binding of Mei2 to meiRNA, a noncoding RNA that forms a complex with Mei2 and is indispensable for the progression of meiosis I [92]. The ability of Mei2 to bind to RNA is crucial for its activity to promote meiosis [89]. Altogether, Pat1 kinase downregulates Mei2 in both activity and stability (Figure 12.6A).

In cells passing through conjugation under nitrogen starvation, expression of *mat1-Pi* and *mat1-Mi* is stimulated by the mating pheromone signaling via the Byr2–Byr1–Spk1 MAP kinase cascade, which also boosts expression of *ste11*. Coexpression of *mat1-Pi* and *mat1-Mi* in a resulting

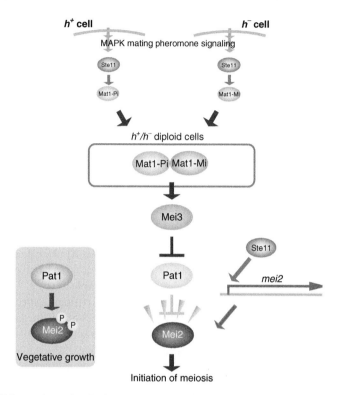

Fɪɢ. 12.5. A scheme for the initiation of meiosis in fission yeast. Diploid cells starved for nitrogen produce both Mat1-Pi and Mat1-Mi, which cooperatively induce expression of *mei3*. Mei3 is an inhibitor of Pat1 kinase, which phosphorylates and inactivates Mei2. Thus, diploid cells under nitrogen starvation accumulate unphosphorylated Mei2, which has the ability to provoke meiosis.

diploid cell allows expression of Mei3, which is a pseudosubstrate of Pat1 kinase [83]. It can bind to the catalytic site of Pat1, but has no residue to be phosphorylated. Thus, Mei3 functions as an inhibitor that masks the kinase activity of Pat1 [90] (Figure 12.5). In this way diploid cells accumulate unphosphorylated Mei2 under nitrogen starvation, and furthermore, expression of Mei2 is elevated via the function of Ste11 under nitrogen starvation. Unphosphorylated Mei2 exerts the function to switch the cell cycle from mitotic to meiotic. It has been observed that Mei2 forms a characteristic dot (Mei2 dot) in the nucleus during prophase of meiosis I [93, 94]. This Mei2 dot is associated with the chromosome II at the *sme2* gene, which encodes meiRNA [95]. It appears that nascent meiRNA

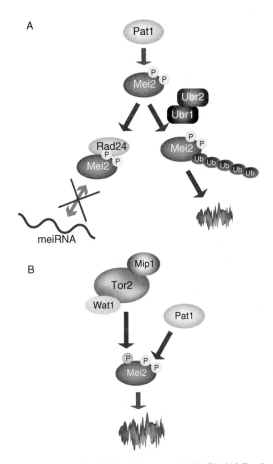

Fig. 12.6. Mei2 is a substrate of both Pat1 kinase and TORC1. (A) Dual regulation of Mei2 by Pat1 kinase. During vegetative growth, Pat1 phosphorylates Mei2. Phosphorylated Mei2 is degraded rapidly by proteasomes. In addition, Rad24 (14-3-3) binds preferentially to phosphorylated Mei2 and prevents it from binding to meiRNA. (B) Regulation of Mei2 by TORC1. In addition to Pat1, TORC1 apparently phosphorylates Mei2 during vegetative growth. This phosphorylation also accelerates degradation of Mei2.

transcribed from *sme2* forms a complex with Mei2 and is bound to the chromosome. The Mei2 dot is essential for the cell to initiate meiosis I.

Evidence suggests that Mei2 may carry out multiple molecular functions to complete meiosis. A single function of Mei2 that has been clearly elucidated to date is the inactivation of the DSR–Mmi1 system, which operates in mitotically growing cells to eliminate undesired meiosis-specific transcripts [96]. Transcripts of a number of meiosis-specific genes have been shown to contain a region termed DSR (determinant of selective removal), which is a marker for the respective transcript to be subjected to selective elimination in mitotic cells. Mmi1, a YTH-family protein, binds to the DSR region and directs the transcript to destruction by nuclear exosomes. Upon entry to meiosis, the nuclear Mei2 dot entraps Mmi1 proteins and sequesters them throughout prophase I, thereby allowing meiosis-specific mRNAs to be expressed stably and perform their function [3].

B.   INTERACTION OF MEI2 WITH TORC1

As discussed above, Mip1, the raptor ortholog in fission yeast, binds to Tor2 and composes TORC1 [13, 33, 35]. A truncated form of Mip1, missing the N-terminal region, was isolated as a high-copy suppressor of the ectopic meiosis induced by the expression of unphosphorylated Mei2, namely Mei2-SATA, when the identity of raptor was not yet established. It has been demonstrated that Mip1 physically interacts with Mei2 and Ste11 [97]. It has been also shown recently that Tor2 physically interacts with Mei2 and Ste11 [13]. Furthermore, it is reported that the *Arabidopsis* homolog of Mei2, AML1, interacts with the *Arabidopsis* homolog of raptor, AtRaptor1B [98].

In wild-type fission yeast cells, the level of Mei2 protein is very low during the mitotic cell cycle. However, we have observed that Mei2 is considerably stabilized in *tor2-ts* mutant cells shifted to the restrictive temperature. Furthermore, *in vitro* kinase assay has indicated that Tor2 directly phosphorylates Mei2. Mei2-SATA could also be phoshorylated by Tor2, suggesting that Tor2 and Pat1 phosphorylate Mei2 on different sites (our unpublished results). Thus, it appears that both TORC1 and Pat1 kinase phosphorylate Mei2 and contribute to destabilize this meiotic inducer under the nutritional conditions favorable for vegetative growth (Figure 12.6B).

ACKNOWLEDGMENTS

We thank Dr. Akira Yamashita for a number of helpful comments during the preparation of this manuscript. Our studies cited in this chapter were supported by Grants-in-Aid for Scientific Research from the Ministry of Education, Culture, Sports, Science, and Technology (MEXT) of Japan.

References

1. Yamamoto, M., Imai, Y., and Watanabe, Y. (1997). Mating and sporulation in *Shizosaccharomyces pombe*. In The Molecular and Cellular Biology of the Yeast Saccharomyces, J. Pringle, J. Broach and E. Jones, (eds.), pp. 1037–1106. Cold Spring Harbor Laboratory Press, Cold Spring Harbor, NY.
2. Otsubo, Y., and Yamamato, M. (2008). TOR signaling in fission yeast. *Crit Rev Biochem Mol Biol* 43.277–283.
3. Harigaya, Y., and Yamamoto, M. (2007). Molecular mechanisms underlying the mitosis-meiosis decision. *Chromosome Res* 15:523–537.
4. MacNeil, S.A., and Nurse, P. (1997). Cell cycle control in fission yeast. In The Molecular and Cellular Biology of the yeast Saccharomyces, J. Pringle, J. Broach and E. Jones, (eds.), pp. 697–763. Cold Spring Harbor Laboratory Press, Cold Spring Harbor, NY.
5. Hartwell, L.H. (1974). *Saccharomyces cerevisiae* cell cycle. *Bacteriol Rev* 38:164–198.
6. Pardee, A.B., Dubrow, R., Hamlin, J.L., and Kletzien, R.F. (1978). Animal cell cycle. *Annu Rev Biochem* 47:715–750.
7. Nurse, P., and Bissett, Y. (1981). Gene required in G1 for commitment to cell cycle and in G2 for control of mitosis in fission yeast. *Nature* 292:558–560.
8. Correa-Bordes, J., and Nurse, P. (1995). p25^{rum1} orders S phase and mitosis by acting as an inhibitor of the p34^{cdc2} mitotic kinase. *Cell* 83:1001–1009.
9. Yamaguchi, S., Murakami, H., and Okayama, H. (1997). A WD repeat protein controls the cell cycle and differentiation by negatively regulating Cdc2/B-type cyclin complexes. *Mol Biol Cell* 8:2475–2486.
10. Kitamura, K., Maekawa, H., and Shimoda, C. (1998). Fission yeast Ste9, a homolog of Hct1/Cdh1 and Fizzy-related, is a novel negative regulator of cell cycle progression during G1-phase. *Mol Biol Cell* 9:1065–1080.
11. Blanco, M.A., Sanchez-Diaz, A., de Prada, J.M., and Moreno, S. (2000). APC$^{ste9/srw1}$ promotes degradation of mitotic cyclins in G_1 and is inhibited by cdc2 phosphorylation. *EMBO J* 19:3945–3955.
12. Kominami, K., Seth-Smith, H., and Toda, T. (1998). Apc10 and Ste9/Srw1, two regulators of the APC-cyclosome, as well as the CDK inhibitor Rum1 are required for G1 cell-cycle arrest in fission yeast. *EMBO J* 17:5388–5399.
13. Alvarez, B., and Moreno, S. (2006). Fission yeast Tor2 promotes cell growth and represses cell differentiation. *J Cell Sci* 119:4475–4485.
14. Ikeda, K., Morigasaki, S., Tatebe, H., Tamanoi, F., and Shiozaki, K. (2008). Fission yeast TOR complex 2 activates the AGC-family Gad8 kinase essential for stress resistance and cell cycle control. *Cell Cycle* 7:358–364.
15. Aono, T., Yanai, H., Miki, F., Davey, J., and Shimoda, C. (1994). Mating pheromone-induced expression of the *mat1-Pm* gene of *Schizosaccharomyces pombe*: identification of signalling components and characterization of upstream controlling elements. *Yeast* 10:757–770.
16. Sugimoto, A., Iino, Y., Maeda, T., Watanabe, Y., and Yamamoto, M. (1991). *Schizosaccharomyces pombe ste11$^+$* encodes a transcription factor with an HMG motif that is a critical regulator of sexual development. *Genes Dev* 5:1990–1999.
17. Maeda, T., Mochizuki, N., and Yamamoto, M. (1990). Adenylyl cyclase is dispensable for vegetative cell growth in the fission yeast *Schizosaccharomyces pombe*. *Proc Natl Acad Sci USA* 87:7814–7818.
18. Mochizuki, N., and Yamamoto, M. (1992). Reduction in the intracellular cAMP level triggers initiation of sexual development in fission yeast. *Mol Gen Genet* 233:17–24.

19. Isshiki, T., Mochizuki, N., Maeda, T., and Yamamoto, M. (1992). Characterization of a fission yeast gene, *gpa2*, that encodes a G alpha subunit involved in the monitoring of nutrition. *Genes Dev* 6:2455–2462.
20. Welton, R.M., and Hoffman, C.S. (2000). Glucose monitoring in fission yeast via the gpa2 Galpha, the git5 Gbeta and the git3 putative glucose receptor. *Genetics* 156:513–521.
21. Hoffman, C.S. (2005). Glucose sensing via the protein kinase A pathway in *Schizosaccharomyces pombe*. *Biochem Soc Trans* 33:257–260.
22. Kawamukai, M., Ferguson, K., Wigler, M., and Young, D. (1991). Genetic and biochemical analysis of the adenylyl cyclase of *Schizosaccharomyces pombe*. *Cell Regul* 2:155–164.
23. Maeda, T., Watanabe, Y., Kunitomo, H., and Yamamoto, M. (1994). Cloning of the *pka1* gene encoding the catalytic subunit of the cAMP-dependent protein kinase in *Schizosaccharomyces pombe*. *J Biol Chem* 269:9632–9637.
24. Kunitomo, H., Higuchi, T., Iino, Y., and Yamamoto, M. (2000). A zinc-finger protein, Rst2p, regulates transcription of the fission yeast *ste11*[+] gene, which encodes a pivotal transcription factor for sexual development. *Mol Biol Cell* 11:3205–3217.
25. Higuchi, T., Watanabe, Y., and Yamamoto, M. (2002). Protein kinase A regulates sexual development and gluconeogenesis through phosphorylation of the Zn finger transcriptional activator Rst2p in fission yeast. *Mol Cell Biol* 22:1–11.
26. Vassarotti, A., and Friesen, J.D. (1985). Isolation of the fructose-1, 6-bisphosphatase gene of the yeast *Schizosaccharomyces pombe*. Evidence for transcriptional regulation. *J Biol Chem* 260:6348–6353.
27. Hoffman, C.S., and Winston, F. (1991). Glucose repression of transcription of the *Schizosaccharomyces pombe fbp1* gene occurs by a cAMP signaling pathway. *Genes Dev* 5:561–571.
28. Hoffman, C.S., and Winston, F. (1990). Isolation and characterization of mutants constitutive for expression of the *fbp1* gene of *Schizosaccharomyces pombe*. *Genetics* 124:807–816.
29. Kawai, M., *et al.* (2001). Fission yeast *tor1* functions in response to various stresses including nitrogen starvation, high osmolarity, and high temperature. *Curr Genet* 39:166–174.
30. Weisman, R., and Choder, M. (2001). The fission yeast TOR homolog, *tor1*[+], is required for the response to starvation and other stresses via a conserved serine. *J Biol Chem* 276:7027–7032.
31. Uritani, M., Hidaka, H., Hotta, Y., Ueno, M., Ushimaru, T., and Toda, T. (2006). Fission yeast Tor2 links nitrogen signals to cell proliferation and acts downstream of the Rheb GTPase. *Genes Cells* 11:1367–1379.
32. Weisman, R., Roitburg, I., Schonbrun, M., Harari, R., and Kupiec, M. (2007). Opposite effects of *tor1* and *tor2* on nitrogen starvation responses in fission yeast. *Genetics* 175:1153–1162.
33. Matsuo, T., Otsubo, Y., Urano, J., Tamanoi, F., and Yamamoto, M. (2007). Loss of the TOR kinase Tor2 mimics nitrogen starvation and activates the sexual development pathway in fission yeast. *Mol Cell Biol* 27:3154–3164.
34. Urano, J., Sato, T., Matsuo, T., Otsubo, Y., Yamamoto, M., and Tamanoi, F. (2007). Point mutations in TOR confer Rheb-independent growth in fission yeast and nutrient-independent mammalian TOR signaling in mammalian cells. *Proc Natl Acad Sci USA* 104:3514–3519.
35. Hayashi, T., *et al.* (2007). Rapamycin sensitivity of the *Schizosaccharomyces pombe tor2* mutant and organization of two highly phosphorylated TOR complexes by specific and common subunits. *Genes Cells* 12:1357–1370.

36. Roberg, K.J., Bickel, S., Rowley, N., and Kaiser, C.A. (1997). Control of amino acid permease sorting in the late secretory pathway of *Saccharomyces cerevisiae* by *SEC13*, *LST4*, *LST7* and *LST8*. *Genetics* 147:1569–1584.

37. Ochotorena, I.L., *et al.* (2001). Conserved Wat1/Pop3 WD-repeat protein of fission yeast secures genome stability through microtubule integrity and may be involved in mRNA maturation. *J Cell Sci* 114:2911–2920.

38. Kemp, J.T., Balasubramanian, M.K., and Gould, K.L. (1997). A *wat1* mutant of fission yeast is defective in cell morphology. *Mol Gen Genet* 254:127–138.

39. Matsumoto, S., Bandyopadhyay, A., Kwiatkowski, D.J., Maitra, U., and Matsumoto, T. (2002). Role of the Tsc1-Tsc2 complex in signaling and transport across the cell membrane in the fission yeast *Schizosaccharomyces pombe*. *Genetics* 161:1053–1063.

40. van Slegtenhorst, M., Carr, E., Stoyanova, R., Kruger, W.D., and Henske, E.P. (2004). *Tsc1$^+$* and *tsc2$^+$* regulate arginine uptake and metabolism in *Schizosaccharomyces pombe*. *J Biol Chem* 279:12706–12713.

41. Urano, J., *et al.* (2005). Identification of novel single amino acid changes that result in hyperactivation of the unique GTPase, Rheb, in fission yeast. *Mol Microbiol* 58:1074–1086.

42. Mach, K.E., Furge, K.A., and Albright, C.F. (2000). Loss of Rhb1, a Rheb-related GTPase in fission yeast, causes growth arrest with a terminal phenotype similar to that caused by nitrogen starvation. *Genetics* 155:611–622.

43. Yang, W., Tabancay, A.P., Jr., Urano, J., and Tamanoi, F. (2001). Failure to farnesylate Rheb protein contributes to the enrichment of G0/G1 phase cells in the *Schizosaccharomyces pombe* farnesyltransferase mutant. *Mol Microbiol* 41:1339–1347.

44. Weisman, R., Choder, M., and Koltin, Y. (1997). Rapamycin specifically interferes with the developmental response of fission yeast to starvation. *J Bacteriol* 179:6325–6334.

45. Weisman, R. (2004). The fission yeast TOR proteins and the rapamycin response: an unexpected tale. *Curr Top Microbiol Immunol* 279:85–95.

46. Weisman, R., Roitburg, I., Nahari, T., and Kupiec, M. (2005). Regulation of leucine uptake by *tor1$^+$* in *Schizosaccharomyces pombe* is sensitive to rapamycin. *Genetics* 169:539–550.

47. Petersen, J., and Nurse, P. (2007). TOR signalling regulates mitotic commitment through the stress MAP kinase pathway and the Polo and Cdc2 kinases. *Nat Cell Biol* 9:1263–1272.

48. Urban, J., *et al.* (2007). Sch9 is a major target of TORC1 in *Saccharomyces cerevisiae*. *Mol Cell* 26:663–674.

49. Toda, T., Cameron, S., Sass, P., Zoller, M., and Wigler, M. (1987). Three different genes in *S. cerevisiae* encode the catalytic subunits of the cAMP-dependent protein kinase. *Cell* 50:277–287.

50. Toda, T., Cameron, S., Sass, P., and Wigler, M. (1988). *SCH9*, a gene of *Saccharomyces cerevisiae* that encodes a protein distinct from, but functionally and structurally related to, cAMP-dependent protein kinase catalytic subunits. *Genes Dev* 2:517–527.

51. Jin, M., *et al.* (1995). *sck1*, a high copy number suppressor of defects in the cAMP-dependent protein kinase pathway in fission yeast, encodes a protein homologous to the *Saccharomyces cerevisiae* SCH9 kinase. *Genetics* 140:457–467.

52. Fujita, M., and Yamamoto, M. (1998). *S. pombe sck2$^+$*, a second homologue of *S. cerevisiae SCH9* in fission yeast, encodes a putative protein kinase closely related to PKA in function. *Curr Genet* 33:248–254.

53. Mukai, H., *et al.* (1995). Identification of *Schizosaccharomyces pombe* gene *psk1$^+$*, encoding a novel putative serine/threonine protein kinase, whose mutation conferred resistance to phenylarsine oxide. *Gene* 166:155–159.

54. Jacinto, E., and Lorberg, A. (2008). TOR regulation of AGC kinases in yeast and mammals. *Biochem J* 410:19–37.
55. Shiozaki, K., and Russell, P. (1995). Cell-cycle control linked to extracellular environment by MAP kinase pathway in fission yeast. *Nature* 378:739–743.
56. Shiozaki, K., Shiozaki, M., and Russell, P. (1997). Mcs4 mitotic catastrophe suppressor regulates the fission yeast cell cycle through the Wik1-Wis1-Spc1 kinase cascade. *Mol Biol Cell* 8:409–419.
57. Samejima, I., Mackie, S., and Fantes, P.A. (1997). Multiple modes of activation of the stress-responsive MAP kinase pathway in fission yeast. *EMBO J* 16:6162–6170.
58. Samejima, I., Mackie, S., Warbrick, E., Weisman, R., and Fantes, P.A. (1998). The fission yeast mitotic regulator *win1*⁺ encodes an MAP kinase kinase kinase that phosphorylates and activates Wis1 MAP kinase kinase in response to high osmolarity. *Mol Biol Cell* 9:2325–2335.
59. Gaits, F., Degols, G., Shiozaki, K., and Russell, P. (1998). Phosphorylation and association with the transcription factor Atf1 regulate localization of Spc1/Sty1 stress-activated kinase in fission yeast. *Genes Dev* 12:1464–1473.
60. Shiozaki, K., and Russell, P. (1996). Conjugation, meiosis, and the osmotic stress response are regulated by Spc1 kinase through Atf1 transcription factor in fission yeast. *Genes Dev* 10:2276–2288.
61. Wilkinson, M.G., *et al.* (1996). The Atf1 transcription factor is a target for the Sty1 stress-activated MAP kinase pathway in fission yeast. *Genes Dev* 10:2289–2301.
62. Kanoh, J., Watanabe, Y., Ohsugi, M., Iino, Y., and Yamamoto, M. (1996). *Schizosaccharomyces pombe gad7*⁺ encodes a phosphoprotein with a bZIP domain, which is required for proper G1 arrest and gene expression under nitrogen starvation. *Genes Cells* 1:391–408.
63. Watanabe, Y., and Yamamoto, M. (1996). *Schizosaccharomyces pombe pcr1*⁺ encodes a CREB/ATF protein involved in regulation of gene expression for sexual development. *Mol Cell Biol* 16:704–711.
64. Takeda, T., Toda, T., Kominami, K., Kohnosu, A., Yanagida, M., and Jones, N. (1995). *Schizosaccharomyces pombe atf1*⁺ encodes a transcription factor required for sexual development and entry into stationary phase. *EMBO J* 14:6193–6208.
65. Warbrick, E., and Fantes, P.A. (1991). The wis1 protein kinase is a dosage-dependent regulator of mitosis in *Schizosaccharomyces pombe*. *EMBO J* 10:4291–4299.
66. Matsuo, T., Kubo, Y., Watanabe, Y., and Yamamoto, M. (2003). *Schizosaccharomyces pombe* AGC family kinase Gad8p forms a conserved signaling module with TOR and PDK1-like kinases. *EMBO J* 22:3073–3083.
67. Wilkinson, M.G., *et al.* (1999). Sin1: an evolutionarily conserved component of the eukaryotic SAPK pathway. *EMBO J* 18:4210–4221.
68. Hilti, N., Baumann, D., Schweingruber, A.M., Bigler, P., and Schweingruber, M.E. (1999). Gene *ste20* controls amiloride sensitivity and fertility in *Schizosaccharomyces pombe*. *Curr Genet* 35:585–592.
69. Millar, J.B., Buck, V., and Wilkinson, M.G. (1995). Pyp1 and Pyp2 PTPases dephosphorylate an osmosensing MAP kinase controlling cell size at division in fission yeast. *Genes Dev* 9:2117–2130.
70. Hartmuth, S., and Petersen, J. (2009). Fission yeast Tor1 functions as part of TORC1 to control mitotic entry through the stress MAPK pathway following nutrient stress. *J Cell Sci* 122:1737–1746.
71. Kelly, M., Burke, J., Smith, M., Klar, A., and Beach, D. (1988). Four mating-type genes control sexual differentiation in the fission yeast. *EMBO J* 7:1537–1547.

72. Kjaerulff, S., Dooijes, D., Clevers, H., and Nielsen, O. (1997). Cell differentiation by interaction of two HMG-box proteins: Mat1-Mc activates M cell-specific genes in *S. pombe* by recruiting the ubiquitous transcription factor Ste11 to weak binding sites. *EMBO J* 16:4021–4033.

73. Nielsen, O., Friis, T., and Kjaerulff, S. (1996). The *Schizosaccharomyces pombe map1* gene encodes an SRF/MCM1-related protein required for P-cell specific gene expression. *Mol Gen Genet* 253:387–392.

74. Yabana, N., and Yamamoto, M. (1996). *Schizosaccharomyces pombe map1⁺* encodes a MADS-box-family protein required for cell-type-specific gene expression. *Mol Cell Biol* 16:3420–3428.

75. Imai, Y., and Yamamoto, M. (1994). The fission yeast mating pheromone P-factor: its molecular structure, gene structure, and ability to induce gene expression and G1 arrest in the mating partner. *Genes Dev* 8:328–338.

76. Davey, J. (1992). Mating pheromones of the fission yeast *Schizosaccharomyces pombe*: purification and structural characterization of M-factor and isolation and analysis of two genes encoding the pheromone. *EMBO J* 11:951–960.

77. Kitamura, K., and Shimoda, C. (1991). The *Schizosaccharomyces pombe mam2* gene encodes a putative pheromone receptor which has a significant homology with the *Saccharomyces cerevisiae* Ste2 protein. *EMBO J* 10:3743–3751.

78. Tanaka, K., Davey, J., Imai, Y., and Yamamoto, M. (1993). *Schizosaccharomyces pombe map3⁺* encodes the putative M-factor receptor. *Mol Cell Biol* 13:80–88.

79. Nadin-Davis, S.A., and Nasim, A. (1988). A gene which encodes a predicted protein kinase can restore some functions of the *ras* gene in fission yeast. *EMBO J* 7:985–993.

80. Nadin-Davis, S.A., and Nasim, A. (1990). *Schizosaccharomyces pombe ras1* and *byr1* are functionally related genes of the *ste* family that affect starvation-induced transcription of mating-type genes. *Mol Cell Biol* 10:549–560.

81. Toda, T., Shimanuki, M., and Yanagida, M. (1991). Fission yeast genes that confer resistance to staurosporine encode an AP-1-like transcription factor and a protein kinase related to the mammalian ERK1/MAP2 and budding yeast FUS3 and KSS1 kinases. *Genes Dev* 5:60–73.

82. Gotoh, Y., Nishida, E., Shimanuki, M., Toda, T., Imai, Y., and Yamamoto, M. (1993). *Schizosaccharomyces pombe* Spk1 is a tyrosine-phosphorylated protein functionally related to *Xenopus* mitogen-activated protein kinase. *Mol Cell Biol* 13:6427–6434.

83. Willer, M., Hoffmann, L., Styrkarsdottir, U., Egel, R., Davey, J., and Nielsen, O. (1995). Two-step activation of meiosis by the *mat1* locus in *Schizosaccharomyces pombe*. *Mol Cell Biol* 15:4964–4970.

84. Wang, Y., Xu, H.P., Riggs, M., Rodgers, L., and Wigler, M. (1991). *byr2*, a *Schizosaccharomyces pombe* gene encoding a protein kinase capable of partial suppression of the *ras1* mutant phenotype. *Mol Cell Biol* 11:3554–3563.

85. Nielsen, O., Davey, J., and Egel, R. (1992). The *ras1* function of *Schizosaccharomyces pombe* mediates pheromone-induced transcription. *EMBO J* 11:1391–1395.

86. Hughes, D.A., Fukui, Y., and Yamamoto, M. (1990). Homologous activators of *ras* in fission and budding yeast. *Nature* 344:355–357.

87. Hughes, D.A., Yabana, N., and Yamamoto, M. (1994). Transcriptional regulation of a Ras nucleotide-exchange factor gene by extracellular signals in fission yeast. *J Cell Sci* 107(Pt 12):3635–3642.

88. Watanabe, Y., Lino, Y., Furuhata, K., Shimoda, C., and Yamamoto, M. (1988). The *S. pombe mei2* gene encoding a crucial molecule for commitment to meiosis is under the regulation of cAMP. *EMBO J* 7:761–767.

89. Watanabe, Y., and Yamamoto, M. (1994). *S. pombe mei2⁺* encodes an RNA-binding protein essential for premeiotic DNA synthesis and meiosis I, which cooperates with a novel RNA species meiRNA. *Cell* 78:487–498.

90. Li, P., and McLeod, M. (1996). Molecular mimicry in development: identification of *ste11⁺* as a substrate and *mei3⁺* as a pseudosubstrate inhibitor of *ran1⁺* kinase. *Cell* 87:869–880.

91. Kitamura, K., *et al.* (2001). Phosphorylation of Mei2 and Ste11 by Pat1 kinase inhibits sexual differentiation via ubiquitin proteolysis and 14-3-3 protein in fission yeast. *Dev Cell* 1:389–399.

92. Sato, M., Watanabe, Y., Akiyoshi, Y., and Yamamoto, M. (2002). 14-3-3 protein interferes with the binding of RNA to the phosphorylated form of fission yeast meiotic regulator Mei2p. *Curr Biol* 12:141–145.

93. Watanabe, Y., Shinozaki-Yabana, S., Chikashige, Y., Hiraoka, Y., and Yamamoto, M. (1997). Phosphorylation of RNA-binding protein controls cell cycle switch from mitotic to meiotic in fission yeast. *Nature* 386:187–190.

94. Yamashita, A., Watanabe, Y., Nukina, N., and Yamamoto, M. (1998). RNA-assisted nuclear transport of the meiotic regulator Mei2p in fission yeast. *Cell* 95:115–123.

95. Shimada, T., Yamashita, A., and Yamamoto, M. (2003). The fission yeast meiotic regulator Mei2p forms a dot structure in the horse-tail nucleus in association with the *sme2* locus on chromosome II. *Mol Biol Cell* 14:2461–2469.

96. Harigaya, Y., *et al.* (2006). Selective elimination of messenger RNA prevents an incidence of untimely meiosis. *Nature* 442:45–50.

97. Shinozaki-Yabana, S., Watanabe, Y., and Yamamoto, M. (2000). Novel WD-repeat protein Mip1p facilitates function of the meiotic regulator Mei2p in fission yeast. *Mol Cell Biol* 20:1234–1242.

98. Anderson, G.H., and Hanson, M.R. (2005). The Arabidopsis Mei2 homologue AML1 binds AtRaptor1B, the plant homologue of a major regulator of eukaryotic cell growth. *BMC Plant Biol* 5:2.

13

Fission Yeast TOR and Rapamycin

RONIT WEISMAN[a,b]

[a]*Department of Natural and Life Sciences*
The Open University of Israel
Raanana
Israel

[b]*Department of Molecular Microbiology and Biotechnology*
Tel Aviv University
Tel Aviv
Israel

I. Abstract

Recent studies of target of rapamycin (TOR) in fission yeast revealed that in this organism, like in budding yeast and human cells, the TORC1 complex containing TOR together with a raptor-like protein, plays a central role in regulating growth, while inhibiting starvation responses. Disruption of TORC1 in fission yeast results in a phenotype very similar to that of wild-type cells starved for nitrogen, suggesting that TORC1 may regulate growth in response to nitrogen availability. The TORC2 complex in fission yeast contains TOR together with a rictor-like protein. In fission yeast, this complex is not essential under normal growth conditions but is required for survival under stress and for starvation responses. More recent studies demonstrate that TORC2 in fission yeast also has a profound role in gene silencing, telomere length maintenance and DNA damage response. Most interestingly, rapamycin does not inhibit the essential role of TORC1 or most of the cellular functions of TORC2. Yet, accumulation of data suggests that rapamycin can inhibit either TORC1- or TORC2-dependent functions under certain nutritional growth conditions and/or in the presence of loss-of-function mutations in TORC1 or its regulators. Understanding

THE ENZYMES, Vol. XXVII
251
ISSN NO: 1874-6047
DOI: 10.1016/S1874-6047(10)27013-0

the determinants that render cells sensitive to rapamycin is critical for understanding the mode of action of rapamycin and may also extend our understanding of potential interactions between TOR containing complexes.

II. Introduction

One of my first observations, which incited me into exploring the target of rapamycin (TOR) pathway in fission yeast, was that rapamycin did not inhibit vegetative growth in this organism [1]. This was in sharp contrast to the growth inhibitory effect of rapamycin on budding yeast [2]. The question that immediately comes to mind is: what renders fission yeast resistant to rapamycin?

Resistance to rapamycin in fission yeast appears less of a surprise considering the response to the drug of many higher eukaryotic cell types. A common use of rapamycin is as an immunosuppressive drug following transplant surgeries. However, rapamycin does not inhibit proliferation of T lymphocytes that have already entered the cell cycle, but rather blocks resting T cells from entering the cell cycle [3]. The proliferation of certain cancerous cells is also highly sensitive to rapamycin, a finding that led to the use of rapamycin as an anticancer drug (reviewed in Ref. [4]). Which molecular mechanisms underlie rapamycin resistance/sensitivity in various cell types? Can we learn from fission yeast about mechanisms that render cells resistant to the drug?

The budding yeast *Saccharomyces cerevisiae* and the fission yeast *Schizosaccharomyces pombe* are distantly related. Indeed, the evolutionary distance between these two yeasts is as far apart from each other as from higher eukaryotes [5]. Thus, processes that are conserved between budding and fission yeasts are often also conserved in higher eukaryotes. Previous studies of the two yeasts have taught us that cellular processes that at first sight appeared substantially different, later proved to share common and conserved molecular mechanisms. For example, the regulation of the cell cycle first appeared fundamentally dissimilar in the two yeasts, partly owing to the fact that under optimal growth conditions, most of the control over cell cycle progression occurs at the G_1/S transition in budding yeast, while the G_2/M transition is the major control point in fission yeast. Further studies revealed that the same key regulators, in particular the master regulator CDK (cyclin-dependent kinase) are shared between the two yeasts and are conserved in higher eukaryotes [6, 7]. Indeed, it was the difference between the two yeasts that propelled many cell cycle studies, leading to a better understanding of the evolutionary conserved mechanisms [8]. Would this also be the case for the studies of TOR pathways and rapamycin response?

Recent progress in understanding TOR signaling suggests that this may well be the case. At first, the TOR signaling pathway appeared very different in the two yeasts. One of the reasons for this apparent difference was that the first TOR homolog characterized in fission yeast was Tor1, which was subsequently identified as part of TORC2. Later, identification and characterization of Tor2 (which forms TORC1) revealed that in fission yeast, like in budding yeast, TOR plays a major role in regulating growth. Studies of the FRB (FKBP12-rapamycin binding) domains of either Tor1 or Tor2 in fission yeast demonstrated that these domains can bind rapamycin in the presence of FKBP12, indicating that the basic structural requirements for rapamycin-mediated inhibition exist. We and others identified several genetic backgrounds, in which rapamycin inhibits the growth of fission yeast cells. Although currently we understand little about the determinants that affect rapamycin sensitivity, characterization of rapamycin-sensitive mutants is likely to help us understand the rapamycin response.

III. TORC1 is a Major Regulator of Cellular Growth

A. DISRUPTION OF TORC1 MIMICS RESPONSE TO NITROGEN STARVATION

Fission yeast contains two TOR homologs, Tor1 and Tor2 [9]. These were identified through sequence comparison analyses of data originating from the fission yeast genome sequencing project [10]. Tor2 mainly associates with Mip1 (similar to mammalian raptor and budding yeast Kog1) to form TOR complex 1 (TORC1). Tor1 mainly associates with Ste20 (similar to mammalian rictor and budding yeast Avo3) and with Sin1 (similar to mammalian Sin1 and budding yeast Avo1) to form TOR complex 2 (TORC2) [11, 12]. TORC1 and TORC2 in fission yeast include additional conserved components. The detailed composition of the TOR complexes is discussed by M. Yanagida in chapter 14.

$tor2^+$ is an essential gene. Depletion of $tor2^+$ using a heterologous repressible promoter [13] or temperature sensitive alleles ($tor2$-ts) [11, 12, 14, 15] resulted in a phenotype that highly resembled wild-type cells starved for nitrogen. This was the first indication that TOR in fission yeast plays a major role in controlling cellular growth in response to nutrient availability.

Since Tor2 physically associates with a raptor-like protein, Mip1, the structure and function of the TORC1 complex is conserved in fission yeast.

Fission yeast cells grow normally as haploid, rod-shaped cells, which divide by medial fission, giving rise to two daughter cells of equal size. Upon either nitrogen, or carbon starvation, cells exit the logarithmic phase and can opt for two alternative pathways: entrance into quiescence (stationary phase) or sexual development. Either option results in cells (or spores) that can better maintain viability for long periods and/or under harsh conditions, compared with vegetative growing cells. Entrance into sexual development requires the presence of the two opposite mating-type cells, h^+ and h^-, which can mate to form zygotes that subsequently undergo meiosis and sporulation. The nitrogen and carbon starvation responses differ in several aspects. Suffice here to say that the nitrogen-starvation response induces cells to enter mitosis at a reduced cell size, resulting in small-sized rounded stationary cells that accumulate at the G_1 phase of the cell cycle. In contrast, no mitosis advancement occurs when cells are starved for carbon, and under such conditions cells arrest their growth at the G_2 phase of the cell cycle [16]. Whether starved for nitrogen or carbon, mating occurs between two opposite mating-type cells at the G_1 phase of the cell cycle.

Like nitrogen-starved cells, cells depleted of Tor2 divide a few times, accumulate at the G_1 phase of the cell cycle and arrest their growth as small and round cells [11–15]. Since yeast cells normally grow in nature in the presence of abundant carbon but under nitrogen-limiting conditions, nitrogen availability is likely to be a major cue for growth control. Thus, Tor2 may play a major role under natural growth conditions [17].

Cells depleted of Tor2 showed a reduction in gene expression of ribosomal proteins and increased expression of nitrogen-starvation specific genes, such as regulators of sexual development, autophagy, membrane transporters, and amino acid permeases [12–15]. In accordance with this transcription profile, depletion of Tor2 induced autophagy [15] and sexual development [12–15]. Overexpression of $tor2^+$ or hyperactive $tor2$ alleles inhibited mating and delayed cell-size adaptation in response to nitrogen starvation [14, 22]. Thus, Tor2 has a dual role as a positive regulator of growth and as a negative regulator of starvation responses (Figure 13.1). Overexpression of $tor2^+$ also rendered cells resistant to canavanine, a toxic analog of arginine, and was deleterious when combined with an auxotrophic mutation to leucine [13]. These findings suggest that overexpression of Tor2 inhibits amino acid uptake. The role of Tor2 in regulating amino acid uptake is particularly relevant for our discussion of the response to rapamycin, and will be further discussed in the next section.

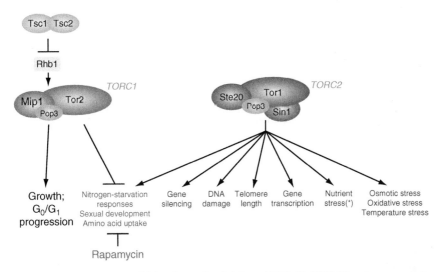

FIG. 13.1. Schematic model for the mode of action of TORC1, TORC2, and rapamycin in fission yeast. Only TORC1 complexes containing Tor2 and TORC2 complexes containing Tor1 are described, however, this may not represent the full repertoire of TOR containing complexes. It is notable that the cellular functions that are oppositely regulated by the two TOR complexes are sensitive to rapamycin. (*): Response to nutrient stress refers to the ability of the cells to properly enter and maintain stationary phase physiology following nutrient depletion.

B.   REGULATION OF NITROGEN STARVATION RESPONSES AND AMINO ACID UPTAKE BY THE TSC–RHB1–TORC1 MODULE

Fission yeast provides a particularly valuable system for studying TOR pathways, since the regulation of TORC1 by the TSC–Rheb module is conserved in this organism [18]. Here, I will focus on the features of the TSC–Rheb module which are relevant for our discussion of the rapamycin response.

The fission yeast Tsc1 and Tsc2 proteins form a complex [19, 20], in which Tsc2 acts as GTPase activating protein (GAP) for the small GTPase Rhb1 (Rheb homolog) [21–23]. Rhb1 associates with Tor2 and positively regulates its activity [15, 22]. Accordingly, depletion of Rhb1 displays a nitrogen-starvation phenotype [24, 25] and activated Tor2 mutant alleles can compensate for the loss of function of Rhb1 [26].

While loss of function of Tor2 or Rhb1 mimic nitrogen starvation, deletion of $tsc1^+$ or $tsc2^+$ resulted in viable cells that exhibited reduced amino acid uptake, low concentration of inner amino acid pools, inability to induce several nitrogen-starvation response genes and partial sterility [19–21]. The defect in amino acid uptake was demonstrated in tsc mutant cells by several methods: direct measurement of uptake of radioactively labeled leucine or arginine [19, 20], resistance to canavanine, thialysine or ethionine, which are the toxic analogs of arginine, lysine, and methionine, respectively [20, 22, 23] or slow growth when combined with auxotrophic mutations, in particular, auxotrophy for leucine [19, 20].

Several studies indicated that the *tsc* genes regulated amino acid uptake via negatively regulating Rhb1, including the finding that partial loss of function of *rhb1*$^+$ (*rhb1*$^{G63D/S165N}$) reversed the canavanine-resistant phenotype of *tsc* mutant cells (49), while hyperactive *rhb1* mutants conferred resistance to canavanine [22]. Overexpression of *tor2*$^+$ conferred canavanine and thialysine resistance and inhibited the growth of cells auxotrophic for leucine [13]. In contrast, downregulation of Tor2, using *tor2-ts* alleles, rescued the adenine uptake defect in Δ*tsc2* mutant cells [12]. Collectively, these findings are in line with the suggestion that the defect in amino acid uptake in *tsc* mutant cells results from overactivation of Tor2 via Rhb1 (Figure 13.1). It is possible that hyperactivation of Tor2 mimics conditions in which nitrogen is abundant, leading to the downregulation of amino acid import. However, the cue that is sensed by TOR or TSC–Rhb1 is currently unknown.

The phenotype of inefficient amino acid uptake in *tsc* mutant cells is associated with the downregulation of the expression of a group of amino acid permeases, *isp5*$^+$, 7G5.06, and c869.10 [20]. The transcripts of these permeases are strongly upregulated by a shift to medium with a poor nitrogen source (proline), suggesting that they may act as general amino acid permeases [27]. *isp5*$^+$ and c869.10 were also identified by global transcriptional profiling as genes whose transcripts were upregulated by depletion of Tor2 [12]. We found that overexpression of Tor2 downregulated the expression of 7G5.06 and c869.10 (M. Schonbrun and R. Weisman, unpublished data). Thus, the TSC complex and Tor2 oppositely regulate the transcription of the same amino acid permeases. The TSC–Rheb module also regulates amino acid uptake posttranscriptionally, as mislocalization of amino acid permeases was demonstrated in *tsc* mutant cells [19, 28].

Intriguingly, a recent study demonstrated that a *tor2-ts* mutation conferred resistance to canavanine [29], suggesting that downregulation of Tor2, as well as its upregulation, may lead to reduced uptake of amino acids. It is currently not clear whether a common underlying mechanism underlies canavanine resistance in *tor2-ts* or overexpression of *tor2*$^+$. In this respect, it is interesting to note that either upregulation of TORC1 by mutations in TSC1/2 or downregulation of TORC1 by mutations in the Birt-Hogg-Dube protein (folliculin) may lead to similar development of tumors [30].

Finally, although a clear role for Rhb1 was identified in controlling Tor2, Rhb1 may also exert Tor2-independent functions. Accordingly, hyperactivated *tor2* alleles, which rescued the lethal phenotype of deletion of *rhb1*$^+$, did not compensate for the sensitivity to osmotic stress, high temperature or thialysine in the double mutant $\Delta rhb1\ tor2^{act}$ cells. Thus, Rhb1 may also regulate amino acid uptake or survival under stress conditions in a Tor2-independent manner [26]. The possibility that Rhb1 may regulate other proteins beside Tor2 has also been suggested by [29], who isolated the $rhb1^{Q52RI76F}$ allele that conferred canavanine resistance but, unlike other constitutive activated $rhb1$ allele ($rhb1^{V17A}$), did not show a defect in response to nitrogen starvation. No other targets for such Rhb1 activity are known at present. An intriguing possibility is Tor1. Tor1 is required for cell survival under osmotic stress and positively regulates amino acid uptake (see below). However, so far, no clear link has been identified between TSC–Rhb1 and Tor1/TORC2.

C. RAPAMYCIN REDUCES AMINO ACID UPTAKE IN WILD-TYPE FISSION YEAST CELLS

Rapamycin does not inhibit vegetative growth in fission yeast; therefore, it is evident that under optimal growth conditions, the drug does not inhibit the essential role of Tor2. Yet, rapamycin inhibited the growth of auxotrophic mutant cells that are dependent on amino acid import for their survival [27]. Like the effect described for *tsc* mutant cells, leucine auxotrophs were particularly sensitive to the growth inhibitory effect of rapamycin. Rapamycin also reduced the efficiency of radioactively labeled leucine uptake [27] and conferred resistance to canavanine or thialysine [13]. Thus, surprisingly, introduction of rapamycin leads to an effect which is similar to activation, rather than inhibition of Tor2 (TORC1).

The slow growth of leucine auxotrophs in the presence of rapamycin was completely rescued by deletion of $fkh1^{+}$, the FKBP12 homolog in fission yeast, indicating that rapamycin inhibits a TOR-dependent function [27]. Further studies suggested that reduction of amino acid uptake is the result of inhibition of Tor1. First, deletion of $tor1^{+}$ resulted in reduced uptake of radioactive-labeled leucine and reduced expression of the $isp5^{+}$, 7G5.06, and c869.10 amino acid permeases [27]. Second, a rapamycin-binding defective allele, $tor1^{S1834E}$, conferred rapamycin resistance to leucine auxotrophs [27]. The sensitivity of auxotroph mutant cells to rapamycin was rescued by replacement of ammonia, which is normally used as the nitrogen source in minimal medium, with proline. Since the presence of proline leads to induction of amino acid uptake [27, 31], the rescue of the rapamycin-sensitive phenotype by using proline as the nitrogen source further supports a defect in amino acid uptake.

We also detected a reduction in the expression of the transcripts of $isp5^{+}$, 7G5.06, and c869.10 when cells were grown in the presence of rapamycin in minimal or proline medium [27]. In contrast to our findings, it was reported that 3 hours after a shift into rapamycin-containing medium, the transcription of *isp5*+, 7G5.06, and c869.10 was upregulated [32]. This discrepancy may be reconciled if there is a dynamic change in expression of amino acid permeases in response to rapamycin, in which there is first an induction in expression, followed by downregulation to levels that are lower compared to untreated cells. A tempting explanation is that rapamycin first inhibits TORC1, leading to upregulation of expression of amino acid permeases. Following a long period of exposure to the drug, the TORC2 complex is also inhibited, resulting in a decrease in expression of amino acid permeases. In mammalian cells, rapamycin inhibits TORC1 upon short exposure to the drug, but upon long exposure also causes disassembly and inhibition of TORC2 [33]. A similar mechanism may operate in fission yeast.

Combining the deletion of *tor1*+ (Δ*tor1*) with the Δ*tsc1* or Δ*tsc2* mutations resulted in a further reduction in amino acid uptake and amino acid permease gene expression, suggesting that Tor1 acts in parallel with Tsc1/2 in regulating amino acid uptake [13]. However, *tor1*+ and *tsc1/2*+ show a complex genetic interaction pattern. Cell size is oppositely regulated by Tor1 and Tsc1/2. Δ*tor1* cells are elongated; in contrast, *tsc1* or *tsc2* mutants are slightly smaller than wild-type cells [13]. The Δ*tor1* mutation reversed the short cell size phenotype of *tsc1/2* mutants [13]. Also, mutation in *tsc* partially suppressed the sensitivity of Δ*tor1* to a variety of stresses [15]. Further studies are required to determine functional interactions between the TSC complex, Rhb1 and Tor1.

In conclusion, rapamycin appears to inhibit amino acid uptake via inhibiting a Tor1-dependent activity. It remains to be determined whether Tor1 exerts this function as part of TORC2. It will also be important to find out whether the apparent opposite roles of Tor1 and Tor2 in regulating amino acid uptake reflects an interdependent regulation between these two complexes, which may be disrupted by rapamycin.

IV. TORC2 is Required for Responses to Starvation, Survival Under Stress Conditions, Chromatin-Mediated Functions, DNA Damage Response and Maintenance of Telomere Length

A. Tor1 and Tor2 Oppositely Regulate Responses to Nitrogen Starvation

The first TOR homolog to be characterized in fission yeast was Tor1. This TOR homolog was subsequently shown to mainly associate with Ste20, Sin1, and Wat1 to form a TORC2-like complex [11, 12]. *tor1*+ is not essential for cell viability; however, disruption of *tor1*$^+$ resulted in pleiotropic defects. Δ*tor1* mutant cells show slightly elongated cell morphology, sensitivity to osmotic, oxidative or temperature stress conditions, inability to enter sexual development or acquire stationary phase physiology, and decrease in amino acid uptake [9, 27, 34, 35]. More recently, we found that Tor1 is also required for cell survival under DNA-damaging conditions, gene silencing and regulation of telomere length [36]. The elongated cell morphology, stress sensitivity, defects in response to starvation and sensitivity to DNA damages are shared between Δ*tor1* and deletion of other components of TORC2, supporting the notion that Tor1 mainly acts as part of a TORC2 complex [12, 34, 36]. These defects are also shared with mutant cells lacking the AGC kinase Gad8 [12, 36]. TORC2 activates Gad8 via its phosphorylation, in line with the suggestion that Gad8 is a major downstream target of TORC2 in fission yeast [34, 37]. Whether the cellular roles of Tor1 in regulating amino acid uptake or gene silencing are also regulated when Tor1 associates with the subunits of TORC2 and/or via Gad8, is yet to be established. One notable cellular function which is not attributed to TORC2 in fission yeast is the regulation of the actin cytoskeleton [12, 15, 34], a function that is attributed to TORC2 in budding yeast and mammalian cells [38–40].

The wide variety of cellular functions in which Tor1 is involved is perplexing. Unlike the "classical" role of TOR as a major regulator of cellular growth, Tor1 (TORC2) is not required under optimal conditions. Moreover, in certain aspects, Tor1 and Tor2 oppositely regulate the same cellular functions (Figure 13.1). Thus, while deletion of *tor2*$^+$ leads to an "always starved for nitrogen" phenotype, deletion of *tor1*$^+$ results in a "never starved for nitrogen" phenotype; cells depleted for Tor2 get shorter and are induced into sexual development on rich medium, while Δ*tor1* cells do not adapt their cell-size in response to nitrogen starvation conditions and are highly sterile. Loss-of-function mutants of *tor1*$^+$ or *tor2*$^+$ also show opposite phenotypes at the molecular level. Disruption of Tor1 leads to reduced expression of nitrogen-starvation induced genes, while disruption of Tor2 results in the induction of these genes. These include genes encoding amino acid permeases and genes required for sexual development [12, 27, 36]. The CDK inhibitor, *rum1*$^+$, which is required for G_1 arrest in response to nitrogen starvation, is also oppositely regulated by Tor2 and Tor1 [14, 34].

Combining the Δ*tor1* and Δ*tor2* or *tor2-ts* mutations resulted in nonviable double mutant cells. The double mutant cells arrested their growth at the G_2 phase and were sterile, similar to single Δ*tor1* mutant cells [13, 15]. Thus, Tor1 may act either downstream of Tor2 or independently. The Δ*tor1* Δ*tor2* double mutant cells arrested their growth at an intermediate cell-size, compared with the respective single mutants, which suggests that Tor2 requires Tor1 for divisions at a small cell size. The finding that Tor1 and Tor2 regulate the same cellular processes argues for a close functional relationship.

B. TORC2 AFFECTS CHROMATIN-MEDIATED FUNCTIONS AND DNA DAMAGE OR STRESS RESPONSES

A genome-wide expression profiling of $\Delta tor1$ cells revealed changes in the expression of many genes. An extensive overlap with genes that were upregulated in chromatin-structure mutants was observed [36]. These included mutations in the histone deacetylases $clr3^+$ ($\Delta clr3$) or $clr6^+$ ($clr6-1$) [41] or in the subunit of the chromatin-remodeling RSC complex, $rsc58^+$ ($\Delta rsc58$) [42]. Genes that are upregulated in $\Delta tor1$ cells included repeated genes and genes that were clustered at subtelomeric regions, suggesting that Tor1 affects gene silencing at heterochromatic regions. Consistent with this suggestion, cells deleted for $tor1^+$ derepressed the expression of a reporter gene ($ade6^+$) that was inserted into the mating-type locus, a region that is characterized by a heterochromatic structure [36].

$clr6^+$ is an essential gene, and the $clr6-1$ hypomorphic allele has been extensively analyzed. Single $clr6-1$ mutants or $clr6-1$ $\Delta clr3$ double mutant cells show a variety of defects that include elongated telomeres and sensitivity to osmotic stress, DNA damaging condition or drugs that destabilize microtubules [41, 43]. These defects are also shared by $\Delta tor1$ mutant cells or deletion of other components of the TORC2, or deletion of $gad8^+$ [36]. Telomere length analysis showed that the telomeres of $\Delta tor1$, $\Delta ste20$, or $\Delta gad8$ mutant cells are elongated by ~150 bp compared to wild type. This elongation is similar to that observed in $clr6-1$ single mutant cells [41]. Mutations in specific components of TORC2 or deletion of $gad8^+$ also rendered the cells sensitive to thiabendazole, a drug that destabilizes microtubules [36]. Taken together, these findings suggest that TORC2 has a profound effect on heterochromatic regions: the mating-type locus, the telomeres, and possibly also the centromeres.

TORC2–Gad8 is also required for cell survival under DNA damaging conditions. Cells deleted for $tor1^+$, $sin1^+$, $ste20^+$, or $gad8^+$ showed great sensitivity to hydroxyurea (HU), which induces replication stress by depleting dNTP pools, or to methyl-methane sulfonate (MMS), a DNA alkylating agent [36]. Recently, we also found that mutations in TORC2 rendered cells highly sensitive to camptothecin (M. Schonbrun and R. Weisman, unpublished data), further supporting a general role for TORC2 in survival under DNA damaging conditions. Cells disrupted for TORC2 showed greater sensitivity to DNA damaging conditions compared with histone deacetylase mutants, indicating that the putative defect in chromatin structure alone cannot explain the strong sensitivity to DNA damaging conditions in TORC2 mutants.

In response to DNA replication stress or DNA damage, cells inhibit cell cycle progression, repair the damage and then reenter the cell cycle. This set of actions is orchestrated by the checkpoint kinase Rad3 (similar to mammalian ATR). Like TOR, Rad3 is a member of the phosphatydilinositol kinase-related kinases. Rad3 is responsible for the activation of Chk1 in response to DNA damage or Cds1 (similar to mammalian Chk2) in response to replication stress. Upon activation by Rad3, Chk1 or Cds1 inhibit entrance into mitosis by regulating the activity of the cyclin-dependent kinase Cdc2 (CDK). This is achieved by determining the status of tyrosine-15 phophorylation on Cdc2 via dual regulation of the phophatase Cdc25 and the kinase Wee1 (reviewed in Ref. [44]). The sensitivity of TORC2 mutant cells to either HU or MMS was comparable to that of Δrad3 mutant cells; however, the mechanism that underlies this sensitivity is distinct. Cells lacking Rad3 failed to activate Chk1 or Cds1; consequently, cells fail to inhibit entrance into mitosis and rapidly lose their viability in the presence of the DNA damage [44]. In contrast, cells lacking Tor1 activated Chk1 in response to MMS and delayed progression into mitosis in the presence of DNA replication stress [36]. In response to HU, Δtor1 cells accumulated with highly phosphorylated (inactive) Cdc2 kinase, consistent with our suggestion that Tor1 is not required for mitotic inhibition in the presence of unreplicated chromosomes. In wild-type cells, Cdc2 is dephosphorylated ~100 min after release from HU arrest [36, 45]. In contrast, upon release of Δtor1 mutant cells from HU, Cdc2 remained in its phosphorylated (inactive) form and cells failed to reenter the cell cycle [36]. Our data thus suggest that unlike Rad3, Tor1 is required for a later stage in response to DNA damage, either for repair (or the stabilization of the stalled replication forks in the presence of HU) or for reentering the cell cycle upon completion of the repair. In this respect, it is interesting to note that cells defective in histone deacetylase activity (clr6-1) also failed to reestablish growth following removal of HU, possibly due to failure to inactivate Cds1 [46]. Whether a similar mechanism occurs in Δtor1 mutant cell is currently being investigated in our laboratory.

Treatment of wild-type cells with rapamycin did not affect survival under osmotic, oxidative, or temperature stress conditions. Also, rapamycin did not affect telomere length, gene silencing, or survival in the presence of HU or MMS. Thus, most of the cellular functions of TORC2–Gad8 are not inhibited by rapamycin [36].

C. Tor1 Regulates Mitotic Entry

Cells deleted for *tor1⁺*, *ste20⁺*, *sin1⁺*, or *gad8⁺* are slightly elongated [9, 12, 34, 35]. Recently, it was also reported that overexpression of *tor1⁺* can lead to a short cell phenotype [47]. Since the timing of entrance into mitosis in fission yeast is the main transition point that operates to coordinate cell growth with cell division under optimal growth conditions, the elongated morphology of TORC2 mutant cells implies a positive role for TORC2 in regulating entrance into mitosis [34, 36]. Results obtained from combining the Δ*tor1* mutation with a set of other cell cycle mutants suggest that Tor1 positively regulates entrance into mitosis by regulating Tyr-15 phosphorylation of Cdc2, possibly by controlling both the activity of Wee1 and Cdc25 [36].

Surprisingly, decrease in the level of Tor1, rather than its complete deletion, led to advancement of entry into mitosis and to a reduced cell-size phenotype [48], rather than the elongated cells observed in Δ*tor1* [36]. Further studies suggested that Tor1 can act as a negative regulator of mitotic entry via inhibition of the stress activated MAPK Sty1 (also known as Spc1), which is required for Plo1 (Polo-like)-dependent activation of Cdc2 [48]. The role of Tor1 as a modulator of Sty1/MAPK signaling has been recently reviewed [49] and will not be detailed here. More recently, it was suggested that Tor1 negatively regulates entrance into mitosis in response to a shift to poor nitrogen source but not in response to complete withdrawal of nitrogen [47]. The negative regulation of mitosis by Tor1 occurs when Tor1 is associated with Mip1, as part of TORC1, but not when it acts as part of TORC2 [47].

Does Tor2 also regulate entrance into mitosis? Depletion of Tor2 mimicked a complete withdrawal of the nitrogen source. Under such conditions, a transient inhibition of entrance into mitosis occurs, followed by two to three divisions at a reduced cell size. Depletion of Tor2 was also found to induce the levels of Rum1, which is a negative regulator of CDK [14]. Thus, Tor2 (as part of the essential TORC1 complex) may negatively regulate entrance into mitosis, in contrast to the positive role of TORC2 in regulating mitosis [34, 36]. A role for TORC1 as a negative regulator of entrance into mitosis in fission yeast is in line with findings in budding yeast, *Drosophila* and mammalian cells that demonstrate that TORC1 inhibits entrance into mitosis [50–52].

V. The Response to Rapamycin in Fission Yeast

A. The Effects of Rapamycin on Sexual Development

One of our very first observations was that rapamycin reduced the efficiency of the sexual development process [1]. Rapamycin reduced the mating between haploid fission yeast cells, one of the early steps in the sexual development process, under either no-nitrogen or low glucose conditions [1]. Thus, rapamycin does not interfere with the sensing of the particular nutrient which is depleted. If the starvation conditions prevail, the diploid zygotes that are formed by the mating of two haploid cells undergo meiosis and sporulation. Introduction of rapamycin to diploid cells under starvation conditions only slightly reduced the efficiency of meiosis and sporulation [1], indicating that rapamycin is more effective at inhibiting early stages of the sexual development pathway. Rapamycin did not affect entrance into stationary phase, and cells that exit the logarithmic phase in the presence of rapamycin arrested properly and maintained viability comparable to untreated cells [1]. In addition, rapamycin did not interfere with resumption of vegetative growth following nitrogen or carbon starvation (S. Pur and R. Weisman, unpublished data).

If rapamycin inhibited a TOR-dependent function, leading to reduced efficiency of sexual development, then deletion of the fission yeast FKBP12 would be expected to result in a rapamycin-resistance phenotype. However, deletion of $fkh1^+$ resulted in partial sterility and did not affect entrance into stationary phase, similar to treatment with rapamycin [53]. Thus, rapamycin most likely directly inhibits the function of Fkh1. The partial sterile phenotype induced by rapamycin was used to isolate $fkh1$ mutant genes that conferred resistance to the inhibitory effect of rapamycin. This screen identified five amino acids that are located in the rapamycin-binding pocket of Fkh1 and are critical for the effect of rapamycin on sexual development [53].

As described above, Tor1 and Tor2 are also strongly implicated in regulating sexual development. Cells disrupted for $tor1^+$ are highly sterile and thus show a far more severe phenotype compared with $\Delta fkh1$ cells or rapamycin treatment. Cells depleted for Tor2 show derepression of sexual development. Unlike our observation which indicated inhibitory effect of rapamycin on sexual development, it was reported that addition of rapamycin to wild-type cells grown in glutamate medium induced sexual development [48], mimicking a shift from "good" to "poor" nitrogen medium. It is thus possible that rapamycin also affects TOR-dependent regulation of sexual development; however, further studies are required to determine the specific conditions that are required for such an effect and the underlying molecular mechanism.

B. MUTATIONS THAT RENDER THE GROWTH OF FISSION YEAST SENSITIVE TO RAPAMYCIN

The FRB domain of TOR is highly conserved. The FRB of Tor1 [9] or Tor2 (S. Finkelstein and R. Weisman, unpublished data) binds FKBP12 in the presence of rapamycin, as determined by two hybrid assays. Yet, wild-type cells are not sensitive to rapamycin during the growth phase or under stress conditions. As described above, the growth of auxotrophic mutant cells, particularly those auxotrophic for leucine, is sensitive to rapamycin. This sensitivity is associated with reduction in amino acid permeases gene expression, and is mediated by inhibition of Tor1 [27].

Cells auxotrophic for leucine (e.g., *leu1-32* strains) were sensitive to rapamycin only when grown in a minimal (ammonia-based) medium supplemented with leucine (50–75 μg/ml). No growth inhibition by rapamycin was observed when auxotrophic mutant cells were grown in a rich (YE) or minimal medium in which the nitrogen source is proline. In contrast, deletion of *tor1*$^+$ rendered cells sensitive to rapamycin in a rich as well as minimal medium [35, 54]. This rapamycin sensitivity was dependent on the presence of Fkh1 (FKBP12) suggesting that Tor2 is inhibited in this genetic background. Reduction in Tor2 activity by *tor2-ts* mutant alleles [11] also rendered cells sensitive to rapamycin, in rich as well as in minimal medium. These findings may suggest that Tor1 and Tor2 share a common function which is sensitive to rapamycin. Alternatively, the absence of Tor1 may cause a reduction in the level of TORC1 and consequently render cells sensitive to rapamycin. Fragmentary data also suggest a reduction in cell-size following treatment with rapamycin. Either wild-type cells grown in glutamate medium [48] or *tsc* mutant cells [13] show reduction in cell size following rapamycin treatment. This rapamycin-induced reduction in cell size is somewhat reminiscent of the reduction in cell size in mammalian cells following rapamycin treatment [55] and may reflect acceleration of entrance into mitosis and/or inhibition of TORC1 complexes.

Two other mutations that reduce TORC1-dependent activities: deletion of *bhd*$^+$ (the homolog of the Birt-Hogg-Dube disease gene) or the hypomorphic allele of *rhb1*$^+$ (*rhb1*$^{G63D/S165N}$), also rendered cells sensitive to rapamycin during vegetative growth. Growth of the double mutant Δbhd *rhb1*$^{G63D/S165N}$ cells was completely inhibited by rapamycin, and expression of genes encoding amino acid permeases was elevated, compared with each of the single mutants. These findings suggest that Bhd and Rhb1 regulate TORC1 by separate mechanisms [32] and that in the Δbhd *rhb1*$^{G63D/S165N}$ the levels of TORC1 activity is low, rendering cell sensitive to rapamycin.

Most surprisingly, we noted that rapamycin completely inhibited growth of prototrophic $\Delta tsc1$ or $\Delta tsc2$ mutant cells, when the nitrogen source in the medium is proline [13]. The sensitivity of $\Delta tsc1$ or $\Delta tsc2$ cells to rapamycin is rescued by deletion of *fkh1*$^+$, indicating that a TOR-dependent function is inhibited. $\Delta tsc1$ or $\Delta tsc2$ mutant cells did not show rapamycin-sensitivity when the nitrogen source in the medium was ammonia or glutamate [13]. The finding that cells lacking a negative inhibitor of TORC1 are sensitive to rapamycin is surprising, since in such genetic background the activity of

TORC1 is expected to be high. Further studies using *tor1-* and *tor2*-rapamycin-binding defective alleles are required to ascertain this possibility.

VI.   Conclusion and Future Prospective

Fission yeast, like budding yeast has two TOR homologs. In budding yeast either Tor1 or Tor2 can form TORC1, whereas TORC2 contains Tor2 only. In fission yeast Tor1 mainly associates with partner proteins to form TORC2, while Tor2 forms TORC1 [11, 12]. However, Tor1 may also be part of TORC1 [47], while Tor2 may also form complexes with TORC2 subunits [12]. It should be stressed that the major cellular functions of Tor1 and Tor2 are not interchangeable.

The cellular function of TORC1 as a central growth regulator is conserved in fission yeast. Yet, unlike budding yeast and certain mammalian cell types, the essential role of TORC1 is not inhibited by rapamycin under normal growth conditions. The TORC2 complex is required under various stress conditions but is not essential for growth under normal conditions. Our recent results, suggesting that TORC2 affects chromatin structure and participates in DNA damage response, are most intriguing. A defect in chromatin structure may explain several defects associated with disruption of TORC2, including derepression in gene silencing, elongated telomeres and possibly also defects in regulating the cell cycle progression and growth under stress conditions. In budding yeast, TORC1 has been suggested to regulate chromatin structure in a rapamycin-sensitive manner *via* the histone deacetylase Rpd3 [56]. TORC1 in budding yeast has also been suggested to positively regulate ribosomal gene expression via a chromatin mediated mechanism [57]. Our studies in fission yeast suggest that TORC2 is required for gene silencing. Thus, yet again, TORC1 and TORC2 appear to have opposite effects. TORC1 has also been implicated in regulating the response to DNA damage in budding yeast and higher eukaryotes (e.g., see Refs. [58, 59]). However, the relevance of these studies to the role of TORC2 under DNA damage conditions in fission yeast is not clear [60, 62].

Finally TORC1 and TORC2 oppositely regulate response to nitrogen starvation, regulation of amino acid uptake and possibly also entrance into mitosis. Interestingly, amino acid uptake and sexual development are two cellular processes that are inhibited by rapamycin. At least with respect to amino acid uptake, rapamycin appears to have an immediate inhibitory effect on TORC1 and a long-term inhibitory effect on TORC2. These findings together with the observation that sensitivity of *tsc* mutant cells are sensitive to rapamycin suggest a close relationship between TORC1 and TORC2 that awaits further characterization.

ACKNOWLEDGMENTS

I thank Martin Kupiec, Miriam Schonbrun, Dana Laor, and Shiri Pur for critical reading of the manuscript and for enthusiastic and fruitful discussions.

REFERENCES

1. Weisman, R., Choder, M., and Koltin, Y. (1997). Rapamycin specifically interferes with the developmental response of fission yeast to starvation. *J Bacteriol* 179:6325–6334.
2. Heitman, J., Movva, N.R., and Hall, M.N. (1991). Targets for cell cycle arrest by the immunosuppressant rapamycin in yeast. *Science* 253:905–909.
3. Terada, N., Franklin, R.A., Lucas, J.J., Blenis, J., and Gelfand, E.W. (1993). Failure of rapamycin to block proliferation once resting cells have entered the cell cycle despite inactivation of p70 S6 kinase. *J Biol Chem* 268:12062–12068.
4. Sabatini, D.M. (2006). mTOR and cancer: insights into a complex relationship. *Nat Rev Cancer* 6:729–734.
5. Sipiczki, M. (2000). Where does fission yeast sit on the tree of life? *Genome Biol* 1: REVIEWS 1011.
6. Moser, B.A., and Russell, P. (2000). Cell cycle regulation in *Schizosaccharomyces pombe*. *Curr Opin Microbiol* 3:631–636.
7. Nurse, P. (1990). Universal control mechanism regulating onset of M-phase. *Nature* 344:503–508.
8. Forsburg, S.L. (1999). The best yeast? *Trends Genet* 15:340–344.
9. Weisman, R., and Choder, M. (2001). The fission yeast TOR homolog, tor1$^+$, is required for the response to starvation and other stresses via a conserved serine. *J Biol Chem* 276:7027–7032.
10. Wood, V., Gwilliam, R., Rajandream, M.A., Lyne, M., Lyne, R., Stewart, A., Sgouros, J., Peat, N., Hayles, J., Baker, S., Basham, D., Bowman, S., *et al.* (2002). The genome sequence of *Schizosaccharomyces pombe*. *Nature* 415:871–880.
11. Hayashi, T., Hatanaka, M., Nagao, K., Nakaseko, Y., Kanoh, J., Kokubu, A., Ebe, M., and Yanagida, M. (2007). Rapamycin sensitivity of the *Schizosaccharomyces pombe* tor2 mutant and organization of two highly phosphorylated TOR complexes by specific and common subunits. *Genes Cells* 12:1357–1370.
12. Matsuo, T., Otsubo, Y., Urano, J., Tamanoi, F., and Yamamoto, M. (2007). Loss of the TOR kinase Tor2 mimics nitrogen starvation and activates the sexual development pathway in fission yeast. *Mol Cell Biol* 8:3154–3164.
13. Weisman, R., Roitburg, I., Schonbrun, M., Harari, R., and Kupiec, M. (2007). Opposite effects of tor1 and tor2 on nitrogen starvation responses in fission yeast. *Genetics* 175:1153–1162.
14. Alvarez, B., and Moreno, S. (2006). Fission yeast Tor2 promotes cell growth and represses cell differentiation. *J Cell Sci* 119:4475–4485.
15. Uritani, M., Hidaka, H., Hotta, Y., Ueno, M., Ushimaru, T., and Toda, T. (2006). Fission yeast Tor2 links nitrogen signals to cell proliferation and acts downstream of the Rheb GTPase. *Genes Cells* 11:1367–1379.
16. Su, S.S., Tanaka, Y., Samejima, I., Tanaka, K., and Yanagida, M. (1996). A nitrogen starvation-induced dormant G0 state in fission yeast: the establishment from uncommitted G1 state and its delay for return to proliferation. *J Cell Sci* 109(Pt 6):1347–1357.

17. Yanagida, M. (2009). Cellular quiescence: are controlling genes conserved? *Trends in Cell Biol* 12:705–715.
18. Aspuria, P.J., Sato, T., and Tamanoi, F. (2007). The TSC/Rheb/TOR signaling pathway in fission yeast and mammalian cells: temperature sensitive and constitutive active mutants of TOR. *Cell Cycle* 6:1692–1695.
19. Matsumoto, S., Bandyopadhyay, A., Kwiatkowski, D.J., Maitra, U., and Matsumoto, T. (2002). Role of the Tsc1-Tsc2 complex in signaling and transport across the cell membrane in the fission yeast *Schizosaccharomyces pombe*. *Genetics* 161·1053–1063.
20. Van Slegtenhorst, M., Carr, E., Stoyanova, R., Kruger, W., and Henske, E.P. (2004). Tsc1$^+$ and tsc2$^+$ regulate arginine uptake and metabolism in *Schizosaccharomyces pombe*. *J Biol Chem* 279:12706–12713.
21. Nakase, Y., Fukuda, K., Chikashige, Y., Tsutsumi, C., Morita, D., Kawamoto, S., Ohnuki, M., Hiraoka, Y., and Matsumoto, T. (2006). A defect in protein farnesylation suppresses a loss of *Schizosaccharomyces pombe* tsc2$^+$, a homolog of the human gene predisposing tuberous sclerosis complex (TSC). *Genetics* 173:569–578.
22. Urano, J., Comiso, M.J., Guo, L., Aspuria, P.J., Deniskin, R., Tabancay, A.P., Jr., Kato-Stankiewicz, J., and Tamanoi, F. (2005). Identification of novel single amino acid changes that result in hyperactivation of the unique GTPase, Rheb, in fission yeast. *Mol Microbiol* 58:1074–1086.
23. van Slegtenhorst, M., Mustafa, A., and Henske, E.P. (2005). Pas1, a G1 cyclin, regulates amino acid uptake and rescues a delay in G1 arrest in Tsc1 and Tsc2 mutants in *Schizosaccharomyces pombe*. *Hum Mol Genet* 14:2851–2858.
24. Mach, K.E., Furge, K.A., and Albright, C.F. (2000). Loss of Rhb1, a Rheb-related GTPase in fission yeast, causes growth arrest with a terminal phenotype similar to that caused by nitrogen starvation. *Genetics* 155:611–622.
25. Yang, W., Tabancay, A.P., Jr., Urano, J., and Tamanoi, F. (2001). Failure to farnesylate Rheb protein contributes to the enrichment of G0/G1 phase cells in the *Schizosaccharomyces pombe* farnesyltransferase mutant. *Mol Microbiol* 41:1339–1347.
26. Urano, J., Sato, T., Matsuo, T., Otsubo, Y., Yamamoto, M., and Tamanoi, F. (2007). Point mutations in TOR confer Rheb-independent growth in fission yeast and nutrient-independent mammalian TOR signaling in mammalian cells. *Proc Natl Acad Sci USA* 104:3514–3519.
27. Weisman, R., Roitburg, I., Nahari, T., and Kupiec, M. (2005). Regulation of leucine uptake by tor1$^+$ in *Schizosaccharomyces pombe* is sensitive to rapamycin. *Genetics* 169:539–550.
28. Aspuria, P.J., and Tamanoi, F. (2008). The Tsc/Rheb signaling pathway controls basic amino acid uptake via the Cat1 permease in fission yeast. *Mol Genet Genomics* 279:441–450.
29. Murai, T., Nakase, Y., Fukuda, K., Chikashige, Y., Tsutsumi, C., Hiraoka, Y., and Matsumoto, T. (2009). Distinctive responses to nitrogen starvation in the dominant active mutants of the fission yeast Rheb GTPase. *Genetics* 183:517–527.
30. Hartman, T.R., Nicolas, E., Klein-Szanto, A., Al-Saleem, T., Cash, T.P., Simon, M.C., and Henske, E.P. (2009). The role of the Birt-Hogg-Dube protein in mTOR activation and renal tumorigenesis. *Oncogene* 28:1594–1604.
31. Karagiannis, J., Saleki, R., and Young, P.G. (1999). The pub1 E3 ubiquitin ligase negatively regulates leucine uptake in response to NH(4)(+) in fission yeast. *Curr Genet* 35:593–601.
32. van Slegtenhorst, M., Khabibullin, D., Hartman, T.R., Nicolas, E., Kruger, W.D., and Henske, E.P. (2007). The Birt-Hogg-Dube and tuberous sclerosis complex homologs have opposing roles in amino acid homeostasis in *Schizosaccharomyces pombe*. *J Biol Chem* 282:24583–24590.

33. Sarbassov dos, D., Ali, S.M., Sengupta, S., Sheen, J.H., Hsu, P.P., Bagley, A.F., Markhard, A.L., and Sabatini, D.M. (2006). Prolonged rapamycin treatment inhibits mTORC2 assembly and Akt/PKB. *Mol Cell* 22:159–168.
34. Ikeda, K., Morigasaki, S., Tatebe, H., Tamanoi, F., and Shiozaki, K. (2008). Fission yeast TOR complex 2 activates the AGC-family Gad8 kinase essential for stress resistance and cell cycle control. *Cell Cycle* 7:358–364.
35. Kawai, M., Nakashima, A., Ueno, M., Ushimaru, T., Aiba, K., Doi, H., and Uritani, M. (2001). Fission yeast tor1 functions in response to various stresses including nitrogen starvation, high osmolarity, and high temperature. *Curr Genet* 39:166–174.
36. Schonbrun, M., Laor, D., Lopez-Maury, L., Bahler, J., Kupiec, M., and Weisman, R. (2009). TOR complex 2 controls gene silencing, telomere length maintenance, and survival under DNA-damaging conditions. *Mol Cell Biol* 29:4584–4594.
37. Matsuo, T., Kubo, Y., Watanabe, Y., and Yamamoto, M. (2003). *Schizosaccharomyces pombe* AGC family kinase Gad8p forms a conserved signaling module with TOR and PDK1-like kinases. *EMBO J* 22:3073–3083.
38. Jacinto, E., Loewith, R., Schmidt, A., Lin, S., Ruegg, M.A., Hall, A., and Hall, M.N. (2004). Mammalian TOR complex 2 controls the actin cytoskeleton and is rapamycin insensitive. *Nat Cell Biol* 6:1122–1128.
39. Sarbassov, D.D., Ali, S.M., Kim, D.H., Guertin, D.A., Latek, R.R., Erdjument-Bromage, H., Tempst, P., and Sabatini, D.M. (2004). Rictor, a novel binding partner of mTOR, defines a rapamycin-insensitive and raptor-independent pathway that regulates the cytoskeleton. *Curr Biol* 14:1296–1302.
40. Schmidt, A., Kunz, J., and Hall, M.N. (1996). TOR2 is required for organization of the actin cytoskeleton in yeast. *Proc Natl Acad Sci USA* 93:13780–13785.
41. Hansen, K.R., Burns, G., Mata, J., Volpe, T.A., Martienssen, R.A., Bahler, J., and Thon, G. (2005). Global effects on gene expression in fission yeast by silencing and RNA interference machineries. *Mol Cell Biol* 25:590–601.
42. Monahan, B.J., Villen, J., Marguerat, S., Bahler, J., Gygi, S.P., and Winston, F. (2008). Fission yeast SWI/SNF and RSC complexes show compositional and functional differences from budding yeast. *Nat Struct Mol Biol* 15:873–880.
43. Nicolas, E., Yamada, T., Cam, H.P., Fitzgerald, P.C., Kobayashi, R., and Grewal, S.I. (2007). Distinct roles of HDAC complexes in promoter silencing, antisense suppression and DNA damage protection. *Nat Struct Mol Biol* 14:372–380.
44. Carr, A.M. (2002). DNA structure dependent checkpoints as regulators of DNA repair. *DNA Repair (Amst)* 1:983–994.
45. Rhind, N., and Russell, P. (1998). Tyrosine phosphorylation of cdc2 is required for the replication checkpoint in *Schizosaccharomyces pombe*. *Mol Cell Biol* 18:3782–3787.
46. Kunoh, T., Habu, T., and Matsumoto, T. (2008). Involvement of fission yeast Clr6-HDAC in regulation of the checkpoint kinase Cds1. *Nucleic Acids Res* 36:3311–3319.
47. Hartmuth, S., and Petersen, J. (2009). Fission yeast Tor1 functions as part of TORC1 to control mitotic entry through the stress MAPK pathway following nutrient stress. *J Cell Sci* 122:1737–1746.
48. Petersen, J., and Nurse, P. (2007). TOR signalling regulates mitotic commitment through the stress MAP kinase pathway and the Polo and Cdc2 kinases. *Nat Cell Biol* 9:1263–1272.
49. Petersen, J. (2009). TOR signalling regulates mitotic commitment through stress-activated MAPK and Polo kinase in response to nutrient stress. *Biochem Soc Trans* 37:273–277.
50. Nakashima, A., Maruki, Y., Imamura, Y., Kondo, C., Kawamata, T., Kawanishi, I., Takata, H., Matsuura, A., Lee, K.S., Kikkawa, U., Ohsumi, Y., Yonezawa, K., *et al.* (2008). The yeast Tor signaling pathway is involved in G2/M transition via polo-kinase. *PLoS ONE* 3:e2223.

51. Smith, E.M., and Proud, C.G. (2008). cdc2-cyclin B regulates eEF2 kinase activity in a cell cycle- and amino acid-dependent manner. *EMBO J* 27:1005–1016.
52. Wu, M.Y., Cully, M., Andersen, D., and Leevers, S.J. (2007). Insulin delays the progression of *Drosophila* cells through G2/M by activating the dTOR/dRaptor complex. *EMBO J* 26:371–379.
53. Weisman, R., Finkelstein, S., and Choder, M. (2001). Rapamycin blocks sexual development in fission yeast through inhibition of the cellular function of an FKBP12 homolog. *J Biol Chem* 276:24736–24742.
54. Weisman, R. (2004). The fission yeast TOR proteins and the rapamycin response: an unexpected tale. *Curr Top Microbiol Immunol* 279:85–95.
55. Fingar, D.C., Salama, S., Tsou, C., Harlow, E., and Blenis, J. (2002). Mammalian cell size is controlled by mTOR and its downstream targets S6K1 and 4EBP1/eIF4E. *Genes Dev* 16:1472–1487.
56. Rohde, J.R., and Cardenas, M.E. (2003). The tor pathway regulates gene expression by linking nutrient sensing to histone acetylation. *Mol Cell Biol* 23:629–635.
57. Tsang, C.K., Bertram, P.G., Ai, W., Drenan, R., and Zheng, X.F. (2003). Chromatin-mediated regulation of nucleolar structure and RNA Pol I localization by TOR. *EMBO J* 22:6045–6056.
58. Reiling, J.H., and Sabatini, D.M. (2006). Stress and mTORture signaling. *Oncogene* 25:6373–6383.
59. Shen, C., Lancaster, C.S., Shi, B., Guo, H., Thimmaiah, P., and Bjornsti, M.A. (2007). TOR signaling is a determinant of cell survival in response to DNA damage. *Mol Cell Biol* 27:7007–7017.
60. Lee, C.H., Inoki, K., Karbowniczek, M., Petroulakis, E., Sonenberg, N., Henske, E.P., and Guan, K.L. (2007). Constitutive mTOR activation in TSC mutants sensitizes cells to energy starvation and genomic damage via p53. *EMBO J* 26:4812–4823.
61. Budanov, A.V., and Karin, M. (2008). P53 target genes sestin1 and sestrin2 connect genotoxic stress and mTOR signaling. *Cell* 134:451–460.
62. Bandhakavi, S., Kim, Y.M., Ro, S.H., Xie, H., Onsongo, G., Jun, C.B., Kim, D.H., and Griffin, T.J. (2010). Quantitative nuclear proteomics identifies mTOR regulation of DNA damage response. *Mol Cell Proteomics* 9:403–414.

14

Structure of TOR Complexes in Fission Yeast

JUNKO KANOH[a] • MITSUHIRO YANAGIDA[b,c]

[a]*Institute for Protein Research*
Osaka University
Osaka, Japan

[b]*Graduate School of Biostudies*
Kyoto University
Kyoto, Japan

[c]*The G0 Cell Unit*
Okinawa Institute of Science and Technology Promotion Corporation
Okinawa, Japan

I. Abstract

Fission yeast, *Schizosaccharomyces pombe*, is an excellent model system for analysis of target of rapamycin (TOR). Fission yeast has Tor1 and Tor2 (homologs of mTOR), Tsc1 and Tsc2 (homologs of mammalian TSC1 and TSC2, respectively), and Rhb1 (a homolog of mammalian Rheb) proteins. In addition, fission yeast has two TORCs, namely TORC1 and TORC2, and the similar TORC components to those in other organisms. Thus, it appears that fission yeast possesses the regulatory mechanisms for TORs that are conserved among eukaryotes. This contrasts to budding yeast, *Saccharomyces cerevisiae*, which does not have *TSC* genes. Furthermore, Tel2 protein associates with TOR kinases, as in mammals. Therefore, fission yeast will provide us a number of useful information about the regulations and functions of TORCs in eukaryotes.

THE ENZYMES, Vol. XXVII
© 2010 Elsevier Inc. All rights reserved.

271

ISSN NO: 1874-6047
DOI: 10.1016/S1874-6047(10)27014-2

II. *S. pombe* TOR Kinases

A. STRUCTURE OF *S. POMBE* TOR KINASES

Fission yeast has two TOR kinases, Tor1 and Tor2. They were identified in the *S. pombe* genome database as TOR homologs [1, 2]. The *tor1*$^+$ and *tor2*$^+$ genes encode 2335- and 2337-amino acid proteins, respectively. Their amino acid sequences share 52% overall identity, and each shares 42–44% identity with human or budding yeast TOR kinases.

Tor1 and Tor2 belong to the PIKK (phosphoinositide 3-kinase-related kinase) family [3]. This family includes Rad3 (an ATR homolog), Tel1 (an ATM homolog), Tra1 and Tra2 (TRRAP: transformation/transcription domain-associated protein homologs) in fission yeast. In addition, mammals have DNA-PKcs (DNA-dependent protein kinase catalytic subunit) and SMG-1 (suppressor of morphogenesis in genitalia-1) as members of the PIKK family. PIKKs contain several structural domains in common; the HEAT (huntingtin, elongation factor 3, a subunit of protein phosphatase 2A and TOR1) repeats, FAT (FRAP, ATM and TRRAP), kinase catalytic and FATC (FRAP, ATM and TRRAP, C-terminus) domains [4, 5]. Additionally, TOR kinases contain the FRB (FKBP12/rapamycin-binding) domain located between the FAT and kinase domains (Figure 14.1).

The HEAT repeats are tandem repeated sequences (40–54 repeats) of approximately 50 amino acids, and span the large N-terminal region of a PIKK protein [5]. Each repeat is composed of a pair of interacting antiparallel helices linked by loops [6]. It is suggested that the HEAT repeat domain mediates protein–protein interactions. In fact, raptor (a subunit of TORC1 in mammals) binds to the HEAT repeat domain of mTOR [7], and ATRIP (a binding partner of ATR in mammals) binds to the N-terminal HEAT repeat region of ATR [8].

Although the kinase domains of PIKKs share significant homology to those of PI (phosphoinositide) 3-kinases, PIKKs phosphorylate proteins, rather than lipids. TRRAP proteins, however, lack the conserved amino acids required for ATP-binding and catalytic activity for phosphate transfer in their kinase domains, so they do not possess kinase activity [9–11]. Among the PIKK proteins, ATR, ATM, DNA-PKcs, and SMG-1 preferentially phosphorylate serine (Ser) or threonine (Thr) residues followed by glutamine (Gln) [12, 13]. In contrast, TOR kinases phosphorylate Ser/Thr residues with no clear consensus sequence motif.

The FAT domain is located C-terminal to the HEAT repeat domain, and spans approximately 500 amino acids [4]. This domain is unique to the PIKK family proteins and is not found in other proteins. The FAT domain is always accompanied by the FATC domain (approximately 35 amino

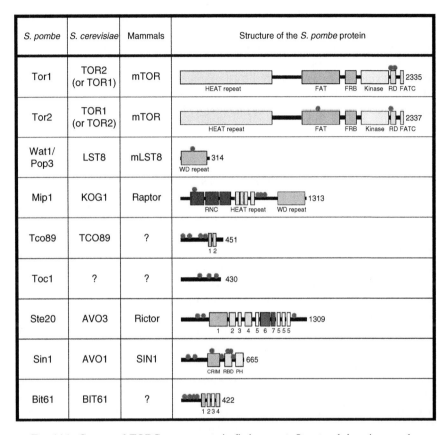

FIG. 14.1. Conserved TORC components in fission yeast. Structural domains are shown. See text for the explanation of each domain. Red circles indicate phosphorylation sites identified by mass spectroscopic analyses [32]. Numbers on the right side of the protein structures indicate the total numbers of amino acids.

acids), which is located at the C-terminal end. Functions of the FAT and FATC domains are largely unknown, but it is speculated that these domains fold together and facilitate the activation of the kinase domain, which is located between the FAT and FATC domains. In fact, mutation in the FATC domain of mTOR abolishes its kinase activity, indicating that at least the FATC domain is critical for the mTOR function [14, 15].

TOR kinases have the FRB domain between the FAT and kinase domains. Rapamycin, associated with the peptidyl prolyl isomerase

FKBP12, binds to the FRB domain and inhibits TOR, although rapamycin does not inhibit all the kinase-dependent functions of TOR [16]. Unexpectedly, mutation in the FRB domain diminishes the activity of Tor1 in fission yeast [1]. Furthermore, the function of Fkh1 (an FKBP12 homolog) in sexual development is inhibited by rapamycin in this organism [17]. The FRB domain binds to phosphatidic acid in mammalian cells, which possibly mediates localization of TOR at membrane [18–20].

B.   FUNCTIONS OF *S. POMBE* TOR KINASES

S. pombe TOR kinases, Tor1 and Tor2, play different roles. Tor1 is not essential for the normal cell growth, because the *tor1*+-deleted strain is viable. Tor1 is, however, required for entry into stationary phase and sexual development upon nutrient starvation, for amino acid uptake, and for survival under various stress conditions, such as high osmolarity, oxidative stress, low or high temperature and DNA replication block [1, 2, 21–23]. Tor1 is also required for the maintenance of telomere DNA length and heterochromatin at centromeres, telomeres, and the mating-type locus [23]. Furthermore, Tor1 promotes mitotic entry by regulating Cdc2 activity [23–26]. Tor1 executes these multiple functions through phosphorylation and activation of the AGC (protein kinase A/protein kinase G/ protein kinase C) family protein kinase Gad8, a putative homolog of mammalian AKT/PKB [23, 26, 27]. Gad8 is the only protein that has been demonstrated to be a direct target of TOR kinases in fission yeast. On the other hand, other proteins, such as S6K (S6 kinase, an AGC family kinase) and 4EBP (eukaryotic initiation factor 4E-binding protein), have been shown to be the downstream targets of TOR kinases in budding yeast and mammals [28, 29]. Therefore, it is possible that the fission yeast TOR kinases also regulate some of those homologs.

Tor2 is essential for cell growth, while Tor1 is not. The *tor2*+-deleted spores germinate, but do not undergo further cell divisions [1]. Functions of Tor2 have been investigated by analyzing the multiple *tor2* mutants. Inactivation of Tor2 results in accumulation of cells with a 1C (G1) DNA content, which mimics nitrogen starvation (the condition inducing sexual development) [22, 30–33]. In fact, the *tor2* ts (temperature sensitive) mutants undergo meiosis and sporulation in the presence of nitrogen at the restrictive temperatures [22, 30, 33]. Thus, Tor1 and Tor2 have opposite effects in sexual development. In addition, a *tor2* ts mutant, *tor2-L2048S*, is hypersensitive to rapamycin [32]. Further analyses are required to clarify the unknown regulations and functions of fission yeast TOR kinases.

III. *S. pombe* TORC1

As in other organisms, the fission yeast TOR kinases function in two distinct complexes, TORC1 and TORC2 (Figure 14.2). Recent works revealed that fission yeast has the similar components of TORC1/2 to those in other organisms (Figure 14.1) [30, 32, 33].

A. TOR KINASE IN TORC1

Tor2 is the major TOR kinase in TORC1 in fission yeast [30, 32, 33]. Minority of Tor1 proteins, however, associates with a TORC1 component (Mip1) especially in minimal media, not in nutrient-rich media, to control mitotic entry upon nutrient stress [25, 33]. It is proposed that the Tor2-containing TORC1 regulates G1/S cell cycle progression and sexual development, whereas the Tor1-containing TORC1 regulates G2/M progression [25]. Budding yeast TOR1 and TOR2 also associate with TORC1

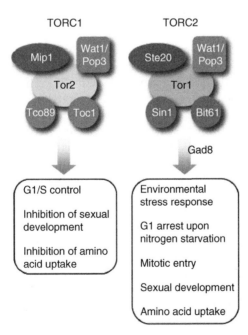

FIG. 14.2. Functions of TORCs in fission yeast. Major TORC components and the functions of each TORC are shown.

components [34], and the TOR1-containing TORC1 and the TOR2-containing TORC1 seem to have distinct roles [35]. It is unknown whether other organisms also have more than one sorts of TORC1.

B. Mip1

Mip1 (Mei2-interacting protein 1) is a homolog of mammalian raptor and budding yeast KOG1, subunits of TORC1. Truncated form of Mip1 was originally identified as a high-copy suppressor of ectopic meiosis induced by the activated Mei2, an RNA-binding protein required for meiosis [36]. The $mip1^+$ gene encodes a protein of 1313 amino acids, which contains the RNC (raptor N-terminal conserved), HEAT repeat and WD40 repeat domains [7, 34, 36]. Mip1 is essential for cell growth, and $mip1^+$ deletion results in accumulation of cells with a G1 DNA content [36]. These phenotypes are similar to those of the $tor2^+$-deleted strain. Mip1 physically associates with Mei2 as well as TOR kinases [36].

C. Wat1/Pop3

Wat1/Pop3 (watermelon-shaped 1/polyploidy 3) is a homolog of mammalian and budding yeast LST8. The $wat1$ mutant was originally identified as a mutant defective in localization of F-actin [37]. The $pop3$ mutant was identified as a sterile mutant defective in the maintenance of genome ploidy [38]. The $wat1/pop3^+$ gene encodes a protein of 314 amino acids, which is composed almost entirely of seven WD40 repeats [37, 38]. Wat1/Pop3 binds to both Tor1 and Tor2 [30, 32, 33]. It is thus believed that Wat1/Pop3 is the common subunit of TORC1 and TORC2. In mammals and budding yeast, LST8 binds to the kinase domain of TOR kinase to stimulate the kinase activity [39, 40]. Unexpectedly, the $wat1/pop3^+$-deleted strain is viable, while deletion of $LST8$ in budding yeast is lethal [37, 41]. The $wat1/pop3$ mutant is defective in sexual development, and is hypersensitive to osmotic stress and high temperature [26, 38]. These phenotypes are similar to those of the mutants of TORC2 (see below). Similarly, analyses of mLST8-knockout cells indicate that mLST8 is required for mTORC2 but not mTORC1 functions in mammals [42]. It appears that neither TORC1 nor TORC2 are directly involved in actin organization, suggesting that Wat1/Pop3 has a TOR-independent function in actin regulation [22, 33]. Further analyses are required to clarify the physiological functions of Wat1/Pop3.

D.   Tco89

Tco89 is a homolog of budding yeast TCO89, a subunit of TORC1. Fission yeast Tco89 was identified by the mass spectroscopic analysis of Tor2-interacting proteins [32]. The *tco89*$^+$ gene encodes a protein of 451 amino acids. Homologs are found only in fungi, and these proteins share two conserved domains [32]. The physiological functions of Tco89 are currently unknown.

E.   Toc1

Toc1 (Tor complex 1) was also identified by the mass spectroscopic analysis of Tor2-interacting proteins [32]. The *toc1*$^+$ gene encodes a protein of 430 amino acids. A clear homolog is found only in *Schizosaccharomyces japonicus* [32]. The physiological functions of Toc1 are currently unknown.

IV.   *S. pombe* TORC2

TORC2 regulates the cell survival upon various environmental stresses, induction of sexual development and activation of amino acid uptake. Deletion strains of each TORC2 component show similar phenotypes. They are viable but sterile and sensitive to various forms of stress, and show a cell cycle delay at G2/M.

A.   TOR KINASE IN TORC2

Tor1 is the major TOR kinase in TORC2 in fission yeast [30, 32, 33]. Minority of Tor2 proteins, however, associates with the TORC2 components [33], although physiological significance of Tor2 in TORC2 is unknown. In the absence of Tor1 or other TORC2 components, the AGC-kinase Gad8 is not phosphorylated and activated [26, 27]. Therefore, it is likely that Gad8 is the downstream target of TORC2 but not TORC1. Recent study, however, suggested the possibility that the Tor1-containing TORC1 (see above) also regulates Gad8 kinase to control G2/M cell cycle progression [25]. Further analysis is required to clarify the regulation of Gad8 kinase.

B.   STE20

Ste20 (sterile 20) is a homolog of mammalian rictor and budding yeast AVO3, subunits of TORC2. The *ste20* mutant was originally identified as an amiloride-resistant and sterile mutant [43]. The *ste20*$^+$ gene encodes a protein of 1309 amino acids. Ste20 and its homologs in other organisms share common domain architectures [44] (Figure 14.1). Deletion of *ste20*+ is viable, and the *ste20* mutant shows the similar phenotypes to those of the *tor1* or *gad8* mutants, such as a delay in cell cycle progression, a defect in G1 arrest upon nitrogen starvation and in sexual development, hypersensitivity to osmotic stress and high temperature [26, 33, 43].

C. Wat1/Pop3

Wat1/Pop3 is a common subunit of TORC1 and TORC2. The features of this protein are described above.

D. Sin1

Sin1 (SAPK-interacting protein 1) is a homolog of mammalian SIN1 and budding yeast AVO1, subunits of TORC2. The *sin1*+ gene encodes a protein of 665 amino acids. Sin1 homologs share three common domains, the CRIM (conserved region in middle of all proteins in the Sin1 family), RBD (Ras-binding domain), and PH (Pleckstrin homology) domains [45]. Sin1 was originally identified as a protein that interacts with Spc1/Sty1, a stress-activated MAPK (SAPK) [46]. It was, however, reported that Sin1 does not play an important role in the Spc1/Sty1 MAPK pathway [26]. Deletion of *sin1*+ is viable, and shows a cell-cycle delay, hypersensitivity to osmotic stress and high temperature, and a defect in sexual development, like the *tor1* or *gad8* mutants [26, 46].

E. Bit61

Bit61 is a homolog of budding yeast BIT61, a component of TORC2. Fission yeast Bit61 was identified by the mass spectroscopic analysis of Tor1-interacting proteins [32]. The *bit61*+ gene encodes a protein of 422 amino acids. Bit61 belongs to the HbrB/Bit61 family, and the family members share four conserved domains [32]. There is another HbrB/Bit61 family protein in fission yeast (SPAC6B12.03c), but it appears not to interact with Tor1 or Tor2 [32]. The physiological functions of Bit61 are currently unknown.

V.   Phosphorylation of TORC Components

Recent study reported the identification of phosphorylation sites in TORC components in the nonstressed condition (Figure 14.1)[32]. It appears that all the TORC components are phosphorylated at serine or threonine residue(s). Thus, TORCs seem to be regulated in multiple ways by phosphorylations and dephosphorylations. Many of the locations (not the exact amino acid residues) of the phosphorylated sites are conserved between fission yeast and humans [32, 47]. Therefore, it is likely that eukaryotic TORCs are regulated by phosphorylation in common ways, which may involve changes in protein–protein interactions or protein folding. For example, fission yeast Tor1, Tor2 and human mTOR are commonly phosphorylated at the region C-terminal to the kinase domain. Because this region is considered as a repressor domain (RD in Figure 14.1) for mTOR [48], phosphorylation(s) may regulate the kinase activity of TORs. Currently, it is unknown which kinases phosphorylate the TORC components.

VI. Other TOR-Associated Proteins

In addition to the TORC components described above, three proteins, Tel2 (a homolog of mammalian and budding yeast Tel2), Tti1 (Tel-two-interacting protein 1), and Cka1/Orb5 (casein kinase II) are associated with both Tor1 and Tor2 [32]. Tel2 is conserved among eukaryotes and belongs to the Tel2/Clk-2/Rad-5 family. Tel2 was originally identified in the genome database as a homolog of Tel2 proteins in other organisms [49]. Tti1 is also conserved among eukaryotes, and it appears that Tel2 and Tti1 form a stable complex *in vivo* [32]. Interestingly, Tel2 associates with all the PIKK family proteins in fission yeast, not only with Tor1 and Tor2 [32] (Figure 14.3). Mammalian Tel2 also associates with all the PIKKs [50]. Thus, Tel2 is a common partner of PIKKs, although it is unknown whether budding yeast TEL2 associates with TOR1 or TOR2. Fission yeast Tel2 is essential for cell growth, and mammalian Tel2 is also essential for embryonic development and cell growth [49, 50]. Analyses of Tel2/Clk-2/Rad-5 family proteins revealed that Tel2 is involved in variety of cellular activities, such as DNA damage/replication checkpoint, regulation of life span and responses to various forms of stress, implying that Tel2 positively regulates PIKK proteins [51]. It is suggested that mammalian Tel2 regulates the stability of PIKK proteins [50], whereas the TEL2–TEL1 (an ATM homolog) interaction in budding yeast appears to be required for the activation and localization of TEL1 to DNA damage sites, not for the stability of TEL1 [52]. It is currently unknown how Tel2 regulates Tor1 and Tor2 in fission yeast, but recent study proposed the possibility that Tor1 cooperatively functions with other PIKKs, such as Rad3 (an ATR homolog), for the genome integrity [23]. Tel2 may mediate the signal transductions between Tor1 and other PIKKs.

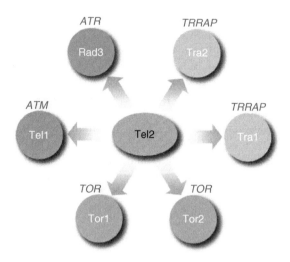

Fig. 14.3. Tel2 associates with all the PIKK family proteins. See text for details.

Cka1/Orb5 is a casein kinase II, essential for cell viability and cell shape [53]. Functional link between Cka1/Orb5 and TORCs is currently unknown. Cka1/Orb5 may phosphorylate some of the TORC components to maintain cell shape or to regulate other essential cellular activities.

VII. Conclusion

As described above, overall structures of fission yeast TORCs are similar to those in budding yeast and mammals. There are, however, some differences among those organisms. First of all, mammals have only one TOR kinase, mTOR, while fission and budding yeasts have two TOR kinases. In fission yeast, Tor2 is the major kinase in TORC1, and Tor1 in TORC2. This is, in a sense, of great advantage to elucidation of distinct functions of a TOR kinase in each TORC.

Mammalian and budding yeast TORC2s regulate actin cytoskeleton, whereas it appears that neither TORC1 nor TORC2 in fission yeast has an important role in actin regulation [22, 33, 54]. On the other hand, fission yeast TORCs regulate sexual development, while budding yeast TORCs appear not to do so. Although some output signals from TORCs are different, the fundamental roles of TORCs, responses to various forms of stress, are highly conserved among eukaryotes.

One of the important aspects of TOR studies is that TOR kinase is regulated by rapamycin. Rapamycin is currently used as immunosuppressant. Therefore, analyses on TORCs are important for medical treatment, not only for the basic studies on cell growth control. Analysis on the multiple rapamycin-sensitive *tor* mutants in fission yeast will clarify the unknown regulations of TORCs by rapamycin.

All together, fission yeast is a very useful model system for TOR study. Because fission yeast is a unicellular organism equipped with relatively simple signal cascades compared with those in mammals, it is easy for us to clarify the fundamental mechanism of each signal cascade regulated by TORC.

REFERENCES

1. Weisman, R., and Choder, M. (2001). The fission yeast TOR homolog, *tor1+*, is required for the response to starvation and other stresses via a conserved serine. *J. Biol. Chem.* 276:7027–7032.
2. Kawai, M., Nakashima, A., Ueno, M., Ushimaru, T., Aiba, K., Doi, H., and Uritani, M. (2001). Fission yeast Tor1 functions in response to various stresses including nitrogen starvation, high osmolarity, and high temperature. *Curr Genet* 39:166–174.
3. Keith, C.T., and Schreiber, S.L. (1995). PIK-related kinases: DNA repair, recombination, and cell cycle checkpoints. *Science* 270:50–51.
4. Bosotti, R., Isacchi, A., and Sonnhammer, E.L.L. (2000). FAT: a novel domain in PIK-related kinases. *Trends Biochem Sci* 25:225–227.
5. Perry, J., and Kleckner, N. (2003). The ATRs, ATMs, and TORs are giant HEAT repeat proteins. *Cell* 112:151–155.
6. Andrade, M.A., Petosa, C., O'Donoghue, S.I., Muller, C.W., and Bork, P. (2001). Comparison of ARM and HEAT protein repeats. *J Mol Biol* 309:1–18.
7. Kim, D.-H., Sarbassov, D.D., Ali, S.M., King, J.E., Latek, R.R., Erdjument-Bromage, H., Tempst, P., and Sabatini, D.M. (2002). mTOR interacts with raptor to form a nutrient-sensitive complex that signals to the cell growth machinery. *Cell* 110:163–175.
8. Ball, H.L., Myers, J.S., and Cortez, D. (2005). ATRIP binding to replication protein A-single-stranded DNA promotes ATR-ATRIP localization but is dispensable for Chk1 phosphorylation. *Mol Biol Cell* 16:2372–2381.
9. Vassilev, A., Yamauchi, J., Kotani, T., Prives, C., Avantaggiati, M.L., Qin, J., and Nakatani, Y. (1998). The 400 kDa subunit of the PCAF histone acetylase complex belongs to the ATM superfamily. *Mol Cell* 2:869–875.
10. Saleh, A., Schieltz, D., Ting, N., McMahon, S.B., Litchfield, D.W., Yates, J.R., III, Lees-Miller, S.P., Cole, M.D., and Brandl, C.J. (1998). Tra1p is a component of the yeast Ada·Spt transcriptional regulatory complexes. *J Biol Chem* 273:26559–26565.
11. McMahon, S.B., Van Buskirk, H.A., Dugan, K.A., Copeland, T.D., and Cole, M.D. (1998). The novel ATM-related protein TRRAP is an essential cofactor for the c-Myc and E2F oncoproteins. *Cell* 94:363–374.
12. Yamashita, A., Ohnishi, T., Kashima, I., Taya, Y., and Ohno, S. (2001). Human SMG-1, a novel phosphatidylinositol 3-kinase-related protein kinase, associates with components of

the mRNA surveillance complex and is involved in the regulation of nonsense-mediated mRNA decay. *Genes Dev* 15:2215–2228.

13. Kim, S.-T., Lim, D.-S., Canman, C.E., and Kastan, M.B. (1999). Substrate specificities and identification of putative substrates of ATM kinase family members. *J Biol Chem* 274:37538–37543.

14. Peterson, R.T., Beal, P.A., Comb, M.J., and Schreiber, S.L. (2000). FKBP12-Rapamycin-associated protein (FRAP) autophosphorylates at serine 2481 under translationally repressive conditions. *J Biol Chem* 275:7416–7423.

15. Takahashi, T., Hara, K., Inoue, H., Kawa, Y., Tokunaga, C., Hidayat, S., Yoshino, K., Kuroda, Y., and Yonezawa, K. (2000). Carboxyl-terminal region conserved among phosphoinositide-kinase-related kinases is indispensable for mTOR function *in vivo* and *in vitro*. *Genes Cells* 5:765–775.

16. Edinger, A.L., Linardic, C.M., Chiang, G.G., Thompson, C.B., and Abraham, R.T. (2003). Differential effects of rapamycin on mammalian target of rapamycin signaling functions in mammalian cells. *Cancer Res* 63:8451–8460.

17. Weisman, R., Finkelstein, S., and Choder, M. (2001). Rapamycin blocks sexual development in fission yeast through inhibition of the cellular function of an FKBP12 homolog. *J Biol Chem* 276:24736–24742.

18. Avila-Flores, A., Santos, T., Rincon, E., and Merida, I. (2005). Modulation of the mammalian target of rapamycin pathway by diacylglycerol kinase-produced phosphatidic acid. *J Biol Chem* 280:10091–10099.

19. Fang, Y., Vilella-Bach, M., Bachmann, R., Flanigan, A., and Chen, J. (2001). Phosphatidic acid-mediated mitogenic activation of mTOR signaling. *Science* 294:1942–1945.

20. Veverka, V., Crabbe, T., Bird, I., Lennie, G., Muskett, F.W., Taylor, R.J., and Carr, M.D. (2008). Structural characterization of the interation of mTOR with phosphatidic acid and a novel class of inhibitor: compelling evidence for a central role of the FRB domain in small molecule-mediated regulation of mTOR. *Oncogene* 27:585–595.

21. Weisman, R., Roitburg, I., Nahari, T., and Kupiec, M. (2005). Regulation of leucine uptake by *tor1+* in *Schizosaccharomyces pombe* is sensitive to rapamycin. *Genetics* 169:539–550.

22. Uritani, M., Hidaka, H., Hotta, Y., Ueno, M., Ushimaru, T., and Toda, T. (2006). Fission yeast Tor2 links nitrogen signals to cell proliferation and acts downstream of the Rheb GTPase. *Genes Cells* 11:1367–1379.

23. Schonbrun, M., Laor, D., Lopez-Maury, L., Bahler, J., Kupiec, M., and Weisman, R. (2009). TOR complex 2 controls gene silencing, telomere length maintenance, and survival under DNA-damaging conditions. *Mol Cell Biol* 29:4584–4594.

24. Petersen, J., and Nurse, P. (2007). TOR signalling regulates mitotic commitment through the stress MAP kinase pathway and the Polo and Cdc2 kinases. *Nat Cell Biol* 9:1263–1272.

25. Hartmuth, S., and Petersen, J. (2009). Fission yeast Tor1 functions as part of TORC1 to control mitotic entry through the stress MAPK pathway following nutrient stress. *J Cell Sci* 122:1737–1746.

26. Ikeda, K., Morigasaki, S., Tatebe, H., Tamaoi, F., and Shiozaki, K. (2008). Fission yeast TOR complex 2 activates the AGC-family Gad8 kinase essential for stress resistance and cell cycle control. *Cell Cycle* 7:358–364.

27. Matsuo, T., Kubo, Y., Watanabe, Y., and Yamamoto, M. (2003). *Schizosaccharomyces pombe* AGC family kinase Gad8p forms a conserved signaling module with TOR and PDK1-like kinases. *EMBO J* 22:3073–3083.

28. Inoki, K., Ouyang, H., Li, Y., and Guan, K.-L. (2005). Signaling by target of rapamycin proteins in cell growth control. *Microbiol Mol Biol Rev* 69:79–100.

29. Jacinto, E., and Lorberg, A. (2008). TOR regulation of AGC kinases in yeast and mammals. *Biochem J* 410:19–37.
30. Alvarez, B., and Moreno, S. (2006). Fission yeast Tor2 promotes cell growth and represses cell differentiation. *J Cell Sci* 119:4475–4485.
31. Weisman, R., Roitburg, I., Schonbrun, M., Harari, R., and Kupiec, M. (2006). Opposite effects of Tor1 and Tor2 on nitrogen starvation responses in fission yeast. *Genetics* 175:1153 1162.
32. Hayashi, T., Hatanaka, M., Nagao, K., Nakaseko, Y., Kanoh, J., Kokubu, A , Ebe, M., and Yanagida, M. (2007). Rapamycin sensitivity of the *Schizosaccharomyces pombe tor2* mutant and organization of two highly phosphorylated TOR complexes by specific and common subunits. *Genes Cells* 12:1357–1370.
33. Matsuo, T., Otsubo, Y., Urano, J., Tamaoi, F., and Yamamoto, M. (2007). Loss of the TOR kinase Tor2 mimics nitrogen starvation and activates the sexual development pathway in fission yeast. *Mol Cell Biol* 27:3154–3164.
34. Loewith, R., Jacinto, E., Wullschleger, S., Lorberg, A., Crespo, J.L., Bonenfant, D., Oppliger, W., Janoe, P., and Hall, M.N. (2002). Two TOR complexes, only one of which is rapamycin sensitive, have distinct roles in cell growth control. *Mol Cell* 10:457–468.
35. Xie, M.W., Jin, F., Hwang, H., Hwang, S., Anand, V., Duncan, M.C., and Huang, J. (2005). Insights into TOR function and rapamycin response: chemical genomic profiling by using a high-density cell array method. *Proc Natl Acad Sci USA* 102:7215–7220.
36. Shinozaki-Yabana, S., Watanabe, Y., and Yamamoto, M. (2000). Novel WD-repeat protein Mip1p facilitates function of the meiotic regulator Mei2p in fission yeast. *Mol Cell Biol* 20:1234–1242.
37. Kemp, J.T., Balasubramanian, M.K., and Gould, K.L. (1997). A *wat1* mutant of fission yeast is defective in cell morphology. *Mol Gen Genet* 254:127–138.
38. Ochotorena, I.L., Hirata, D., Kominami, K.-i., Patashkin, J., Sahin, F., Wentz-Hunter, K., Gould, K.L., Sato, K., Yoshida, Y., Vardy, L., and Toda, T. (2001). Conserved Wat1/Pop3 WD-repeat protein of fission yeast secures genome stability through microtubule integrity and may be involved in mRNA maturation. *J Cell Sci* 114:2911–2920.
39. Kim, D.-H., Sarbassov, D.D., Ali, S.M., Latek, R.R., Guntur, K.V., Erdjument-Bromage, H., Tempst, P., and Sabatini, D.M. (2003). GβL, a positive regulator of the rapamycin-sensitive pathway required for the nutrient-sensitive interaction between raptor and mTOR. *Mol Cell* 11:895–904.
40. Wullschleger, S., Loewith, R., Oppliger, W., and Hall, M.N. (2005). Molecular organization of target of rapamycin complex 2. *J Biol Chem* 280:30697–30704.
41. Roberg, K.J., Bickel, S., Rowley, N., and Kaiser, C.A. (1997). Control of amino acid permiase sorting in the late secretory pathway of *Saccharomyces cerevisiae* by SEC13, LST4, LST7 and LST8. *Genetics* 147:1569–1584.
42. Guertin, D.A., Stevens, D.M., Thoreen, C.C., Burds, A.A., Kalaany, N.Y., Moffat, J., Brown, M., Fitzgerald, K.J., and Sabatini, D.M. (2006). Ablation in mice of the mTORC components raptor, rictor, or mLST8 reveals that mTORC2 is required for signaling to Akt-FOXO andPKCα, but not S6K1. *Dev Cell* 11:859–871.
43. Hilti, N., Baumann, D., Schweingruber, A.-M., Bigler, P., and Schweingruber, M.E. (1999). Gene *ste20* controls amiloride sensitivity and fertility in *Schizosaccharomyces pombe*. *Curr Genet* 35:585–592.
44. Sarbassov, D.D., Ali, S.M., Kim, D.-H., Guertin, D.A., Latek, R.R., Erdjument-Bromage, H., Tempst, P., and Sabatini, D.M. (2004). Rictor, a novel binding partner of mTOR, defines a rapamycin-insensitive and Raptor-independent pathway that regulates the cytoskeleton. *Curr Biol* 14:1296–1302.

45. Schroder, W.A., Buck, M., Cloonan, N., Hancock, J.F., Suhrbier, A., Sculley, T., and Bushell, G. (2007). Human Sin1 contains Ras-binding and pleckstrin homology domains and suppress Ras signalling. *Cell Signal* 19:1279–1289.
46. Wilkinson, M.G., Pino, T.S., Tournier, S., Buck, V., Martin, H., Christiansen, J., Wilkinson, D.G., and Millar, J.B.A. (1999). Sin1: an evolutionarily conserved component of the eukaryotic SAPK pathway. *EMBO J* 18:4210–4221.
47. Olsen, J.V., Blagoev, B., Gnad, F., Macek, B., Kumar, C., Mortensen, P., and Mann, M. (2006). Global, in vivo, and site-specific phosphorylation dynamics in signaling networks. *Cell* 127:635–648.
48. Sekulic, A., Hudson, C.C., Homme, J.L., Yin, P., Otterness, D.M., Karnitz, L.M., and Abraham, R.T. (2000). A direct linkage between the phosphoinositide 3-kinase-AKT signaling pathway and the mammalian target of rapamycin in mitogen-stimulated and transformed cells. *Cancer Res* 60:3504–3513.
49. Shikata, M., Ishikawa, F., and Kanoh, J. (2007). Tel2 is required for activation of the Mrc1-mediated replication checkpoint. *J Biol Chem* 282:5346–5355.
50. Takai, H., Wang, R.C., Takai, K.K., Yang, H., and de Lange, T. (2007). Tel2 regulates the stability of PI3K-related protein kinases. *Cell* 131:1248–1259.
51. Kanoh, J., and Yanagida, M. (2007). Tel2: a common partner of PIK-related kinases and a link between DNA checkpoint and nutritional response? *Genes Cells* 12:1301–1304.
52. Anderson, C.M., Korkin, D., Smith, D.L., Makovets, S., Seidel, J.J., Sali, A., and Blackburn, E.H. (2008). Tel2 mediates activation and licalization of ATM/Tel1 kinase to a double-strand break. *Genes Dev* 22:854–859.
53. Snell, V., and Nurse, P. (1994). Genetic analysis of cell morphogenesis in fission yeast—a role for casein kinase II in the establishment of polarized growth. *EMBO J* 13:2066–2074.
54. Wullschleger, S., Loewith, R., and Hall, M.N. (2006). TOR signaling in growth and metabolism. *Cell* 124:471–484.

15

The TOR Complex and Signaling Pathway in Plants

MANON MOREAU[a] • RODNAY SORMANI[a] • BENOIT MENAND[b]
BRUCE VEIT[c] • CHRISTOPHE ROBAGLIA[b] •
CHRISTIAN MEYER[a]

[a]Institut Jean-Pierre Bourgin, UMR 1318, INRA
Versailles, France

[b]Laboratoire de Génétique et Biophysique des Plantes CNRS-CEA-Université de la
Mediterranée Faculté des Sciences de Luminy
Marseille, France

[c]AgResearch Grasslands
Tennent Drive, Palmerston North, New Zealand

I. Abstract

The mechanisms of plant growth and development are specific and have been shaped by their evolution as immobile, autotrophic and multicellular organisms. Indeed, plant growth is highly plastic, relatively undetermined, and strongly influenced by external conditions. Signaling pathways must therefore constantly link environmental inputs to growth and development. Among the known eukaryotic pathways, target of rapamycin (TOR) signaling, a central component of the perception and transduction of exogenous signals, has been shown to link cell and organism growth to the environment. But little is so far known on this signaling pathway in photosynthetic organisms. In this review, we summarize recent studies that address the nature of TOR complexes and signaling in plants, highlighting similarities and differences with respect to other eukaryotic kingdoms.

THE ENZYMES, Vol. XXVII
285
ISSN NO:1874-6047
DOI: 10.1016/S1874-6047(10)27015-4

II. Introduction

Many facets of plant biology are governed by the fact that they are unable to move, and hence have to face adverse conditions, but also that they are autotrophic organisms which rely on complex photosynthetic processes to synthetize organic carbon compounds. Although all eukaryotes rely on several core ancestral mechanisms to support growth, significant and specific variations to these mechanisms can be observed within individual groups like plants, which moreover have evolved as multicellular organisms independently of animals. Multicellular organisms usually start as single cells which grow, divide, and develop into different organs. Unlike animals where the development plan is determined (i.e., growth is limited and the organization plan is conserved between individuals) and generally occurs in the embryo, plants start as simple organisms and develop new organs, which connect to older parts, throughout their life to elaborate complex and typically adaptive architectures. Other differences in growth control could be expected to arise from unique sets of selection pressures. For plants, some of these would reflect adaptations to a sessile lifestyle, thus providing the means to adapt to the prevailing, but sometimes changeable environment. Indeed, the plasticity of their development patterns allow plants to respond to environmental signals such as variations in nutrition, light, temperature, water, etc. (for a review, see Refs. [1-3]). For instance, nitrogen starvation stimulates the production of new roots to forage N-rich soil patches while phosphate deficiencies result in the appearance of new specialized roots able to better take up this nutrient [4].

In the context of this pattern of indeterminate growth, many aspects of plant development are exemplified in the life cycle of a single organ. A leaf for example, is first born as a group of uncommitted cells derived from the shoot apical meristem which are recruited on the flanks of the shoot apex. These cells grow, differentiate and eventually, as the continuous production of new organs generally implies a concomitant loss of older ones, senesce and die. During plant development, cell growth occurs both during cell proliferation, involving cytosolic growth which depends mostly on the synthesis of ribosomes and new proteins, and during cell expansion, where a large size is reached mainly through vacuole and cell wall expansion [5, 6]. This "low-cost" growth occurs for instance in the hypocotyl of etiolated seedlings which need to reach the light above ground. To support continuous growth, plants maintain niches of totipotent, quiescent stem cells in structures called meristems which continuously produce new cells and new organs for the plant adaptative development [1]. In both leaf and root meristematic regions, growth of cells occur mainly by the accumulation of cytosolic material and is coupled to cell division with a high rate of protein synthesis.

The process of cell elongation happens only later either in the upper part of the elongating root or as the newly formed leaf elongates. So far little is known on the genetic and molecular determinants of plant growth despite its importance for crop yield and biomass production [3].

Plant metabolism and nutritional signaling is also clearly different from that of animals. Indeed, after a short period of heterotrophic growth supported by the use of starch or lipids and storage proteins deposited in seed endosperm or embryonic cotyledons, parts of the young seedlings switch to (photo)autotrophic growth. This profound metabolic change depends on a developmental checkpoint that senses the quality of the environment and is dependent on light signaling [7]. As in all other eukaryotic organisms, glucose plays a role as a central regulatory molecule in plants but, unlike heterotrophic cells, accumulation of glucose in actively photosynthesizing plant cells represses carbon accumulation, switching down many metabolic pathways [8]. There is also a great variety in plant lifespan: annual or biennial live few months only, these plants usually flower and get seeds only once in their plant life but perennials can live and produce seeds many years. The maximal life span for perennial plants largely exceed that of animals with records of 4600 years for the bristlecone pine tree (*Pinus longaeva*) and 3200 years for the giant sequoia (*Sequoia gigantea*). Age over 10,000 years have been recorded for some clonal plants like Huckleberry (*Gaylussacia brachycerium*) [9]. This suggests a specific regulation of plant lifespan.

The foregoing discussion emphasizes how plant life depends on a close coupling of many environmental inputs, and how these factors must somehow be coordinated with complex adaptive growth responses. The TOR kinase appears as a good candidate to perform these connections. In the following sections, we consider evidence for how the TOR signaling pathway may have been adapted to fit specific aspects and requirements of plant life and metabolism.

III.   Plant Homologs of the TOR Complex Proteins

A.   IS PLANT TOR A TARGET OF RAPAMYCIN?

The growth of most land plants seems unaffected by the presence of rapamycin, even at high concentrations [10]. This includes the model plant *Arabidopsis* (*Arabidopsis thaliana*), but also wheat, rice and tobacco. Conversely, there have been conflicting reports for rapamycin sensitivity of maize embryonic axes through inhibition of TOR ([11] and references therein). Nevertheless it seems that, unlike most fungi and animals, multicellular plants are resistant to the inhibitory effects of rapamycin. In contrast, the unicellular green alga *Chlamydomonas reinhardtii* is highly sensitive to rapamycin [10, 12]. The lack of rapamycin effects in land plants could be explained by a reduced rapamycin uptake by plant cells, but it appeared that another explanation, confirmed later by other groups, was the inability of plant FKBP12 homologs to bind rapamycin efficiently [13]. This was supported by a yeast three-hybrid analysis showing that *Arabidopsis* FKBP12, as well as other closely related FKBPs, were unable to bind the TOR FRB domain in the presence of rapamycin [10, 14, 15]. This property was ascribed to amino acid changes in the *Arabidopsis* and plant FKBP12 sequences which may have occurred during their evolution. These mutations affect residues known to interact with rapamycin [12, 15]. Conversely, the TOR FRB domain was shown to be able to interact with either yeast [10] or human [14] FKBP12 proteins but solely in the presence of rapamycin. Furthermore the ectopic expression of yeast FKBP12 protein in *Arabidopsis* could restore some rapamycin sensitivity in the transformed plants [15]. Therefore, although most land plants seem resistant to rapamycin, the plant TOR is still a *bona fide* target of rapamycin. This suggests that the conservation of the ability of the TOR FRB domain to bind the rapamycin/FKBP12 complex is linked to a functional role of this domain which was retained during evolution, like the binding of a small molecule or of a protein that could regulate TOR activity. For instance, phosphatidic acid, a small molecule produced by phospholipase D activity and present in both plants and animal cells, was shown to bind the TOR FRB domain [16]. Interestingly, *Arabidopsis* plants overexpressing the phospholipase Dalpha3 have higher levels of AtTor transcripts and of phosphorylated S6 kinase protein than wild-type (WT) plants under hyperosmotic stress [17].

B.   IDENTIFICATION OF PLANT HOMOLOGS OF PROTEINS BELONGING
      TO THE TOR COMPLEXES

1.   TOR

Genes coding for putative orthologs of the TOR protein have now been identified in many plants whose genome has been completely sequenced, including *Arabidopsis*, rice, maize, poplar, mosses and several red and green algae. As in most other eukaryotes, it seems that TOR is encoded by a single copy gene in these plants. This singularity is observed even in plants such as *Arabidopsis*, whose genomes were duplicated relatively recently during evolution, suggesting a possible selection process that specifically eliminates duplicated Tor genes.

The overall *Arabidopsis* TOR protein sequence (AtTOR, At1g50030) is relatively well conserved with the TOR protein sequence from yeast and animals (around 40% of sequence identity) with some domains being better conserved. For instance, the kinase and FATC domains show, respectively, 75% and 74% of identity with the same domains of the human TOR kinase whereas the FRB domain is less conserved (58% of identity). Domains containing the HEAT repeats and the FAT motif are also found in the plant TOR sequences, suggesting that protein interacting with them, like Tel2 [18], might also be conserved. The AtTor gene appeared to be expressed in all tissues with a maximum in developing seeds and siliques (the *Arabidopsis* fruit; Figure 15.1). In contrast, GUS activity derived from a translational TOR–GUS fusion protein was mainly found in young, growing tissues like root tips or emerging leaves (Figure 15.2). The TOR-linked GUS activity was found to be enhanced by the plant growth-promoting hormone auxin in roots [6].

2. RAPTOR

RAPTOR was first identified as a protein partner (Mip1p) of Mei2p in fission yeast [19]. There are two possible RAPTOR proteins in *Arabidopsis* which bear strong similarities to the yeast KOG1 protein as well as with the human RAPTOR protein. All the domains found in KOG1/RAPTOR were also present in the plant sequences. The genes encoding these two proteins were named AtRaptor 1 and AtRaptor 2 by Deprost *et al.* [20] and AtRaptor 1b and 1a by Anderson *et al.* [21] which correspond, respectively, to the At3g08850 and At5g01770 *Arabidopsis* genes. In this review the former gene will be called AtRaptor3g and the latter AtRaptor5g. Interestingly, two genes encoding proteins with similar sequences and sizes were found in the rice genome (genes 9640.t00092 and 9639.t00087 on chromosome 12 and 11, respectively, in the TIGR rice genome project,

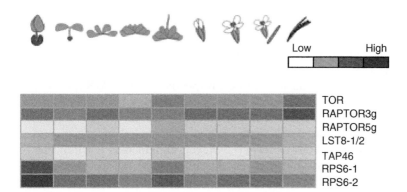

FIG. 15.1. Expression pattern of *Arabidopsis* genes involved in the formation of the TORC1 complex or in the TOR signaling pathway in various plant organs or developmental stages. Microarray expression data are from Genevestigator (https://www.genevestigator.com [51]). RPS6: ribosomal protein S6.

Fig. 15.2. Expression pattern of the translational AtTOR–GUS fusion protein in *Arabidopsis*. GUS activity is visible in young emerging leaves, in leaf conductive tissues and in root tips ([10]; Sormani *et al.*, submitted).

http://www.tigr.org). Conversely, only a single RAPTOR-like protein, as in other eukaryotes, was identified in the genome of the unicellular red alga *Cyanidioschyzon merolae* (LocusCMH109C, http://merolae.biol.s.u-tokyo. ac.jp/) or in green algae. Alignment of AtRAPTOR3g and 5g with RAP-TOR protein sequences from other organisms shows a higher degree of sequence identity among higher plant sequences as well as with the yeast KOG1 (43% and 41% identities, respectively) and the human RAPTOR proteins (41% and 38% identities, respectively). The AtRaptor3g gene is ubiquitously expressed and at a much more higher level than AtRaptor5g (Figure 15.1). The same seems true for the second Raptor gene in other plants and may correspond to the evolution in land plants of a second, more specialized RAPTOR isoform (C. Meyer, unpublished results). The RAP-TOR protein was shown to be phosphorylated by an AMP-dependent kinase (AMPK) activity in animal cells on two conserved Ser residues [22]. One of them, Ser792, is conserved in the plant RAPTOR protein sequences. Since plants contain homologs of AMPK (called SnrK, Snf1-related kinases [23]) which are known to mediate a large part of stress responses in *Arabidopsis* [24], it is tempting to hypothesize that the action of plant SnrKs could be mediated by the inactivation of the TORC1 complex upon phosphorylation of RAPTOR. It is noteworthy that a double SnrK1 mutant in the moss *Physcomitrella patens* showed premature senescence and was unable to grow when a normal day–night cycle was applied, although it grew normally in continuous light [25].

3. LST8

As for the Raptor genes, there are two genes in *Arabidopsis* potentially coding for homologs of LST8/GbetaL proteins (At3g18140 and At2g22040 for respectively AtLST8-1 and AtLST8-2 genes) one of which (AtLST8-1) is also expressed ubiquitously and at a much higher level than the other one (Figure 15.1; Moreau *et al.*, submitted). In *Chlamydomonas* the LST8 protein (CrLST8), as well as CrTOR, were found in high-molecular mass complexes that are associated with internal membranes [26]. The same report showed that CrLST8 protein was also localized around the nucleus and near peri-basal bodies, close to the flagella. The fact that the introduction of the CrLST8 coding sequence in yeast complemented *lst8* mutations strongly suggests that LST8 from photosynthetic organisms is a functional homolog of the yeast protein. The AtLST8-1 protein can also fully restore growth in *lst8* mutants (Moreau *et al.*, submitted). Moreover, mutations of the CrLST8 proteins that led to a complete loss of the ability of these proteins to complement yeast mutants also abolished the interaction with the yeast TOR2 protein [26].

4. The Plant TORC2 Complex

There are no clear homologs in the plant genomes of the TOR complex 2 (TORC2) specific components like AVO1/hSIN1 or AVO3/RICTOR but these proteins present a low degree of similarities which may have hindered their identification in plants and algae. Nevertheless, as stated above, the CrLST8 is probably able to associate with both TOR complexes in yeast as it fully complements the *lst8* mutant phenotype [26].

5. The Plant TORC1 Complex

Conversely, there is strong evidence for the existence of a TORC1 complex in both land plants and algae. Indeed, Mahfouz *et al.* [14] used transient overexpression of tagged protein in tobacco leaves to demonstrate an interaction between the AtRAPTOR3g protein and the AtTOR HEAT repeats. In *Chlamydomonas*, the CrLST8 protein was shown to copurify with the CrTOR protein and to interact, as in other eukaryotes, with its kinase domain [26]. Taken together, these data all support the assumption that a conserved TORC1 protein complex is formed in plants and algae with the TOR, RAPTOR and LST8 partners.

IV.   Components of the TOR Signaling Pathway in Plants

A.   S6K/TAP42

The *Arabidopsis* genome contains two homologs of S6 kinase (AtS6K1 and AtS6K2, At3g08720 and At3g08730, respectively). Even if their N- and C-terminal protein sequences are quite divergent from that of animal S6K, the kinase domain itself is highly conserved between these organisms. This results in the absence of the putative TOR signaling (TOS) motif in the N-terminus of plant S6K and of the autoinhibitory loop at the C-terminus. Nevertheless, the major and regulatory phosphorylation sites identified in human S6K (Thr229, Ser371 Thr389 and Thr226, Ser370, Thr388, for, respectively, hS6K1 and hS6K2) are conserved in the *Arabidopsis* sequence (Ser290, Ser431, Thr449 and Ser296, Ser437, Thr455 for, respectively, AtS6K1 and AtS6K2). The activity of AtS6K has been shown to be inhibited by dephosphorylation [27, 28]. The phosphorylation of Thr388/389 is widely used in animal and human studies as a marker for the TOR kinase activity. Since this site, and the motif surrounding it, is conserved in plants, a similar test has been used in plants and antibodies specific for phosphorylated human Thr389 to recognize a band in *Arabidopsis* whose intensity seems to be modulated by osmotic stress and probably correlated with AtTOR activity ([17]; Sormani *et al.*, submitted). This would represent an invaluable tool for measuring the *in vivo* TOR kinase activity in plants and algae. In *Arabidopsis* the S6K activity was also found to be activated by auxin, which, as stated above, seems to augment the amount of TOR protein [29]. Very recently, it was shown that the overexpression of Lily S6K1 resulted in male sterility in transformed *Arabidopsis* and the authors suggested that this was the consequence of an enhanced translation of regulatory genes containing putative 5'TOP sequences in the 5' UTR of the corresponding mRNAs [30]. Furthermore it has been reported that, like in animal cells, RAPTOR interacts *in vivo* with S6 kinase and regulates its activity towards the ribosomal S6 protein [14, 31]. All these data suggests that S6K are targets of TOR in *Arabidopsis* and probably in other plants.

In yeast cells, the activity of the TOR kinase seems to be mainly channeled through two major substrates which are the Sch9 kinase (a functional homolog of S6K) and the PP2A phosphatase interacting complex TAP42/TIP41 [32]. *Arabidopsis* contains homologs of both TAP42 (named AtTAP46 [33]) and TIP41. AtTAP46 was isolated by a two-hybrid screen using the catalytic subunit of PP2A as a bait and its mRNA was shown to be strongly induced by a cold stress [33].

B. MEI2-LIKE PROTEINS

Studies in *Schizosaccharomyces pombe* point towards an unusual class of RNA binding protein, termed mei2-like, that may represent a novel class of TOR substrate, which is found in plants and certain fungi, but not animals. This model can be traced to the first functional characterization of RAPTOR. In this pioneering work, the protein that would later become known as RAPTOR was instead termed Mei2 Interacting Protein (Mip 1), based on biochemical and genetic evidence [19]. Previous studies had shown that Mei2 promotes meiotic development, in part through a complex program that involves its protein binding to specific nascent noncoding transcripts in the nucleus. Mutational analysis of Mip1 confirmed its involvement in meiotic development, but also showed an essential role in mitotic growth similar to that seen for RAPTOR in other systems. In relation to TOR, more recent studies have documented its physical interaction with mei2p and have suggested a model in which TOR mediated phosphorylation of mei2p inhibits its meiosis promoting activity [34].

In contrast to their single occurrence in *S. pombe* and their total absence in animals, mei2-like genes are well represented in plants, with two major clades of genes found in all plant genomes that have been fully sequenced [35]. One of these contains genes with high level, both structural and functional, affinities with mei2. One *Arabidopsis* member of this clade, AML1 (*Arabidopsis* mei2-like) was first isolated in a heterologous functional analysis by its ability to bypass a *S. pombe* mating type mutant that was blocked for meiosis. More recent work has shown members of this clade are most highly expressed in reproductive tissues and play essential roles in promoting both meiotic and gametophytic development [36]. A second, smaller clade, sometimes termed TERMINAL EAR-like (TEL) after the founding member gene from maize, differs from the first clade in terms of its distinct domain organization, as well as its more vegetative expression, especially in meristematic tissues that maintain the capacity for further cell division and growth [37]. Loss-of-function mutants for these genes in maize and rice show accelerated organ and tissue maturation, suggesting that the activity of these genes may help maintain cells in a proliferative state [37].

Whether these plant mei2-likes represent targets of TOR phosphorylation remains unclear, though parallels with interactions seen in *S. pombe* are suggestive. Interactions between AML1 and AtRAPTOR3g have been shown by yeast two-hybrid, though evidence for in-planta interactions is still lacking [38]. Similarly, the meiotic promoting activities of AML class genes could be likened to those of mei2 of *S. pombe*, in which TOR activity has been proposed to inhibit mei2 activity, thereby delaying a terminal meiotic differentiation pathway. For TEL genes, however, where activity is associated with maintaining growth potential, TOR activity would be more simply modeled as promoting TEL activity. Further, biochemical and genetic analysis of mei2-likes will be required to resolve to what degree they are subject to TOR regulation, and how their distinct activities might mediate TOR related growth regulation.

C.   LKB1/PDK1/AMPK Homologs

There is various evidence that possible TOR substrates and regulators such as AGC kinases, phosphoinositide-3 (PI3) kinases and the 3-phosphoi-nositide-dependent protein kinase 1 (PDK1) are conserved in plants [31, 39, 40]. It was shown that *Arabidopsis* PDK1 phosphorylates AtS6K1 and 2 by, respectively, Otterhag *et al.* [31] and Mahfouz *et al.* [14]. More recently, two *Arabidopsis* kinases (At3g45240 and At5g04510 for, respectively, SnAK1/ GRIK1 and SnAK2/GRIK2) were isolated which complement the yeast *elm1 sak1 tos3* triple mutant affected in proteins regulating the Snf1 kinase [41] or by a two-hybrid screen using the SnrK1 protein, a plant conserved homolog of AMPK/Snf1, as a bait [42]. These kinases are phosphorylating a peptide derived from *Arabidopsis* SnrK1 [41], and were shown to regulate SnrK1 activity by phosphorylating its activation loop [42]. As noted above the plant SnrK1 kinase may regulate directly the plant TORC1 kinase activity through phosphorylation of RAPTOR [22]. The SnAK/GRIK protein kinases could thus be functional homologs of the animal LKB1 kinase, that is known to also regulate TOR activity in animal cells through regulation of AMPK [43, 44]. Conversely, *Arabidopsis* seems devoid of any obvious Rheb or TSC1/TSC2 homologs.

V.   Genetic Analysis of the Plant TOR Signaling Pathway: A Green Growth facTOR?

In addition to previous lines of analyses supporting the existence of a TOR signaling pathway in plants, a number of more recent genetically focused studies have provided additional insight into the functional activities of specific components.

A. RIBOSOMAL S6 PROTEIN AND S6K

Downregulation of the *Arabidopsis* S6 proteins (AtRPS6-1 and AtRPS6-2 coded by, respectively, At4g31700 and At5g10360) by the production of an antisense RNA caused various alterations in leaf and flower positions as well as a bushy phenotype [45]. So far no null AtS6K mutants have hitherto been generated in *Arabidopsis*. Indeed, the close localization of the two AtS6K genes in tandem has precluded the identification of double insertion mutants.

B. RAPTOR

Two independent studies have assessed the functional significance of RAPTOR in plants through analyses of loss-of-function mutants in *Arabidopsis* [20, 21]. No obvious loss-of-function phenotypes were observed for AtRaptor5g, the more weakly expressed of the two AtRaptor genes. By contrast, clear growth related phenotypes could be associated with disruption of the more strongly expressed AtRaptor3g gene. Early embryonic arrest was ascribed to homozygous mutant individuals [20], but under certain growth conditions, some mutant individuals survived to complete embryogenesis [21]. During subsequent vegetative growth, these mutants grew more slowly, had an altered root morphology, formed more branches and flowered later than the WT or heterozygous plants. Mutants in which both AtRaptor genes were disrupted had more limited growth potential, with no double homozygous mutants recovered beyond early seedling stages [21].

Taken together with previously described biochemical evidence, the similarities in the phenotypes associated with plant mutants of AtRaptor and AtTor, are consistent with both animal and fungal based models in which RAPTOR acts as a scaffold to target TOR kinase activity to specific substrates. Some of these, such as S6K or TAP42/TIP41, seem to have homologs in all eukaryotes.

C. TOR

The resistance of land plants to rapamycin has clearly posed a handicap and delayed the onset of genetic, biochemical, and molecular studies on the TOR gene and signaling pathway. Moreover, the disruption of the AtTor gene by T-DNA insertions proved also to be embryo lethal. Homozygous embryos from AtTor heterozygous mutants were stopped at the globular stage after which cell division in plant embryos starts to be accompanied by cell growth [10]. Indeed, postfertilization mitosis occurs without much concomitant cell growth until the embryo reaches the globular stage. It thus seems that the *Arabidopsis* TOR activity is needed for cell growth rather than for cell division, at least during postfertilization mitosis. Interestingly, the development of seed endosperms, a triploid tissue formed by the Angiosperm double fertilization process, was arrested at an earlier stage in homozygous AtTor mutant, with only around 50 nuclei and no cellularization of the endosperm syncitium. The fact that all T-DNA insertion mutants in the AtTor coding sequence were embryo-lethal [46] prompted us to search for hypomorphic mutations by characterizing insertions upstream or downstream of this sequence, which may have had milder effects, or by silencing the AtTor expression using RNAi. Strong promoters are present in the T-DNA sequences used for the disruption of *Arabidopsis* genes which can result in an enhanced transcription of the target genes. Indeed, *Arabidopsis* lines overexpressing the AtTor gene were isolated with T-DNA insertions in the 5'UTR sequence of the gene. These insertion mutants were viable at the homozygous state. In parallel, plants harboring a construct aiming at silencing the AtTor gene were obtained and a few independent *Arabidopsis* lines with a reduced AtTor expression were obtained [46]. In both cases, the level of AtTor expression was very well correlated with the size of the plants and also with the amount of seed produced (Figure 15.3). Conversely, cell elongation did not seem to be affected since there were no differences in the length of etiolated hypocotyls [46]. This represents one of the few studies which examine the impact of variations in TOR expression on the overall size of a multicellular organism. Furthermore, it was found by the same authors that AtTor overexpressing plants were more resistant to osmotic and salt stress. The expression of the *Arabidopsis* homolog of human EBP1 (ErbB-3 epidermal growth factor receptor binding protein), which regulates translation and ribosome assembly, was correlated to variations in the level of AtTor [46]. Interestingly, it was also reported that elevated or decreased levels of AtEBP1 resulted in corresponding changes in organ growth [47]. This suggests that AtEBP1 could act downstream of AtTOR to control the translation machinery and ultimately organ growth.

    To circumvent the difficulty of completely silencing the expression of an essential gene like AtTor, ethanol-inducible RNAi lines were obtained that allow a conditional silencing of the AtTor gene [46]. When the expression of AtTor was abolished by ethanol induction, plant's growth was arrested and senescence-linked markers (genes and metabolites) became upregulated. Similarly, *Chlamydomonas* cells treated with rapamycin stopped growth and the size of the vacuole increased [12]. This is reminiscent of the effects observed upon induction of autophagic processes in yeast or

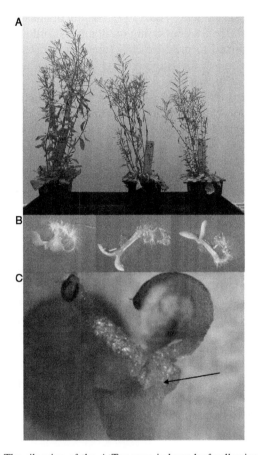

FIG. 15.3. (A) The silencing of the AtTor gene induces leaf yellowing and symptoms of early senescence in *Arabidopsis* plants grown in the greenhouse and (B) stops seedling development at an early stage (plants grown *in vitro* on an ethanol-containing solid medium). Two independent ethanol-silenced lines are shown on the right compared to a control, ethanol-treated, *Ler Arabidopsis* line on the left (see Ref. [46] for details). (C) Phenotype of a seedling from an AtTor RNAi line after 1 week of silencing induction on ethanol. The hypocotyl cells appear distorted and disorganized cell proliferation occurs around the shoot apical meristem (arrow shows callus formation).

algae. Indeed, autophagy genes (ATG) and mechanisms seem also present and conserved in algae and land plants (for a review, see Ref. [48]). Moreover in the presence of ethanol, germinating plantlets from AtTor silenced lines were halted at a postgerminative checkpoint (Figure 15.3) that is used by the emerging plants to sense if the environment is favorable for the establishment of the seedlings, and is mainly controlled by the phytohormone abscisic acid (ABA) [7]. When plants were grown for a longer time on an ethanol-containing medium that induced silencing, they did not die but started to produce calli (unorganized tissues) on both the hypocotyl and the apical meristem zone (Figure 15.3), which suggests that cell positional identity was lost.

Protein synthesis as a whole, including the production of ribosomes, mobilizes a large part of the cellular resources and is an essential component of cytosolic growth during cell proliferation. The translational control of given mRNAs is also an important regulatory step of gene expression. The TOR pathway was found to act at the level of translation initiation and also in the regulation of the synthesis of ribosomal components [43]. *Arabidopsis* plants silenced for the AtTor expression displayed a significant reduction in polysome abundance [46]. The same result was observed after rapamycin treatment of *Arabidopsis* lines expressing the yeast FKBP12 protein [15]. The fact that transgenic *Arabidopsis* plants expressing a yeast FKBP12 protein showed only a modest reduction of growth compared to AtTor mutants or silenced lines could be due to either a low binding efficiency of the FKBP12/rapamycin complex to the AtTOR FRB domain or to the fact that rapamycin-dependent readouts of AtTOR are restricted to a smaller subset in plants than in other eukaryotes. Indeed, it seems that a complete inactivation of TORC1 activity is more drastic and has more effects than rapamycin treatment [49, 50].

Collectively these results suggest that the TOR activity (probably that of the TORC1 complex) is needed to maintain mRNA translation and plant growth and probably to restrain autophagy and senescence (see Figure 15.4 for a tentative model of TOR complex structure and role in plants).

VI. Conclusion

Multicellular plants usually cannot escape their environment and its changes. Moreover, as autotrophic organisms, they have very diverse and complex metabolic pathways. Therefore their development is highly constrained by environmental cues and, given the central role of the TOR signaling pathway in modulating cell/organ growth and metabolism, it is not surprising that this pathway was recruited during the evolution of plants to suit their needs and hence may be a central player of the adaptation to ever changing outside conditions. On one hand, recent results support the hypothesis that a conserved TORC1 complex, with the TOR, RAPTOR, and LST8 proteins, exists in all photosynthetic organisms, from unicellular green algae to land plants, with an important role in regulating growth, mRNA translation and metabolism. Similarly major TOR substrates

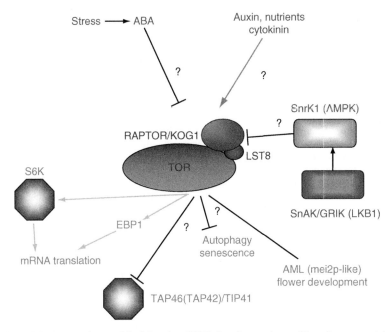

FIG. 15.4. A tentative model of the plant TOR signaling pathway. Homologous proteins in other organisms are indicated in parentheses.

including the S6K and the TAP42/46 proteins are clearly present. On the other hand, the existence of a TORC2 complex or of various regulators of TOR activity, like the TSC complex or the Rheb protein, remains so far to be proven in plants.

ACKNOWLEDGMENTS

This work was partly supported by grants from Agence Nationale de la Recherche to C. M., R. S., and C. R. (ANR Blanc06-3-135436), and to B. M. (ANR-08-JCJC-0054-01). M. M. was supported by a joint grant from INRA and CEA.

REFERENCES

1. Ingram, G., and Waites, R. (2006). Keeping it together: co-ordinating plant growth. *Curr Opin Plant Biol* 9:12–20.
2. Wolters, H., and Jürgens, G. (2009). Survival of the flexible: hormonal growth control and adaptation in plant development. *Nat Rev Genet* 10:305–317.

3. Gonzalez, N., Beemster, G., and Inzé, D. (2009). David and Goliath: what can the tiny weed *Arabidopsis* teach us to improve biomass production in crops? *Curr Opin Plant Biol* 12:157–164.
4. Desnos, T. (2008). Root branching responses to phosphate and nitrate. *Curr Opin Plant Biol* 11:82–87.
5. Menand, B., Meyer, C., and Robaglia, C. (2004). Plant growth and the TOR pathway. *Curr Top Microbiol Immunol* 279:97–113.
6. Robaglia, C., *et al.* (2004). Plant growth: the translational connection. *Biochem Soc Trans* 32:581–584.
7. Lopez-Molina, L., Mongrand, S., and Chua, N. (2001). A postgermination developmental arrest checkpoint is mediated by abscisic acid and requires the ABI5 transcription factor in *Arabidopsis*. *Proc Natl Acad Sci USA* 98:4782–4787.
8. Rook, F., Hadingham, S., Li, Y., and Bevan, M. (2006). Sugar and ABA response pathways and the control of gene expression. *Plant Cell Environ* 29:426–434.
9. Noodén, L. (1988). The phenomena of senescence and aging. In Senescence and Ageing in Plants, L.D. Noodén and A.C. Leopold (eds.), pp. 1–50. Academic Press, San Diego, CA.
10. Menand, B., Desnos, T., Nussaume, L., Berger, F., Bouchez, D., Meyer, C., and Robaglia, C. (2002). Expression and disruption of the *Arabidopsis* TOR (target of rapamycin) gene. *Proc Natl Acad Sci USA* 99:6422–6427.
11. Agredano-Moreno, L., Reyes de la Cruz, H., Martínez-Castilla, L., and Sánchez de Jiménez, E. (2007). Distinctive expression and functional regulation of the maize (*Zea mays* L.) TOR kinase ortholog. *Mol Biosyst* 3:794–802.
12. Crespo, J., Díaz-Troya, S., and Florencio, F. (2005). Inhibition of target of rapamycin signaling by rapamycin in the unicellular green alga *Chlamydomonas reinhardtii*. *Plant Physiol* 139:1736–1749.
13. Xu, Q., Liang, S., Kudla, J., and Luan, S. (1998). Molecular characterization of a plant FKBP12 that does not mediate action of FK506 and rapamycin. *Plant J* 15:511–519.
14. Mahfouz, M., Kim, S., Delauney, A., and Verma, D. (2006). *Arabidopsis* TARGET OF RAPAMYCIN interacts with RAPTOR, which regulates the activity of S6 kinase in response to osmotic stress signals. *Plant Cell* 18:477–490.
15. Sormani, R., *et al.* (2007). *Saccharomyces cerevisiae* FKBP12 binds *Arabidopsis thaliana* TOR and its expression in plants leads to rapamycin susceptibility. *BMC Plant Biol* 7:26.
16. Veverka, V., *et al.* (2008). Structural characterization of the interaction of mTOR with phosphatidic acid and a novel class of inhibitor: compelling evidence for a central role of the FRB domain in small molecule-mediated regulation of mTOR. *Oncogene* 27:585–595.
17. Hong, Y., Pan, X., Welti, R., and Wang, X. (2008). Phospholipase Dalpha3 is involved in the hyperosmotic response in *Arabidopsis*. *Plant Cell* 20:803–816.
18. Takai, H., Wang, R., Takai, K., Yang, H., and de Lange, T. (2007). Tel2 regulates the stability of PI3K-related protein kinases. *Cell* 131:1248–1259.
19. Shinozaki-Yabana, S., Watanabe, Y., and Yamamoto, M. (2000). Novel WD-repeat protein Mip1p facilitates function of the meiotic regulator Mei2p in fission yeast. *Mol Cell Biol* 20:1234–1242.
20. Deprost, D., Truong, H., Robaglia, C., and Meyer, C. (2005). An *Arabidopsis* homolog of RAPTOR/KOG1 is essential for early embryo development. *Biochem Biophys Res Commun* 326:844–850.
21. Anderson, G., Veit, B., and Hanson, M. (2005). The *Arabidopsis* AtRaptor genes are essential for post-embryonic plant growth. *BMC Biol* 3:12.
22. Gwinn, D., *et al.* (2008). AMPK phosphorylation of raptor mediates a metabolic checkpoint. *Mol Cell* 30:214–226.

23. Jossier, M., *et al.* (2009). SnRK1 (SNF1-related kinase 1) has a central role in sugar and ABA signalling in *Arabidopsis thaliana*. *Plant J* 59:316–328.
24. Baena-González, E., and Sheen, J. (2008). Convergent energy and stress signaling. *Trends Plant Sci* 13:474–482.
25. Thelander, M., Olsson, T., and Ronne, H. (2004). Snf1-related protein kinase 1 is needed for growth in a normal day–night light cycle. *EMBO J* 23:1900–1910.
26. Díaz-Troya, S., Florencio, F., and Crespo, J. (2008). Target of rapamycin and LST8 proteins associate with membranes from the endoplasmic reticulum in the unicellular green alga *Chlamydomonas reinhardtii*. *Eukaryot Cell* 7:212–222.
27. Zhang, S., Lawton, M., Hunter, T., and Lamb, C. (1994). atpk1, a novel ribosomal protein kinase gene from *Arabidopsis*. I. Isolation, characterization, and expression. *J Biol Chem* 269:17586–17592.
28. Turck, F., Kozma, S., Thomas, G., and Nagy, F. (1998). A heat-sensitive *Arabidopsis thaliana* kinase substitutes for human p70s6k function in vivo. *Mol Cell Biol* 18:2038–2044.
29. Turck, F., Zilbermann, F., Kozma, S., Thomas, G., and Nagy, F. (2004). Phytohormones participate in an S6 kinase signal transduction pathway in *Arabidopsis*. *Plant Physiol* 134:1527–1535.
30. Tzeng, T., Kong, L., Chen, C., Shaw, C., and Yang, C. (2009). Overexpression of the lily p70(s6k) gene in *Arabidopsis* affects elongation of flower organs and indicates TOR-dependent regulation of AP3, PI and SUP translation. *Plant Cell Physiol* 50:1695–1709.
31. Otterhag, L., *et al.* (2006). *Arabidopsis* PDK1: identification of sites important for activity and downstream phosphorylation of S6 kinase. *Biochimie* 88:11–21.
32. Huber, A., *et al.* (2009). Characterization of the rapamycin-sensitive phosphoproteome reveals that Sch9 is a central coordinator of protein synthesis. *Genes Dev* 23:1929–1943.
33. Harris, D., Myrick, T., and Rundle, S. (1999). The *Arabidopsis* homolog of yeast TAP42 and mammalian alpha4 binds to the catalytic subunit of protein phosphatase 2A and is induced by chilling. *Plant Physiol* 121:609–617.
34. Alvarez, B., and Moreno, S. (2006). Fission yeast Tor2 promotes cell growth and represses cell differentiation. *J Cell Sci* 119:4475–4485.
35. Anderson, G., *et al.* (2004). Diversification of genes encoding mei2 -like RNA binding proteins in plants. *Plant Mol Biol* 54:653–670.
36. Kaur, J., Sebastian, J., and Siddiqi, I. (2006). The *Arabidopsis*-mei2-like genes play a role in meiosis and vegetative growth in *Arabidopsis*. *Plant Cell* 18:545–559.
37. Veit, B., Briggs, S., Schmidt, R., Yanofsky, M., and Hake, S. (1998). Regulation of leaf initiation by the terminal ear1 gene of maize. *Nature* 393:166–168.
38. Anderson, G., and Hanson, M. (2005). The *Arabidopsis* Mei2 homologue AML1 binds AtRaptor1B, the plant homologue of a major regulator of eukaryotic cell growth. *BMC Plant Biol* 5:2.
39. Deak, M., Casamayor, A., Currie, R., Downes, C., and Alessi, D. (1999). Characterisation of a plant 3-phosphoinositide-dependent protein kinase-1 homologue which contains a pleckstrin homology domain. *FEBS Lett* 451:220–226.
40. Bögre, L., Okrész, L., Henriques, R., and Anthony, R. (2003). Growth signalling pathways in *Arabidopsis* and the AGC protein kinases. *Trends Plant Sci* 8:424–431.
41. Hey, S., Mayerhofer, H., Halford, N., and Dickinson, J. (2007). DNA sequences from *Arabidopsis*, which encode protein kinases and function as upstream regulators of Snf1 in yeast. *J Biol Chem* 282:10472–10479.
42. Shen, W., and Hanley-Bowdoin, L. (2006). Geminivirus infection up-regulates the expression of two *Arabidopsis* protein kinases related to yeast SNF1- and mammalian AMPK-activating kinases. *Plant Physiol* 142:1642–1655.

43. Wullschleger, S., Loewith, R., and Hall, M. (2006). TOR signaling in growth and metabolism. *Cell* 124:471–484.

44. Soulard, A., Cohen, A., and Hall, M. (2009). TOR signaling in invertebrates. *Curr Opin Cell Biol* 21:825–836.

45. Morimoto, T., Suzuki, Y., and Yamaguchi, I. (2002). Effects of partial suppression of ribosomal protein S6 on organ formation in *Arabidopsis thaliana*. *Biosci Biotechnol Biochem* 66:2437–2443.

46. Deprost, D., Yao, L., Sormani, R., Moreau, M., Leterreux, G., Nicolaï, M., Bedu, M., Robaglia, C., and Meyer, C. (2007). The *Arabidopsis* TOR kinase links plant growth, yield, stress resistance and mRNA translation. *EMBO Rep* 8:864–870.

47. Horváth, B., *et al.* (2006). EBP1 regulates organ size through cell growth and proliferation in plants. *EMBO J* 25:4909–4920.

48. Díaz-Troya, S., Pérez-Pérez, M., Florencio, F., and Crespo, J. (2008). The role of TOR in autophagy regulation from yeast to plants and mammals. *Autophagy* 4:851–865.

49. Thoreen, C., and Sabatini, D. (2009). Rapamycin inhibits mTORC1, but not completely. *Autophagy* 5:725–726.

50. Thoreen, C., *et al.* (2009). An ATP-competitive mammalian target of rapamycin inhibitor reveals rapamycin-resistant functions of mTORC1. *J Biol Chem* 284:8023–8032.

51. Hruz, T., Laule, O., Szabo, G., Wessendorp, F., Bleuler, S., Oertle, L., Widmayer, P., Gruissem, W., and Zimmermann, P. (2008). Genevestigator V3: a reference expression database for the meta-analysis of transcriptomes. *Adv Bioinform* 2008:(art. no. 420747).

16

Dysregulation of TOR Signaling in Tuberous Sclerosis and Lymphangioleiomyomotosis

JANE YU • ELIZABETH PETRI HENSKE

Division of Pulmonary and Critical Care Medicine
Brigham and Women's Hospital and Harvard Medical School
Boston, Massachusetts, USA

I. Abstract

Activation of the mammalian target of rapamycin complex 1 (mTORC1) has been observed in many human tumors and mTORC1 inhibitors have been approved for the treatment of renal cell carcinoma and other malignancies. The tuberous sclerosis complex (TSC) proteins directly regulate Rheb, a direct activator of TORC1. Therefore, TSC and the related disease, pulmonary lymphangioleiomyomatosis (LAM) have the most direct biochemical link to mTORC1 dysregulation of any known human disease or syndrome, providing a unique window through which the direct impact of mTORC1 activation in humans can be viewed. Yet, despite the direct links between TSC and mTORC1, the relationship between mTORC1 activation and the unusual clinical manifestations of TSC remains largely mysterious. Furthermore, TORC1 inhibition has only a partial impact on tumors in TSC patients, with regrowth following discontinuation. In this chapter, we will review the evidence of mTORC1 activation in LAM and TSC, discuss the impact of mTORC1 inhibition in mouse models of TSC and consider how

DOI: 10.1016/S1874-6047(10)27016-6

these preclinical models can guide the interpretation of the clinical studies of mTORC1 inhibitors in humans with TSC and LAM. We will also consider the growing evidence that at least some of the clinical manifestations of TSC may reflect TORC1-indepenent functions of the TSC proteins. Finally, we will discuss future clinical perspectives for TSC and LAM patients, focusing on therapeutic approaches combining mTORC1 inhibitors and agents targeting survival pathways and/or TSC-dependent/mTORC1-independent pathways.

II. TSC and LAM: Clinical Features

Tuberous sclerosis complex (TSC) is a tumor suppressor gene syndrome in which patients can develop seizures, mental retardation, autism, and tumors in the brain, retina, kidney, heart, and skin [1]. The most frequent tumor types include cerebral cortical tubers, facial angiofibromas, cardiac rhabdomyomas, and renal angiomyolipomas. TSC is caused by germline inactivating mutations in the TSC1 or TSC2 gene. TSC has autosomal dominant inheritance with 95% penetrance, although the clinical manifestations can vary widely even among family members carrying the same germline mutation. TSC has a high new mutation rate, with approximately one-third of all cases represent new germline mutations not present in the parents. Second hit mutations, most often reflected in loss of heterozygosity, have been documented in the majority of the tumors in TSC, including subependymal giant cell astrocytomas, angiomyolipomas, and rhabdomyomas [2–12].

Lymphangioleiomyomatosis (LAM), the pulmonary manifestation of TSC, affects almost exclusively women. Approximately one-third of women with TSC have radiographic evidence of LAM [13, 14]. LAM also occurs in a sporadic form in women who do not have germline TSC gene mutations [3]. The onset of symptoms of LAM is typically in the 20s or early 30s, although cases have been reported in both adolescent girls and postmenopausal women. LAM is characterized by the diffuse proliferation of abnormal smooth muscle cells in both lungs, with emphysema-like destruction of the lung parenchyma [1, 13–20]. In a Mayo Clinic series reported in 1995, LAM was the third most frequent cause of TSC-related death, after renal disease and brain tumors [21].

The first link between the sporadic form of LAM and TSC was in the year 2000, when our group found biallelic TSC2 mutations in angiomyolipomas and pulmonary LAM cells of sporadic LAM patients [3], which was later confirmed in Japanese sporadic LAM patients [22]. Several pieces of genetic evidence indicate that LAM cells metastasize, despite the fact that they appear histologically benign. These include studies of recurrent LAM after lung transplantation which have demonstrated that the recurrent LAM cells are derived from the patient's original LAM cells [10, 23–25].

III.   Evidence of mTOR Activation in TSC and LAM

The protein products of *TSC1* and *TSC2*, hamartin and tuberin, respectively, form heterodimers [26, 27] that inhibit the mammalian target of rapamycin (mTOR) complex 1 (mTORC1) [28–30]. This is achieved via tuberin's evolutionarily conserved GTPase activating domain, which regulates the small GTPase Rheb (Ras homolog enriched in brain) [31–36] (Figure 16.1). mTOR and its substrates, p70 ribosomal protein S6 kinase (p70S6K) and 4EBP1, are components of cellular pathways that regulate protein synthesis, cell size, and cell proliferation [37–39].

The TSC1/TSC2 heterodimer appears to function as a "checkpoint" in the regulation of TORC1, with tuberin (TSC2) and hamartin (TSC1) being directly phosphorylated and regulated in response to growth factors, the cell cycle, and nutrient availability. Kinases that are known to directly phosphorylate tuberin include protein kinase B (Akt) [29, 31, 40–42], p90 ribosomal S6 kinase 1 (RSK1) [43], ERK2 (MAPK) [44], MK2 (which is downstream of p38 MAPK) [45], GSK3β [46], AMP kinase (AMPK) [47], and death-associated protein kinase (DAPK) [48]. Kinases that are known to directly phosphorylate hamartin include cyclin-dependent kinase 1 (CDK1) [49, 50] and IKKβ [51].

As expected because of the role of the TSC1/TSC2 proteins in inhibiting TORC1, hyperphosphorylation of p70S6K and/or its substrate ribosomal protein S6 have been shown by our group and others in cells from TSC and LAM patients [4, 52–55]. Most of these have involved immunohistochemical analyses of archival, paraffin-embedded tissue specimens in which ribosomal protein S6 is hyperphosphorylated relative to adjacent normal cells.

IV.   Evidence That Inhibition of TOR Signaling Inhibits Tumor Formation in Mouse Models

Multiple mouse models of Tsc1 and Tsc2 loss have been developed and used along with the Eker rat model of Tsc2 to examine the efficacy of mTOR inhibitors including rapamycin and its analogs, RAD001, CCI-779, and WAY-129327. However, it is important to note that none of the available mouse models recapitulate the primary tumor types observed in

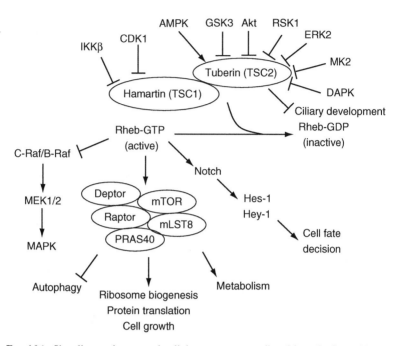

F<small>IG</small>. 16.1. Signaling pathways and cellular processes mediated by tuberin and hamartin. The TSC1 and TSC2 gene products, hamartin and tuberin, form a heterodimer to inactivate the small GTPase Rheb. The active GTP-bound Rheb stimulates mTOR complex 1 which enhances ribosome biogenesis, protein translation, cell metabolism, cell growth, and inhibits autophagy. Independent of mTOR complex 1, tuberin controls ciliary development; and Rheb regulates C-Raf and B-Raf activity and Notch-mediated cell fate decision. Hamartin is inactivated via phosphorylation by CDK1 and IKKβ; and tuberin is activated by AMPK, but inactivated by protein kinases GSK3, Akt, RSK1, ERK2, MK2, and DAPK.

TSC patients (angiomyolipomas, LAM, or rhabdomyomas). The measured outcome of mTOR inhibition in these models varies depending on the model used (Table 16.1) and includes reduction in the size and number of established renal tumors [56–62], reduction in the size of subcutaneous tumors [60, 63–65], improvement in behavioral and neurologic phenotypes [66–69], and enhanced survival [58, 63, 66, 67, 69, 70]. While complete responses are uncommon, a partial effect was observed for every phenotype tested. The most significant tumor outcomes include the complete prevention of subcutaneous tumor growth when rapamycin was given 1 day prior to subcutaneous inoculation of Tsc2-null ELT3 cells [65] and the 98% regression of renal tumors in Eker rats when rapamycin was given for 8 weeks [58]. In other subcutaneous tumor models, rapamycin or CCI-779 was administered in animals bearing established subcutaneous tumors of

TABLE 16.1

Mouse Models Support that Inhibition of mTOR Improves the Symptoms

Model	Treatment	Effects	References
$Tsc2^{Ek/+}$ Eker rat			
Eker rat	Rapamycin (0.1–2.0 mg/kg/day, i.p.)	Decreased renal and pituitary tumor burden	[58, 59]
Eker rat	WAY-129327 (0.5 mg/kg/day, i.p. for 4 months)	Decreased uterine leiomyoma tumor burden	[56]
$Tsc1$ heterozygous mice			
$Tsc1^{+/-}$	Rapamycin (20 mg/kg/day for 4 days, i.p.)	Reduced serum VEGF levels, decreased liver, and kidney tumor burden	[57]
$Tsc2$ heterozygous mice			
$Tsc2^{+/-}$	Rapamycin (8 mg/kg, i.p. daily for 1 month, 16 mg/kg weekly for 5 months, or 8 mg/kg daily for 1 month)	Decreased kidney cystadenoma score	[60]
$Tsc2^{+/-}$	CCI-779 (8 mg/kg/day, i.p.) for 2 months	Decreased kidney cystadenoma score	[61] [64]
	CCI-779 (4 mg/kg, 3 days/week, i.p.) for 3 months		
$Tsc2^{+/-}$ plus ENU	RAD001(10 mg/kg, p.o. 5 days/week)	Decreased kidney cystadenoma score	[62]
$Tsc2^{+/-}$	Rapamycin (1 or 5 mg/kg/day, i.p.)	Improved memory and learning	[66]
Subcutaneous tumor models			
$Tsc2^{-/-} p53^{-/-}$ MEFs in nude mice	CCI-779 (8 mg/kg/day, i.p.)	Suppressed subcutaneous tumor growth	[63]
$Tsc2^{-/-} p53^{-/-}$ MEFs	Rapamycin (8 mg/kg/day)	Suppressed subcutaneous tumor growth	[60]
ELT3 $Tsc2^{Ek/+}$	RAD001 (4 mg/kg, p.o. 5 days)	Inhibited subcutaneous tumor growth	[65]
$Tsc1$ conditional knockout mice			
$Tsc1(\alpha CaMKII)$ CKO	Rapamycin (0.2 mg/kg/day, s.c. for 3 months)	Enhanced survival and reduced brain size	[66]

(Continued)

TABLE 16.1 (*Continued*)

Model	Treatment	Effects	References
Tsc1(GFAP)CKO	Rapamycin (3 mg/kg/day, i.p. for 3 weeks)	Reduced seizures and enhanced survival	[69] [68]
Tsc1(SynI)CKO	Rapamycin (6 mg/kg every other day, i.p.) RAD001 (6 mg/kg every other day, i.p.)	Enhanced survival; improved clasping behavior	[67]
Tsc1(Nse)CKO	Rapamycin (2 mg/kg)	Reduced severity of cystic kidney disease and enhanced survival	[95]
*Tsc1*fl/flmx-1-Cre	Rapamycin (4 mg/kg/every other day, i.p.)	Decreased mitochondrial biogenesis and ROS production; stabilized hematopoiesis	[96]
Rip-*Tsc1*cKO mice	Rapamycin (6 mg/kg every other day, i.p.)	Decreased hyperphagia and obesity	[97]
Pomc-*Tsc1*cKO mice	Rapamycin (6 mg/kg every other day, i.p.)	Decreased hyperphagia and obesity	[97]
Tsc2 conditional knockout mice			
β*Tsc2*$^{-/-}$	Rapamycin (5 mg/kg/day, i.p.)	Normalized glucose levels and decreased hyperinsulinemia	[98]

$Tsc2^{-/-}p53^{-/-}$ MEFs, and the tumors progressed more slowly compared with control animals during the first 2 weeks of treatment. However, despite the continuous treatment with rapamycin or CCI-779, tumor growth accelerated, and by week 6 average tumor volume was similar to the week 2 average tumor volume of control animals [63, 64, 70], indicating that TORC1 inhibitors delays, but does not prevent tumor growth.

A. $Tsc2^{Ek/+}$ EKER RAT

The first evidence of a therapeutic effect of mTOR inhibition in an animal model of TSC was in 2002 in the Eker rat, which carries a germline mutation in exon 30 of the Tsc2 gene [71]. Eker rats develop uterine leiomyomas, renal cysts, and renal cell carcinomas. Kenerson *et al.* [59] treated 12-month-old Eker rats with rapamycin at 0.16, 0.4, and 1 mg/kg/day, i.p. for 3 consecutive days, and compared histology, cell proliferation, and apoptosis in renal tumors with those of vehicle-treated rats. They found that levels of phospho-S6 immunoreactivity were markedly reduced, and PCNA positive cells were reduced from 53% to 8% at 1.0 mg/kg. A dose effect of rapamycin on percent of PCNA positive nuclei, renal tumor size, and cell death was observed. Renal tumor apoptosis, measured using TUNEL staining, was higher in tumors from treated animals, suggesting that rapamycin treatment both decreases proliferation and promotes cell death in Tsc2-null epithelial tumors. Based on these preclinical results, the first clinical trial of rapamycin, using the volume of renal angiomyolipomas as the primary endpoint, was conceived; the results of this trial [72] are discussed later in this chapter.

In a subsequent study by Kenerson *et al.* in 2005, the effect of a longer duration of rapamycin treatment on renal tumors in the Eker rat model was tested [58]. Twelve-month-old Eker rats were treated with rapamycin for 2–8 weeks (0.1–0.2 mg/kg/day, i.p.). An important component of this study was that renal tumor volume was measured using ultrasound imaging, allowing tumor growth within one animal to be monitored. Animals treated with rapamycin at 0.1 mg/kg for 8 weeks exhibited a 98% regression of renal tumor volume (from 0.63 cm^3 pretreatment to 0.05 cm^3). Microscopic examination after 2 weeks or 2 months of rapamycin treatment showed massive necrosis, similar to what was observed in the earlier study in which the animals were treated for only 3 days, and associated with elevated levels of cleaved caspase 3 in renal tumor lysates measured by immunoblotting analysis. In a separate experiment to determine whether rapamycin can prevent tumor development, 2-month-old Eker rats were treated with rapamycin for 2 months (0.1–0.2 mg/kg/day, i.p.), then left untreated for 4 months, and sacrificed at 8 months of age (i.e., 4 months after the last dose of rapamycin). Rapamycin-treated animals did not develop macroscopic renal tumors (≥ 1 mm in diameter) while the controls had an average of 1.25 tumors/rat. Although additional studies are clearly needed, this result may have significant translational implications for TSC patients, since it suggests that a relatively short pulse of rapamycin treatment during development can effectively block tumor progression at later developmental stages. An additional translational implication of this study is that after 4 months off rapamycin treatment, phospho-S6 immunoreactivity was observed in microscopic renal lesions indicating either the initiation of new tumors after rapamycin was discontinued, or the reactivation of TORC1 in microscopic lesions that were "dormant" during therapy. In humans with TSC or LAM treated with rapamycin, regrowth of angiomyolipomas has been consistently observed following discontinuation of therapy, suggesting the reactivation of TORC1, as will be discussed later in this chapter.

A third study in the Eker rat model by Crabtree *et al.* used uterine leiomyoma tumor burdens as the endpoint, and tested the rapamycin analog WAY-129327. Leiomyoma are a later phenotype in the Eker rat model, with incomplete penetrance [73, 74]. Fifteen-month-old Eker rats were treated with WAY-129327 (0.5 mg/kg/day, i.p.) or control for either 2 weeks or 4 months. After 2 weeks, reduced levels of phopsho-S6 were observed in leiomyomas from the treated animals, as expected [56]. Four months of treatment reduced the percentage of animals with a uterine tumor from 77% to 39%, and decreased total tumor numbers from an average of 1.4/animal in the control group to 0.5/animal in the treatment group, but had only a modest effect on the size of the remaining tumors (control 3.9 cm^3 vs. treated 3.1 cm^3, $p < 0.001$). Treated tumors also exhibited a lower proliferation index (3.2%) compared with control (22.4%, $p < 0.01$), while no effect on the apoptotic index was observed.

B. $Tsc1^{+/-}$ Heterozygous Mice

The $Tsc1^{+/-}$ mouse was reported in 2002 by Kwiatkowski *et al.* [75]. The first preclinical trial using these mice was reported in 2003 by El-Hashemite *et al.*, who treated 10-month-old female mice with rapamycin (20 mg/kg/day, i.p.) for 4 days and monitored serum VEGF-A levels and examined histologic sections of the liver and kidneys. VEGF-A levels were about twofold higher in $Tsc1^{+/-}$ mice (99–251 pg/ml) than wild-type mice (50–91 pg/ml) [57]. Four days of rapamycin treatment resulted in a 70% reduction of serum VEGF levels compared to the pretreatment levels (148–594 pg/ml pretreatment vs. 103–132 pg/ml posttreatment). Pathological evaluation and TUNEL staining of liver and kidney sections showed cell death and necrosis in rapamycin-treated mice.

C. $Tsc2^{+/-}$ HETEROZYGOUS MICE

$Tsc2^{+/-}$ mice were first reported in 1999 by Onda et al. [76], and have been used for several preclinical studies. Lee et al. treated $Tsc2^{+/-}$ mice at 5 months of age using three different dose schedules of rapamycin: daily at 8 mg/kg for 1 month, weekly at 16 mg/kg for 5 months, and daily at 8 mg/kg for 1 month. Mice were sacrificed at 13 months of age and the extent of kidney tumors were quantitated using a scoring system developed for this model [61]. Rapamycin treatment caused a 94.5% reduction (untreated 15.00 ± 2.01, rapamycin-treated 0.83 ± 0.18, $p < 0.0001$) in the number of kidney cystadenoma in mice examined by histopathology evaluation [60].

In a separate report, Messina et al. treated $Tsc2^{+/-}$ mice with the rapamycin analog CCI-779 daily at 8 mg/kg for 2 months, beginning at either 6–8 or 10–12 months of age. Animals were sacrificed at the end of the 2 months of treatment, and kidney lesions scored using the quantitative system. CCI-779 treatment at 6–8 months of age caused a 39.7% reduction in the kidney-lesion score (from a score of 10 in the untreated to 6 in the treated, $p < 0.0001$), while treatment at 10–12 months resulted in a 64.4% reduction (from a score of 10 in the untreated to 3.5 in the treated) [61]. In an early study by Lee et al., the effect of CCI-779 with a lower dose and longer duration (4 mg/kg three times per week for 12 weeks) on renal tumor development was examined in 40–52-week-old $Tsc2^{+/-}$ mice. Twelve-week treatment at this lower dose reduced the number of kidney cystadenomas measured by MRI (from 1.1 tumors/kidney in the untreated to 0.3 in the treated, $p = 0.003$), gross count (from 3 tumors/kidney in the untreated to 1.1 in the treated, $p = 0.0001$), and histopathology evaluation (from 8.7 tumors/kidney in the untreated to 1.4 in the treated, an 84% reduction, $p < 0.0001$) [64].

A challenging aspect of the Tsc1 and Tsc2 heterozygous models is the extended time needed to generate animals with evaluable tumors. To decrease this latency, Pollizzi et al. treated $Tsc2^{+/-}$ mice with ENU (60 mg/kg, i.p., a single dose at P9), which has been found to enhance renal tumorigenesis in this model by 50-fold. They found that treatment of 5-month-old animals for 4 weeks with RAD001 (10 mg/kg/day, p.o.) or a dual pan class I PI3K/mTOR catalytic small molecule inhibitor NVP-BEZ235 (45 mg/kg, qd p.o.) caused a 99% reduction of ENU-induced $Tsc2^{+/-}$ kidney tumor scores (gross and microscopic kidney scores) and a 90% reduction of percent of Ki67 positivity [62]. However, when an additional group of animals was similarly treated for 4 weeks and then sacrificed after 8 weeks off treatment, only a 50% reduction of tumor burden was observed, indicating substantial regrowth of tumors after discontinuation of RAD001.

In summary, these studies in animal models in which renal tumors arise either spontaneously or through ENU induction consistently show that mTOR inhibitors reduce tumor burden, although tumors appear to regrow/reinitiate when treatment is discontinued. A limitation of these models is the lengthy latency to develop evaluable tumors, the large number of animals required because of the variable penetrance of the phenotypes, and the lack of kinetic information (how quickly do the tumors regress and how quickly do they regrow) because most studies were done by examining tumors after sacrifice of the animals, rather than with imaging.

D. SUBCUTANEOUS TUMOR MODELS

Several subcutaneous tumor models have been used to evaluate the effect of rapamycin and its analogs in Tsc2-null cells. Lee *et al.* injected $Tsc2^{-/-}p53^{-/-}$ mouse embryonic fibroblasts (MEFs) subcutaneously into nude mice. CCI-779 treatment (8 mg/kg/day, i.p.) was started when the tumor volume was 300 mm^3, and terminated at week 7 postdrug treatment. They found that CCI-779 treated tumors had significant lower volume 2-week posttreatment (control 3074 mm^3 vs. treated 387 mm^3, $p < 0.0001$) [63]. However, tumor growth accelerated after 2 weeks of CCI-779 treatment, and by week 6 on treatment, the average tumor volume reached 3000 mm^3, similar to the week 2 average tumor volume of control animals, indicating that CCI-779 delays but does not prevent tumor development.

In a separate study, Lee *et al.* subcutaneously inoculated $Tsc2^{-/-}p53^{-/-}$ MEFs into nude mice and began rapamycin treatment (8 mg/kg/day, i.p.) when the volume of xenograft tumors reach 150 mm^3. On day 16 of rapamycin treatment, the tumor volume in treated group was lower than the control group (treated 284 mm^3 vs. control 1454 mm^3, $p = 0.0007$) [60]. Treatment was terminated when tumor volume reached 3000 mm^3. Rapamycin treatment slowed but did not prevent tumor growth: by day 42 of rapamycin, the average tumor volume reached 1784 mm^3, similar to the day 26 average tumor volume of placebo-treated animals.

Yu *et al.* tested the effect of RAD001 on tumor development and metastasis in a xenograft model of ELT3 cells, which were derived from a uterine leiomyoma in the Eker rat model of Tsc2 [77] and have been previously shown to carry a somatic "second hit" deletion of the wild-type copy of Tsc2 [78, 79] and to be estrogen responsive *in vitro* and *in vivo* [80]. Estrogen- or placebo-treated CB17-*scid* ovariectomized female mice were inoculated with ELT3 cells subcutaneously, and RAD001 treatment (4 mg/kg/day, p.o.) was started 1-day postcell injection and continued 8–9 weeks. RAD001 treated mice did not form any tumors. In contrast, control mice developed subcutaneous tumors within 5 weeks postcell injection [65], indicating the complete inhibition of tumor initiation by RAD001.

V.   Combinational Therapy in Heterozygous Mouse and Subcutaneous Tumor Models

To access the efficacy of combinational therapy of mTOR inhibitors and other potential agents (Table 16.2), Lee *et al.* tested rapamycin and interferon gamma (IFNγ) in $Tsc2^{+/-}$ mice. Rapamycin (8 mg/kg/day) plus IFNγ (16 mg/kg/week) was not superior to rapamycin alone in reducing the renal tumor scores [70], although in an earlier study using the subcutaneous model, the same combinational treatment appeared to be superior [63], emphasizing the potential importance of models in which tumors arise spontaneously in contrast to those in which the vascular, stromal, and immunologic features of tumorigenesis are vastly different than in subcutaneous models. Pollizzi *et al.* tested the PI3K/Akt and mTOR dual inhibitor NVP-BEZ235 plus RAD001 in the ENU-induced $Tsc2^{+/-}$ kidney tumor model, and found that the combinational treatment was not superior to either agent alone [62]. Lee *et al.* tested sorafenib (VEFG pathway inhibitor), atorvastatin (HMG-CoA reductase inhibitor), and doxycycline (MMP inhibitor) in combination with rapamycin in a $Tsc2^{-/-}p53^{-/-}$ subcutaneous tumor model. They found that rapamycin plus sorafenib treatment was more effective than rapamycin alone in inhibiting early tumor progression [70]. Furthermore, atorvastatin or doxycycline plus rapamycin was not superior to rapamycin alone. Importantly, the tumors regrew during weeks 3–4 despite ongoing therapy treatments in this study [60], indicating the acquisition of the resistance to rapamycin treatment in subcutaneous tumor models.

VI.   Evidence That Inhibition of TOR Signaling Suppresses the Neurologic Manifestation in Mouse Models

One of the most exciting developments in preclinical studies of TSC has been the neurocognitive studies of $Tsc2^{+/-}$ mice and the Eker rats. Ehninger *et al.* reported that $Tsc2^{+/-}$ mice exhibit learning deficits in hippocampal-dependent functions including spatial memory in the Morris water maze, working memory in the eight-arm radial maze, and context discrimination

TABLE 16.2

COMBINATIONAL TREATMENT IN TSC MOUSE MODELS

Model	mTOR inhibitor	Agent 2		Effect	Combinational versus mTOR inhibitor	References
		Drug	Targets			
Tsc2+/− mice	Rapamycin 8 mg/kg, i.p.	IFNγ 20,000 IU, i.p.	VEGF signaling	Effective	Equivalent	[60]
Tsc2+/− mice	RAD001 10 mg/kg, p.o.	NVP-BEZ235 45 mg/kg, p.o. qd	mTOR and Akt	Effective	Equivalent	[62]
Tsc2−/− p53−/− MEFs subcutaneous tumor	Rapamycin 8 mg/kg, i.p.	IFNγ 20,000 IU, i.p.	VEGF signaling	Not tested	Equivalent	[60]
		Sorafenib 60 mg/kg, p.o.	VEGF signaling	Ineffective	More effective than rapamycin	
		Atorvastatin 20 mg/kg, i.p.	HMG-CoA reductase inhibitor	Effective	More effective than rapamycin	
		Doxycycline 10 mg/kg, i.p.	MMP inhibitor	Ineffective	Equivalent	
Tsc2−/− p53−/− MEFs subcutaneous tumor	CCI-779 8 mg/kg, i.p.	IFNγ 20,000 IU, i.p.	VEGF signaling	Ineffective	More effective than CCI-779	[63]

[66]. For example, in the Morris water maze for spatial learning, mice were trained to locate an escape platform hidden under the water surface of a pool. After the trial, mice had to use distal spatial cues to learn to navigate the platform position. Short-term rapamycin treatment (1 or 5 mg/kg/day, i.p.) in $Tsc2^{+/-}$ mice for 5 days improved the percentage time spent in the target quadrant (from 30% vehicle treated to 40% rapamycin treated) and the number of target crossings (from 1.5 vehicle treated to 2.5 rapamycin treated), indicating an improvement in the spatial learning and memory.

In addition to these studies of the heterozygous animals, which are particularly exciting since they may reflect abnormalities that are comparable to those in the heterozygous brain of TSC patient, several groups have studied mTOR inhibitors in conditional knockout models. In the same study of Ehninger et al., mice with conditional deletion of $Tsc1$ in neurons of the postnatal forebrain under the control of the αCaMKII promoter were generated [66]. $Tsc1^{cc}$–αCaMKII-Cre mice exhibited more severe neurologic phenotypes than heterozygous $Tsc2^{+/-}$ mice, including premature death within the first postnatal weeks and severe macroencephaly with 2.5 times higher of brain weight associated with massive neuronal hypertrophy and astrogliosis. Rapamycin treatment (0.2 mg/kg/day, s.c.) beginning at postnatal day 1 and continuing until 3 months of age enhanced the survival of Tsc1^{cc}–αCaMKII-Cre mice (treated 70%, control 4% as expected by Mendelian ratios) and reduced brain weight by 37% (treated 0.6 g, control 0.95 g).

Mice with conditional inactivation of $Tsc1$ in glial fibrillary acidic protein (GFAP)-positive cells ($Tsc1^{GFAP}$CKO) developed histologic abnormalities of the brain (abnormal astrogliosis and disorganized neurons), enlarged brains, seizures beginning at 3 weeks of age and premature death by 4 months [81]. Two rapamycin schedules ("early," beginning at postnatal day 14 prior to the onset of seizures, and "late," beginning at 6 weeks of age) were tested in this model, both at 3 mg/kg/day, i.p. [69]. Five weeks of early rapamycin treatment partially corrected the histological abnormalities including reduced astrogliosis, decreased brain size, and more compact organization to the hippocampal pyramidal neurons. Early rapamycin treatment completely prevented seizure development monitored by video-electroencephalography (EEG) beginning at 1 month of age and enhanced the survival at 16 weeks (from 0% of the controls to 90% of the treated animals) [69]. Later rapamycin treatment in mice with established seizures dramatically reduced seizure frequency. At 3 weeks of treatment, 50% of the rapamycin-treated mice had become seizure free. At 11 weeks, the remaining rapamycin-treated mice had 0.4 ± 0.2 seizures/day compared with 14 ± 4 seizures/day in the control group. In this later-treatment group, rapamycin treatment also enhanced the survival at 17 weeks from 0% of the controls to 100% of the treated animals.

Meikle *et al.* used the synapsin I promoter-driven Cre allele to delete *Tsc1* in differentiating neurons during cortical development [82]. The mutant mice developed several clinical abnormalities, including the development of seizures, and pathologic abnormalities resembling features of TSC, including enlarged and dysplastic cortical neurons and mTOR activation [82]. In a separate study, Meikle *et al.* tested the efficacy of rapamycin and RAD001 using two regimens in this *Tsc1*[null-neuron] model. Rapamycin and RAD001 treatment (6 mg/kg every other day, i.p.) started at P7–P9 and terminated at P100 (continuous), or started at P7–P9, discontinued at P30, and terminated at P100 (on–off). The continuous treatment enhanced the median survival, which was 33 days in the control group versus 80 days in treated group at P100 [67]. The on–off treatment also increased the median survival (control 33 days, rapamycin-treated 79 days, RAD001-treated 77 days), but to a lesser degree compared with the continuous treatment. Treatment with rapamycin or RAD001 improved the behavioral phenotypes including clasping behavior and tremor at P30, P60, and P100 ($p < 0.01$).

VII.   Evidence That Inhibition of TOR Signaling Inhibits Tumor Formation in TSC and LAM

A small number of clinical trials of mTOR inhibitors for the treatment of TSC and LAM patients have been reported (Table 16.3), and others are underway. Bissler *et al.* studied TSC and LAM patients with angiomyolipomas, with angiomyolipoma volume as the primary endpoint. The patients received 12 months of oral rapamycin (Sirolimus), followed by 12 months of follow-up after discontinuation of rapamycin [72]. There was no randomization or placebo control arm in this study. Of 25 men and women enrolled, 20 patients completed 12 months of therapy, 18 patients of which completed the full study including the 12 months of follow-up after therapy was discontinued. Angiomyolipoma volume decreased to 53% of baseline after 12 months of therapy. After discontinuation of Sirolimus at 12 months, angiomyolipoma volume increased, reaching 86% of baseline at the 24-month final endpoint of the trial. In the Bissler *et al.* trial, 18 of the patients had LAM, 12 with TSC-associated LAM, and 6 with sporadic LAM. Among these women, 24-month data was available for 10 patients, and among these, the mean FEV1 increased by 118 ml, the FVC increased by 390 ml, and the RV decreased by 439 ml during the 12 months of rapamycin therapy. At 24 months, the FEV1 was 62 ml above baseline, the FVC

TABLE 16.3

Clinical Use of mTOR Inhibitors for the Treatment of TSC and LAM Patients

Enrollment (no.)	Intervention/ phase	Outcome measures	NCT identifier	References
Renal angiomyolipomas and LAM				
TSC (7) TSC-LAM (12) Sporadic LAM (6)	Sirolimus Phase I/II	Angiomyolipoma volume (MRI); Pulmonary function	NCT00457808	[72]
TSC (7) Sporadic LAM (6)	Sirolimus Phase II	Angiomyolipoma volume (MRI); Pulmonary function	NCT00490789	[83]
Subependymal giant cell astrocytoma				
TSC (5)	Sirolimus	Tumor volume (MRI)		[84]

was 346 ml above baseline, and the RV was 330 ml below baseline, suggesting substantial sustained benefit from rapamycin. Importantly, however, a smaller but similar study in the United Kingdom reported interim results of a 24-month trial of rapamycin for TSC and LAM patients [83], and found similar decreases in angiomyolipoma volume but no evidence of improved FEV1 or FVC in four women with LAM who had completed 12 months of rapamycin. The reasons for these differences are unknown to date, but will be further investigated when the completed data are available from the UK trial and when results are available from the Multicenter International LAM efficacy of Sirolimus (MILES) Trial, the first prospective, randomized clinical trial in LAM, which completed accrual in 2009.

For angiomyolipomas, the approximately 50% regression in size in both trials validates the importance of TORC1 signaling in human TSC. In the interim results from the UK trial, one patient had a 37% reduction in angiomyolipoma volume in just 2 months of treatment. However, the clinical implications of these results are uncertain. Because angiomyolipomas regrew after rapamycin was discontinued, it appears that prolonged or intermittent therapy would be required, with the accompanying common adverse events which include mouth ulcers and hyperlipidemia as well as the potential for infrequent but serious adverse events.

In addition to these two trials in which angiomyolipomas and LAM were primary and secondary endpoints, Franz *et al.* have reported the results of rapamycin treatment of subependymal giant cell astrocytomas (SEGAs) in patients with TSC. Five patients, ranging from 3 to 21 years of age, were treated [84]. The primary outcome was the volume of brain tumors by serial

MRI imaging for 2.5–20 months. The volume of the SEGAs regressed to 50–75% during the course of Sirolimus treatment. In one case, a patient with two tumors discontinued treatment for 4 months, resulting in regrowth of tumors to their pretreatment size, and when rapamycin was resumed the tumors again regressed, further supporting the concept that rapamycin may induce a dormant state of tumor cells in TSC, with regrowth occurring after the discontinuation of treatment.

VIII.   Evidence of TORC1-Independent Phenotypes in TSC

In human studies to date, rapamycin treatment has resulted in partial regression of angiomyolipomas and perhaps a greater response in SEGAs. The underlying reasons for this partial regression could include activation of survival pathways by rapamycin, which may include autophagy [25] and/or the PI3K and MEK/MAPK signaling cascades; incomplete TORC1 inhibition by rapamycin and its analogs; or TORC1-independent functions of the TSC1/TSC2 heterodimer. Currently, Rheb is the only generally accepted target of the TSC1/TSC2 complex, and TORC1 is the only generally accepted target of Rheb. However, we and others have observed functions of TSC2 and Rheb that are clearly rapamycin-insensitive and may prove to be TORC1-independent (Table 16.4). Brugarolos et $al.$ found that elevated VEGF-A secretion in $Tsc2^{-/-}$ MEFs was only partially inhibited by rapamycin treatment [85]. Karbowniczek et $al.$ found that Rheb's inhibition of B-Raf activity [86] and B-Raf and C-Raf heterodimerization [87] are resistant to rapamycin treatment. Gau et $al.$ reported that cell peripheral distribution of actin filament in $Tsc2^{-/-}$ MEFs is not affected by rapamycin treatment [88]. Huang et $al.$ found that the activation of mTORC2 by TSC1–TSC2 complex in $Tsc2^{-/-}p53^{-/-}$ MEFs is not affected by knockdown of mTORC1 components using Raptor siRNA, and by rapamycin treatment [89], indicating a TORC1-independent function of TSC1 and TSC2. Wilson et $al.$ found that renal cysts in $Tsc1^{+/-}$ mice showed low levels of mTOR and S6 phosphorylation [90], and therefore may arise through an mTORC1-independent mechanism. Hartman et $al.$ reported that $Tsc1^{-/-}$ and $Tsc2^{-/-}$ MEFs are more likely to contain primary cilia than wild-type MEFs, and the enhanced ciliary formation in the $Tsc1^{-/-}$ and $Tsc2^{-/-}$ MEFs was not corrected by rapamycin treatment [91]. Zhou et $al.$ found that Rheb-inhibited aggresome formation is not corrected by rapamycin treatment, suggesting TORC1 independence [92]. Lee et $al.$ found that Tsc1/Tsc2-null cells exhibit increased MMP-2 expression and

TABLE 16.4

POSSIBLE TORC1-INDEPENDENT FUNCTIONS OF TSC1/TSC2

Cellular process	Evidence of TORC1 independence	References
VEGF regulation		
Elevated VEGF-A secretion by $Tsc2^{-/-}p53^{-/-}$ MEFs	Partial inhibition by rapamycin	[85]
B-Raf and C-Raf activity		
Inhibition of B-Raf and C-Raf activity and heterodimerization by Rheb	Rapamycin insensitive	[86, 87]
Actin cytoskeleton		
Abnormal distribution of actin filaments in $Tsc2^{-/-}p53^{-/-}$ MEFs	Rapamycin insensitive	[88]
mTORC2 activity		
Activation of mTORC2 by TSC2	Rapamycin and raptor-siRNA insensitive	[89]
Renal cyst formation		
Renal cysts formation in $Tsc1^{+/-}$ mice	Low activation of mTOR and S6 in cysts	[90]
Ciliary development		
Enhanced ciliary development in $Tsc1^{-/-}p53^{-/-}$ and $Tsc2^{-/-}p53^{-/-}$ MEFs	Rapamycin insensitive	[91]
Aggresome formation		
Inhibition of aggresome formation by Rheb	Rapamycin insensitive	[92]
Extracellular matrix remodeling		
MMP2 expression and secretion by angiomyolipoma-derived cells	Rapamycin insensitive	[70]
Notch activation		
Rheb-induced Notch cleavage and activity	Rapamycin, Torin 1, and raptor siRNA insensitive	[94]

activity, which are insensitive to rapamycin treatment [70]. Karbowniczek *et al.* found that increased Notch cleavage and activity in Tsc2-null cells is not corrected by rapamycin, by the Tor kinase domain inhibitor Torin 1 [93], or raptor siRNA treatment [94]. Targeting these pathways in combination with TORC1 could have synergistic benefit in angiomyolipoma and LAM cells.

IX. Clinical Questions Not Fully Explained by TORC1 Activation

The clinical manifestations of TSC are among the most disparate of any autosomal dominant disorder. Patients with TSC develop neurologic disease ranging from seizures, intellectual disability, and autism, to tumors with exceptionally unusual features. These include tumor cell autonomous features, such as the differentiation plasticity present in angiomyolipomas and subependymal giant cell astrocytomas, to the surrounding cellular elements such as the aneurysmal large vessels that feed angiomyolipomas, to the behavior of tumor cells within the organism, such as the apparent metastasis of phenotypically benign LAM cells and the spontaneous regression of cardiac rhabdomyomas. In addition, it is quite clear that at least some of the clinical manifestations of TSC are unrecognized or underrecognized, including aspects of the neurocognitive phenotypes such as anxiety and sleep disorders.

As discussed earlier, the increasing literature contains evidence that TSC1/TSC2 and Rheb have TORC1-independent functions, yet the precise underlying mechanisms have not been clearly defined. In addition, animal models that develop angiomyolipomas, LAM, or rhabdomyomas are not available. At this point, therefore, it is unknown whether activation of TORC1 is fundamentally responsible for all of these clinical features of TSC, the role to which dysregulation of other TSC/Rheb-dependent pathways contributes to or modulates each of these clinical manifestations, or how inhibition of TORC1 will impact each of these clinical manifestations. The so-far elusive TORC1-independent functions of TSC1/TSC2 may have cell type or cell lineage specificity that influences differentiation fate, for example, yet have a relatively minor impact on cell proliferation, such as the effect of Rheb toward B-Raf kinase [86, 87]—in this case, Rheb inhibits B-Raf which is unlikely to impact proliferation unless the tumor contains a Rheb-resistant B-Raf mutation. Alternatively, the TORC1-independent functions could influence both cell fate and cell proliferation, as we have observed with Rheb's activation of Notch signaling [94]. In the case of Notch activation, there could be critical cell lineage effects overlying both the proliferative and differentiation phenotype.

X. Clinical Perspectives

The optimal therapies for TSC patients may need to be tailored to the specific manifestation being targeted and the specific developmental stage of the patient. Ideally, the preclinical models will guide the optimization of therapeutic regimens. Preventative therapies, such as seizure prevention, may involve treatment during a specific developmental window, as shown by Zeng *et al.* in the *Tsc1*(GFAP) CKO mouse model[69] and by Meikel *et al.* in the *Tsc1*(SynI)CKO model [67], while prevention of LAM cell metastasis could involve a treatment during a specific stage of tumorigenesis, as shown by Yu *et al.* in the ELT3 cell xenograft model [65]. Treatment of established tumors may require a "pulse" of therapy at a high enough dose to induce apoptosis, as has been observed by Kenerson *et al.* [58] and Pollizzi *et al.* [62]. Treatment with a prolonged maintenance involving daily low dose, weekly high dose, and daily low dose as studied by Lee *et al.* [60] might also be more effective than daily treatment. Finally, combinatorial therapies in which additional pathways are targeted (TORC1-independent pathways regulated by TSC1/TSC2 or Rheb, or survival pathways such as autophagy that are induced by rapamycin) hold promise for triggering the death of angiomyolipoma and LAM cells and avoiding the need for prolonged, even life-long therapy. A key question as preclinical models are chosen to guide the development of TSC therapies is how to balance the use of the various model systems, so as to bring effective therapies to clinical trials as efficiently and rapidly as possible. Xenograft models allow tumor growth to be readily measured and have short latency, providing rapid answers especially when different combinatorial regimens need to be compared, but are performed in animals without intact immune systems and in tumors whose vascular and stromal network may not replicate a human tumor. Tumors in the "endogenous" tumor models (i.e., the heterozygous mouse models) are more difficult to monitor, potentially leading to a much larger number of animals per experimental arm, and may take up to 18–24 months to complete since some of the manifestations have long latency and incomplete penetrance, such as the uterine leiomyomas in the Eker rat model, and still may not replicate the types of tumors in humans with TSC. For example, the epithelial lesions in the kidneys of the $Tsc1^{+/-}$ mice, $Tsc2^{+/-}$ mice and the Eker rat versus the mesenchymal angiomyolipomas in humans. Given the competing needs of speed and cost versus clinical relevance, and the simple fact that none of the models recapitulates the primary tumor phenotypes of TSC, the optimal approach may be to take advantage of all of the available models while supporting studies to identify biological readouts (biomarkers, functional imaging, or other indicators of disease activity) of TSC and LAM in humans, thereby allowing promising therapies to be readily tested in humans.

ACKNOWLEDGMENTS

Research in our laboratory is supported by grants from the Adler Foundation, the LAM Treatment Alliance, The LAM Foundation, the Tuberous Sclerosis Alliance, the National

Institutes of Health (NHLBI and NIDDK), and the Polycystic Kidney Disease Foundation. We thank Dr. Khadijah Hindi for critical review of the manuscript.

REFERENCES

1. Crino, P.B., Nathanson, K.L., and Henske, E.P. (2006). The tuberous sclerosis complex. *N Engl J Med* 355:1345–1356.
2. Carbonara, C., *et al.* (1994). 9q34 loss of heterozygosity in a tuberous sclerosis astrocytoma suggests a growth suppressor-like activity also for the TSC1 gene. *Hum Mol Genet* 3:1829–1832.
3. Carsillo, T., Astrinidis, A., and Henske, E.P. (2000). Mutations in the tuberous sclerosis complex gene TSC2 are a cause of sporadic pulmonary lymphangioleiomyomatosis. *Proc Natl Acad Sci USA* 97:6085–6090.
4. Chan, J.A., *et al.* (2004). Pathogenesis of tuberous sclerosis subependymal giant cell astrocytomas: biallelic inactivation of TSC1 or TSC2 leads to mTOR activation. *J Neuropathol Exp Neurol* 63:1236–1242.
5. Crooks, D.M., *et al.* (2004). Molecular and genetic analysis of disseminated neoplastic cells in lymphangioleiomyomatosis. *Proc Natl Acad Sci USA* 101:17462–17467.
6. Green, A., Johnson, P., and Yates, J. (1994). The tuberous sclerosis gene on chromosome 9q34 acts as a growth suppressor. *Hum Mol Genet* 3:1833–1834.
7. Han, S., *et al.* (2004). Phosphorylation of tuberin as a novel mechanism for somatic inactivation of the tuberous sclerosis complex proteins in brain lesions. *Cancer Res* 64:812–816.
8. Henske, E.P., Neumann, H.P., Scheithauer, B.W., Herbst, E.W., Short, M.P., and Kwiatkowski, D.J. (1995). Loss of heterozygosity in the tuberous sclerosis (TSC2) region of chromosome band 16p13 occurs in sporadic as well as TSC-associated renal angiomyolipomas. *Genes Chromosomes Cancer* 13:295–298.
9. Henske, E.P., *et al.* (1996). Allelic loss is frequent in tuberous sclerosis kidney lesions but rare in brain lesions. *Am J Hum Genet* 59:400–406.
10. Karbowniczek, M., *et al.* (2003). Recurrent lymphangiomyomatosis after transplantation: genetic analyses reveal a metastatic mechanism. *Am J Respir Crit Care Med* 167:976–982.
11. Niida, Y., *et al.* (2001). Survey of somatic mutations in tuberous sclerosis complex (TSC) hamartomas suggests different genetic mechanisms for pathogenesis of TSC lesions. *Am J Hum Genet* 69:493–503.
12. Yu, J., Astrinidis, A., and Henske, E.P. (2001). Chromosome 16 loss of heterozygosity in tuberous sclerosis and sporadic lymphangiomyomatosis. *Am J Respir Crit Care Med* 164:1537–1540.
13. Franz, D.N., *et al.* (2001). Mutational and radiographic analysis of pulmonary disease consistent with lymphangioleiomyomatosis and micronodular pneumocyte hyperplasia in women with tuberous sclerosis. *Am J Respir Crit Care Med* 164:661–668.
14. Moss, J., *et al.* (2001). Prevalence and clinical characteristics of lymphangioleiomyomatosis (LAM) in patients with tuberous sclerosis complex. *Am J Respir Crit Care Med* 164:669–671.
15. Chorianopoulos, D., and Stratakos, G. (2008). Lymphangioleiomyomatosis and tuberous sclerosis complex. *Lung* 186:197–207.
16. Goncharova, E.A., and Krymskaya, V.P. (2008). Pulmonary lymphangioleiomyomatosis (LAM): progress and current challenges. *J Cell Biochem* 103:369–382.

17. Hohman, D.W., Noghrehkar, D., and Ratnayake, S. (2008). Lymphangioleiomyomatosis: a review. *Eur J Intern Med* 19:319–324.
18. Juvet, S.C., McCormack, F.X., Kwiatkowski, D.J., and Downey, G.P. (2007). Molecular pathogenesis of lymphangioleiomyomatosis: lessons learned from orphans. *Am J Respir Cell Mol Biol* 36:398–408.
19. McCormack, F.X. (2008). Lymphangioleiomyomatosis: a clinical update. *Chest* 133:507–516.
20. Sullivan, E.J. (1998). Lymphangioleiomyomatosis: a review. *Chest* 114:1689–1703.
21. Castro, M., Shepherd, C.W., Gomez, M.R., Lie, J.T., and Ryu, J.H. (1995). Pulmonary tuberous sclerosis. *Chest* 107:189–195.
22. Sato, T., *et al.* (2002). Mutation analysis of the TSC1 and TSC2 genes in Japanese patients with pulmonary lymphangioleiomyomatosis. *J Hum Genet* 47:20–28.
23. Bittmann, I., Rolf, B., Amann, G., and Lohrs, U. (2003). Recurrence of lymphangioleiomyomatosis after single lung transplantation: new insights into pathogenesis. *Hum Pathol* 34:95–98.
24. Yu, J., and Henske, E.P. (2009). mTOR activation, lymphangiogenesis, and estrogen-mediated cell survival: the "perfect storm" of pro-metastatic factors in LAM pathogenesis. *LRB,* in press.
25. Yu, J., Parkhitko, A., and Henske, E.P. (2009). mTOR signaling and autophagy: roles in lymphangioleiomyomatosis therapy. *Proc Am Thorac Soc,* in press.
26. Plank, T.L., Yeung, R.S., and Henske, E.P. (1998). Hamartin, the product of the tuberous sclerosis 1 (TSC1) gene, interacts with tuberin and appears to be localized to cytoplasmic vesicles. *Cancer Res* 58:4766–4770.
27. van Slegtenhorst, M., *et al.* (1998). Interaction between hamartin and tuberin, the TSC1 and TSC2 gene products. *Hum Mol Genet* 7:1053–1057.
28. Gao, X., *et al.* (2002). Tsc tumour suppressor proteins antagonize amino-acid-TOR signalling. *Nat Cell Biol* 4:699–704.
29. Inoki, K., Li, Y., Zhu, T., Wu, J., and Guan, K.L. (2002). TSC2 is phosphorylated and inhibited by Akt and suppresses mTOR signalling. *Nat Cell Biol* 4:648–657.
30. Jaeschke, A., *et al.* (2002). Tuberous sclerosis complex tumor suppressor-mediated S6 kinase inhibition by phosphatidylinositide-3-OH kinase is mTOR independent. *J Cell Biol* 159:217–224.
31. Garami, A., *et al.* (2003). Insulin activation of Rheb, a mediator of mTOR/S6K/4E-BP signaling, is inhibited by TSC1 and 2. *Mol Cell* 11:1457–1466.
32. Inoki, K., Li, Y., Xu, T., and Guan, K.L. (2003). Rheb GTPase is a direct target of TSC2 GAP activity and regulates mTOR signaling. *Genes Dev* 17:1829–1834.
33. Saucedo, L.J., Gao, X., Chiarelli, D.A., Li, L., Pan, D., and Edgar, B.A. (2003). Rheb promotes cell growth as a component of the insulin/TOR signalling network. *Nat Cell Biol* 5:566–571.
34. Stocker, H., *et al.* (2003). Rheb is an essential regulator of S6K in controlling cell growth in *Drosophila*. *Nat Cell Biol* 5:559–566.
35. Tee, A.R., Manning, B.D., Roux, P.P., Cantley, L.C., and Blenis, J. (2003). Tuberous sclerosis complex gene products, Tuberin and Hamartin, control mTOR signaling by acting as a GTPase-activating protein complex toward Rheb. *Curr Biol* 13:1259–1268.
36. Zhang, Y., Gao, X., Saucedo, L.J., Ru, B., Edgar, B.A., and Pan, D. (2003). Rheb is a direct target of the tuberous sclerosis tumour suppressor proteins. *Nat Cell Biol* 5:578–581.
37. Blume-Jensen, P., and Hunter, T. (2001). Oncogenic kinase signalling. *Nature* 411:355–365.

38. Kozma, S.C., and Thomas, G. (2002). Regulation of cell size in growth, development and human disease: PI3K, PKB and S6K. *Bioessays* 24:65–71.

39. Shah, O.J., Anthony, J.C., Kimball, S.R., and Jefferson, L.S. (2000). 4E-BP1 and S6K1: translational integration sites for nutritional and hormonal information in muscle. *Am J Physiol Endocrinol Metab* 279:E715–E729.

40. Dan, H.C., *et al.* (2002). Phosphatidylinositol 3-kinase/Akt pathway regulates tuberous sclerosis tumor suppressor complex by phosphorylation of tuberin. *J Biol Chem* 277:35364–35370.

41. Manning, B.D., Tee, A.R., Logsdon, M.N., Blenis, J., and Cantley, L.C. (2002). Identification of the tuberous sclerosis complex-2 tumor suppressor gene product tuberin as a target of the phosphoinositide 3-kinase/akt pathway. *Mol Cell Biol* 10:151–162.

42. Potter, C.J., Pedraza, L.G., and Xu, T. (2002). Akt regulates growth by directly phosphorylating Tsc2. *Nat Cell Biol* 4:658–665.

43. Roux, P.P., Ballif, B.A., Anjum, R., Gygi, S.P., and Blenis, J. (2004). Tumor-promoting phorbol esters and activated Ras inactivate the tuberous sclerosis tumor suppressor complex via p90 ribosomal S6 kinase. *Proc Natl Acad Sci USA* 101:13489–13494.

44. Ma, L., Chen, Z., Erdjument-Bromage, H., Tempst, P., and Pandolfi, P.P. (2005). Phosphorylation and functional inactivation of TSC2 by Erk implications for tuberous sclerosis and cancer pathogenesis. *Cell* 121:179–193.

45. Li, Y., Inoki, K., Vacratsis, P., and Guan, K.L. (2003). The p38 and MK2 kinase cascade phosphorylates tuberin, the tuberous sclerosis 2 gene product, and enhances its interaction with 14-3-3. *J Biol Chem* 278:13663–13671.

46. Inoki, K., *et al.* (2006). TSC2 integrates Wnt and energy signals via a coordinated phosphorylation by AMPK and GSK3 to regulate cell growth. *Cell* 126:955–968.

47. Inoki, K., Zhu, T., and Guan, K.L. (2003). TSC2 mediates cellular energy response to control cell growth and survival. *Cell* 115:577–590.

48. Stevens, C., *et al.* (2009). Peptide combinatorial libraries identify TSC2 as a death-associated protein kinase (DAPK) death domain-binding protein and reveal a stimulatory role for DAPK in mTORC1 signaling. *J Biol Chem* 284:334–344.

49. Astrinidis, A., Senapedis, W., Coleman, T.R., and Henske, E.P. (2003). Cell cycle-regulated phosphorylation of hamartin, the product of the tuberous sclerosis complex 1 gene, by cyclin-dependent kinase 1/cyclin B. *J Biol Chem* 278:51372–51379.

50. Astrinidis, A., Senapedis, W., and Henske, E.P. (2006). Hamartin, the tuberous sclerosis complex 1 gene product, interacts with polo-like kinase 1 in a phosphorylation-dependent manner. *Hum Mol Genet* 15:287–297.

51. Lee, D.F., *et al.* (2007). IKK beta suppression of TSC1 links inflammation and tumor angiogenesis via the mTOR pathway. *Cell* 130:440–455.

52. El-Hashemite, N., Zhang, H., Henske, E.P., and Kwiatkowski, D.J. (2003). Mutation in TSC2 and activation of mammalian target of rapamycin signalling pathway in renal angiomyolipoma. *Lancet* 361:1348–1349.

53. Goncharova, E.A., *et al.* (2002). Tuberin regulates p70 S6 kinase activation and ribosomal protein S6 phosphorylation: a role for the TSC2 tumor suppressor gene in pulmonary lymphangioleiomyomatosis (LAM). *J Biol Chem* 277:30958–30967.

54. Karbowniczek, M., Yu, J., and Henske, E.P. (2003). Renal angiomyolipomas from patients with sporadic lymphangiomyomatosis contain both neoplastic and non-neoplastic vascular structures. *Am J Pathol* 162:491–500.

55. Yu, J., Astrinidis, A., Howard, S., and Henske, E.P. (2003). Estradiol and tamoxifen stimulate lymphangioleiomyomatosis-associated angiomyolipoma cell growth and activate both genomic and non-genomic signaling pathways. *Am J Physiol Lung Cell Mol Physiol* 286:L694–L700.

56. Crabtree, J.S., *et al.* (2009). Comparison of human and rat uterine leiomyomata: identification of a dysregulated mammalian target of rapamycin pathway. *Cancer Res* 69:6171–6178.
57. El-Hashemite, N., Walker, V., Zhang, H., and Kwiatkowski, D.J. (2003). Loss of Tsc1 or Tsc2 induces vascular endothelial growth factor production through mammalian target of rapamycin. *Cancer Res* 63:5173–5177.
58. Kenerson, H., Dundon, T.A., and Yeung, R.S. (2005). Effects of rapamycin in the Eker rat model of tuberous sclerosis complex. *Pediatr Res* 57:67–75.
59. Kenerson, H.L., Aicher, L.D., True, L.D., and Yeung, R.S. (2002). Activated mammalian target of rapamycin pathway in the pathogenesis of tuberous sclerosis complex renal tumors. *Cancer Res* 62:5645–5650.
60. Lee, N., Woodrum, C.L., Nobil, A.M., Rauktys, A.E., Messina, M.P., and Dabora, S.L. (2009). Rapamycin weekly maintenance dosing and the potential efficacy of combination sorafenib plus rapamycin but not atorvastatin or doxycycline in tuberous sclerosis preclinical models. *BMC Pharmacol* 9:8.
61. Messina, M.P., Rauktys, A., Lee, L., and Dabora, S.L. (2007). Tuberous sclerosis preclinical studies: timing of treatment, combination of a rapamycin analog (CCI-779) and interferon-gamma, and comparison of rapamycin to CCI-779. *BMC Pharmacol* 7:14.
62. Pollizzi, K., Malinowska-Kolodziej, I., Stumm, M., Lane, H., and Kwiatkowski, D. (2009). Equivalent benefit of mTORC1 blockade and combined PI3K-mTOR blockade in a mouse model of tuberous sclerosis. *Mol Cancer* 8:38.
63. Lee, L., Sudentas, P., and Dabora, S.L. (2006). Combination of a rapamycin analog (CCI-779) and interferon-gamma is more effective than single agents in treating a mouse model of tuberous sclerosis complex. *Genes Chromosomes Cancer* 45:933–944.
64. Lee, L., *et al.* (2005). Efficacy of a rapamycin analog (CCI-779) and IFN-gamma in tuberous sclerosis mouse models. *Genes Chromosomes Cancer* 42:213–227.
65. Yu, J.J., *et al.* (2009). Estrogen promotes the survival and pulmonary metastasis of tuberin-null cells. *Proc Natl Acad Sci USA* 106:2635–2640.
66. Ehninger, D., *et al.* (2008). Reversal of learning deficits in a Tsc2+/− mouse model of tuberous sclerosis. *Nat Med* 14:843–848.
67. Meikle, L., *et al.* (2008). Response of a neuronal model of tuberous sclerosis to mammalian target of rapamycin (mTOR) inhibitors: effects on mTORC1 and Akt signaling lead to improved survival and function. *J Neurosci* 28:5422–5432.
68. Xu, L., Zeng, L.H., and Wong, M. (2009). Impaired astrocytic gap junction coupling and potassium buffering in a mouse model of tuberous sclerosis complex. *Neurobiol Dis* 34:291–299.
69. Zeng, L.H., Xu, L., Gutmann, D.H., and Wong, M. (2008). Rapamycin prevents epilepsy in a mouse model of tuberous sclerosis complex. *Ann Neurol* 63:444–453.
70. Lee, P.S., *et al.* (2009). Rapamycin-insensitive up-regulation of MMP2 and other genes in TSC2-deficient LAM-like cells. *Am J Respir Cell Mol Biol.*
71. Yeung, R.S., Xiao, G.H., Jin, F., Lee, W.C., Testa, J.R., and Knudson, A.G. (1994). Predisposition to renal carcinoma in the Eker rat is determined by germ-line mutation of the tuberous sclerosis 2 (TSC2) gene. *Proc Natl Acad Sci USA* 91:11413–11416.
72. Bissler, J.J., *et al.* (2008). Sirolimus for angiomyolipoma in tuberous sclerosis complex or lymphangioleiomyomatosis. *N Engl J Med* 358:140–151.
73. Everitt, J.I., Wolf, D.C., Howe, S.R., Goldsworthy, T.L., and Walker, C. (1995). Rodent model of reproductive tract leiomyomata. Clinical and pathological features. *Am J Pathol* 146:1556–1567.
74. Cook, J.D., and Walker, C.L. (2004). The Eker rat: establishing a genetic paradigm linking renal cell carcinoma and uterine leiomyoma. *Curr Mol Med* 4:813–824.

75. Kwiatkowski, D.J., *et al.* (2002). A mouse model of TSC1 reveals sex-dependent lethality from liver hemangiomas, and up-regulation of p70S6 kinase activity in Tsc1 null cells. *Hum Mol Genet* 11:525–534.
76. Onda, H., Lueck, A., Marks, P.W., Warren, H.B., and Kwiatkowski, D.J. (1999). Tsc2(+/−) mice develop tumors in multiple sites that express gelsolin and are influenced by genetic background. *J Clin Invest* 104:687–695.
77. Howe, S.R., Gottardis, M.M., Everitt, J.I., Goldsworthy, T.L., Wolf, D.C., and Walker, C. (1995). Rodent model of reproductive tract leiomyomata. Establishment and characterization of tumor-derived cell lines. *Am J Pathol* 146:1568–1579.
78. Hino, O., *et al.* (1993). Spontaneous and radiation-induced renal tumors in the Eker rat model of dominantly inherited cancer. *Proc Natl Acad Sci USA* 90:327–331.
79. Walker, C., Goldsworthy, T.L., Wolf, D.C., and Everitt, J. (1992). Predisposition to renal cell carcinoma due to alteration of a cancer susceptibility gene. *Science* 255:1693–1695.
80. Howe, S.R., Gottardis, M.M., Everitt, J.I., and Walker, C. (1995). Estrogen stimulation and tamoxifen inhibition of leiomyoma cell growth in vitro and in vivo. *Endocrinology* 136:4996–5003.
81. Uhlmann, E.J., *et al.* (2002). Astrocyte-specific TSC1 conditional knockout mice exhibit abnormal neuronal organization and seizures. *Ann Neurol* 52:285–296.
82. Meikle, L., *et al.* (2007). A mouse model of tuberous sclerosis: neuronal loss of Tsc1 causes dysplastic and ectopic neurons, reduced myelination, seizure activity, and limited survival. *J Neurosci* 27:5546–5558.
83. Davies, D.M., *et al.* (2008). Sirolimus therapy in tuberous sclerosis or sporadic lymphangioleiomyomatosis. *N Engl J Med* 358:200–203.
84. Franz, D.N., *et al.* (2006). Rapamycin causes regression of astrocytomas in tuberous sclerosis complex. *Ann Neurol* 59:490–498.
85. Brugarolas, J.B., Vazquez, F., Reddy, A., Sellers, W.R., and Kaelin, W.G., Jr. (2003). TSC2 regulates VEGF through mTOR-dependent and -independent pathways. *Cancer Cell* 4:147–158.
86. Karbowniczek, M., Cash, T., Cheung, M., Robertson, G.P., Astrinidis, A., and Henske, E.P. (2004). Regulation of B-Raf kinase activity by tuberin and Rheb is mammalian target of rapamycin (mTOR)-independent. *J Biol Chem* 279:29930–29937.
87. Karbowniczek, M., Robertson, G.P., and Henske, E.P. (2006). Rheb inhibits C-raf activity and B-raf/C-raf heterodimerization. *J Biol Chem* 281:25447–25456.
88. Gau, C.L., Kato-Stankiewicz, J., Jiang, C., Miyamoto, S., Guo, L., and Tamanoi, F. (2005). Farnesyltransferase inhibitors reverse altered growth and distribution of actin filaments in Tsc-deficient cells via inhibition of both rapamycin-sensitive and -insensitive pathways. *Mol Cancer Ther* 4:918–926.
89. Huang, J., Dibble, C.C., Matsuzaki, M., and Manning, B.D. (2008). The TSC1-TSC2 complex is required for proper activation of mTOR complex 2. *Mol Cell Biol* 28:4104–4115.
90. Wilson, C., *et al.* (2006). Tsc1 haploinsufficiency without mammalian target of rapamycin activation is sufficient for renal cyst formation in Tsc1+/− mice. *Cancer Res* 66:7934–7938.
91. Hartman, T.R., *et al.* (2009). The tuberous sclerosis proteins regulate formation of the primary cilium via a rapamycin-insensitive and polycystin 1-independent pathway. *Hum Mol Genet* 18:151–163.
92. Zhou, X., Ikenoue, T., Chen, X., Li, L., Inoki, K., and Guan, K.L. (2009). Rheb controls misfolded protein metabolism by inhibiting aggresome formation and autophagy. *Proc Natl Acad Sci USA* 106:8923–8928.

93. Thoreen, C.C., *et al.* (2009). An ATP-competitive mammalian target of rapamycin inhibitor reveals rapamycin-resistant functions of mTORC1. *J Biol Chem* 284:8023–8032.
94. Karbowniczek, M., *et al.* (2010). The evolutionarily conserved TSC/Rheb pathway activates Notch in tuberous sclerosis complex and *Drosophila* external sensory organ development. *J Clin Invest* 120:93–102.
95. Zhou, J., Brugarolas, J., and Parada, L.F. (2009). Loss of Tsc1, but not Pten, in renal tubular cells causes polycystic kidney disease by activating mTORC1. *Hum Mol Genet* 18:4428–4441.
96. Chen, C., Liu, Y., Liu, R., Ikenoue, T., Guan, K.L., and Zheng, P. (2008). TSC-mTOR maintains quiescence and function of hematopoietic stem cells by repressing mitochondrial biogenesis and reactive oxygen species. *J Exp Med* 205:2397–2408.
97. Mori, H., *et al.* (2009). Critical role for hypothalamic mTOR activity in energy balance. *Cell Metab* 9:362–374.
98. Rachdi, L., *et al.* (2008). Disruption of Tsc2 in pancreatic beta cells induces beta cell mass expansion and improved glucose tolerance in a TORC1-dependent manner. *Proc Natl Acad Sci USA* 105:9250–9255.

17

Chemistry and Pharmacology of Rapamycin and Its Derivatives

ROBERT T. ABRAHAM[a] • JAMES J. GIBBONS[a] •
EDMUND I. GRAZIANI[b]

[a]Center for Integrative Biology and Biotherapeutics
Pfizer Biopharmaceuticals, Pearl River
New York, USA

[b]Pfizer Worldwide Medicinal Chemistry
Groton, Connecticut, USA

I. Abstract

Rapamycin is a bacterially derived natural product with a remarkable history as both a chemical probe for studies of cell growth control-related pathways, and a *bona fide* drug with established or predicted clinical activities in a variety of disease settings. Rapamycin was first noted as a potent antifungal and immunosuppressive agent, and studies of this drug's mechanism of action revealed that rapamycin was a surgically precise inhibitor of a novel protein serine–threonine kinase, appropriately termed the target of rapamycin (TOR). Intensive research efforts have revealed that the TOR orthologs function in a highly conserved pathway of eukaryotic cell growth control. This review focuses on the impact of rapamycin exposure on the TOR ortholog (termed mTOR) expressed in mammalian cells. We briefly describe the biosynthesis of rapamycin and the chemical modifications of the parent compound that yielded several of the experimentally useful and/ or clinically active derivatives, collectively termed rapalogs. We then review in some detail the pharmacology of rapamycin, particularly as it

DOI: 10.1016/S1874-6047(10)27017-8

relates to the growth and proliferation of cancer cells. The opportunities and challenges associated with the development of rapalogs as anticancer agents are then discussed. Finally, we briefly review some provocative recent insights into the effects of rapamycin on the immune system and on organismal aging.

II. Introduction

The inhabitants of Easter Island (*Rapa nui*) erected the world-famous stone monuments (*moai*) during a short but intense period of creative artistry and engineering. A much less spectacular but equally noteworthy event in this volcanic island's tumultuous history occurred in 1965, when a Canadian expeditionary team collected a soil sample from the *Rano kau* region [1]. The soil sample wound its way back to the Ayerst Research Laboratories in Canada, where the resident microbes were grown in culture, and the resulting culture supernatants were tested for antimicrobial activities. The culture supernatant from the *Rano kau* sample contained a potent antifungal activity. Ten years after its arrival at the Ayerst labs, the compound responsible for this activity was identified as rapamycin (sirolimus, Figure 17.1), a secreted product of the bacterial strain, *Streptomyces hygroscopicus* [2]. The Easter Islanders were obviously highly motivated to invest so much effort into the erection of the *moai*: the same can be said for a soil bacterium that invests significant amounts of precious metabolic energy in the synthesis of a macrolactone that is not directly involved in bacterial reproduction. However, microbes also gain a reproductive advantage by out-competing their neighbors for limiting supplies of cell mass-building nutrients. Over millions of years, *S. hygroscopicus* created a remarkable molecule that surgically disrupted a nutrient sensing and response mechanism used by the fungal cohabitants of the volcanic soil of Easter Island.

The investigators who first isolated rapamycin may have appreciated its contributions to the survival of *S. hygroscopicus*, but they could not have anticipated the value that this natural product would deliver to both basic and clinical science over the 40-plus years that followed this seminal discovery. Rapamycin was (and still is) a pivotal chemical probe that enabled the characterization of a previously unappreciated and centrally important mechanism of eukaryotic cell growth control. Furthermore, this drug served as the chemical bait that allowed investigators to capture and characterize the target of rapamycin (TOR; also termed mTOR in mammals), a central component of the nutrient-sensing and growth-regulatory machinery in eukaryotic cells [3–6]. In addition to its utility as a research tool for cell biologists, rapamycin

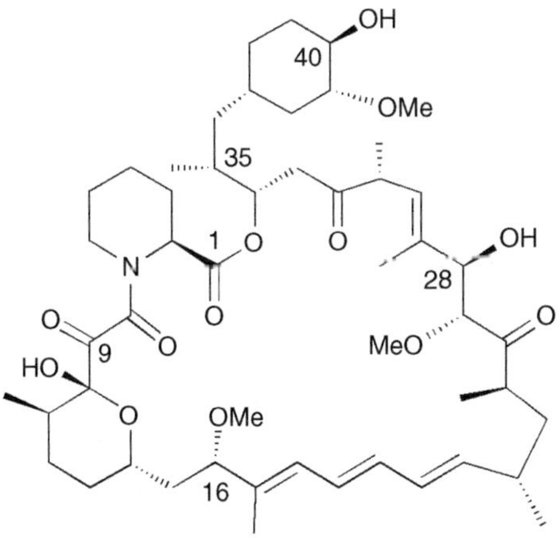

Fig. 17.1. The structure of rapamycin.

and its derivatives (collectively termed "rapalogs") have established clinical activities in transplantation, cardiovascular disease, and oncology, and this list of disease targets will likely grow over the next several years. For full expositions of the history of rapamycin as an experimental and clinical drug, the reader is referred to several earlier reviews [1, 3, 7, 8].

The objectives of the current review are to provide updated information regarding the chemistry and *in vitro* and *in vivo* pharmacology of rapamycin, with a particular focus on the immunological and anticancer activities of this "gift from nature" to science and clinical medicine. Since the identification of mTOR in the mid-1990s, an increasingly broad-based research effort has revealed important new facets of rapamycin pharmacology, and expanding opportunities for the development of rapalogs as therapies for multiple diseases with high unmet clinical needs. Space limitations preclude a comprehensive overview of all of the potential clinical applications of the rapalogs, but we believe that recent lessons learned during the development of the rapalogs in the oncology space will be illustrative of the clinical opportunities and challenges presented by these "first-generation" mTOR inhibitors.

III. Primer on the Mechanism of Action of Rapamycin

At pharmacologically relevant drug concentrations, rapamycin selectively targets a single, conserved component of the eukaryotic proteome. The TOR proteins are members of a family of protein serine–threonine kinases whose catalytic domains bear a clear evolutionary relationship to those of the phosphatidylinositol 3-kinases (PI3Ks) [9]. The PI3K-related kinases (PIKKs) as a group play critical roles in cell growth, proliferation, and stress responses. The TOR proteins reside in two structurally and functionally distinct complexes, termed TOR complexes 1 and 2 (TORC1 and TORC2) [10–12]. In yeast, two *TOR* genes encode highly related Tor proteins that selectively home to TORC1 or TORC2. Surprisingly, meta-zoan cells express only one TOR protein (mTOR in mammals) that can be recruited into either mammalian TORC (mTORC). Remarkably, rapamy-cin selectively and directly inhibits TOR function only in the context of TORC1, although this drug can indirectly suppress the assembly and hence the function of mTORC2 complexes in certain mammalian cell types [13, 14].

In mammalian cells, mTORC1 comprises at least five proteins, including Raptor, PRAS40, LST8 (also termed $G_{\beta}L$), DEPTOR, and mTOR itself [10, 15, 16]. The Raptor subunit plays key roles in the coordination of both signaling inputs into and substrate phosphorylation by mTOR [17, 18]. As detailed in earlier reviews [16, 18, 19], the rapamycin-sensitive functions of mTORC1 are focused mainly on the regulation of mRNA translation and related metabolic events that support cell and tissue mass accumulation. In mammals, mTORC1 represents a nodal hub that coordinates upstream signals related to growth factor and nutrient availability, as well as the cellular bioenergetic state, with anabolic metabolism and cell proliferation. Many, but not all of the signals that govern mTORC1 activity do so indirectly, through the heterodimeric tuberous sclerosis complex (TSC) [16, 20–22]. When supplies of nutrients, bioenergy, and growth factors are low, mTORC1 activity is suppressed, and cells conserve metabolic building blocks and energetic precursors by shifting into a starvation mode. The G_1 phase of the cell cycle is most sensitive to alterations that interfere with the anabolic activities required for mitotic cell division; hence, it is not surprising that inhibition of mTORC1 by rapamycin causes cycling cells to accumulate in G_1 phase [18, 23, 24].

Elegant cellular, biochemical, and structural studies have established the pharmacological mechanism of mTORC1 inhibition by rapamycin [3, 25–27]. These drugs are relatively hydrophobic, and readily cross the plasma membranes of mammalian cells. In the cytoplasm, rapamycin binds to a ubiquitously expressed member of the family of peptidyl-prolyl isomerases (also termed immunophilins), a 12 kDa protein named FK506-binding protein (FKBP) 12. The resulting FKBP12·rapamycin (FRB) complex represents the proximate inhibitor of mTORC1 signaling. However, recent studies suggest that complexation with FKBP12 does not represent an obligate gain-of-function event for rapamycin, as posited in earlier models. Free rapamycin is capable of binding to and inhibiting mTORC1, but the affinity of this interaction is increased by approximately three orders of magnitude in the presence of FKBP12 [28, 29]. Parenthetically, the hydrophobic nature of many rapalogs presumably allows the free drug to accumulate in membrane lipid bilayers to much higher concentrations that those found in the cytoplasm. Thus, cellular exposure to nanomolar concentrations of rapamycin could conceivably lead to micromolar concentrations of drug in membrane-rich compartments, and achieve localized FKBP12-independent suppression of mTORC1 activity. The significance of this potential membrane partitioning of rapalogs is currently uncertain, but could contribute significantly to their overall pharmacological actions, especially when chronically administered at high doses, as is generally the case in the oncology setting.

From a pharmacological perspective, the FKBP12-bound form of rapamycin has received much greater scrutiny than the free drug, and a powerful combination of genetic and structural analyses have shown that the FKBP12·rapalog complex binds directly to mTOR at a conserved stretch of ∼110 amino acids, termed the FRB-binding domain [26]. The FRB domain is positioned amino-terminally relative to the mTOR kinase domain; nonetheless, numerous studies have shown that the immunophilin·drug complex inhibits mTOR (i.e., mTORC1) kinase activity *in vitro*, as well as mTORC1-dependent phosphorylation events in intact cells. Hence, the FKBP12·rapalog complex functions as an allosteric inhibitor of the mTOR catalytic domain, which may explain observations that this complex is a potent but partial antagonist of mTORC1-dependent responses in mammalian cells [14, 30]. Given that mTOR itself is the direct ligand for the FKBP12·rapalog complex, the structural elements that dictate the striking specificity of this complex for mTORC1 relative to mTORC2 remain an intriguing but elusive facet of rapamycin pharmacology.

Numerous sequelae are triggered by exposure of mammalian cells to rapamycin. As stated above, mTORC1 plays well-established roles in the upregulation of protein synthesis in cells that are undergoing hypertrophy or preparing to divide. In general, however, rapamycin does not cause a marked suppression of global mRNA translation at concentrations that maximally suppress the phosphorylation of two key mTORC1 substrates, eIF-4E-binding protein 1 (4E-BP1) and the S6 protein kinases (S6Ks) [19, 29]. Rather, rapamycin exerts selective effects on protein synthesis, with many of the most sensitive translational targets being mRNAs encoding proteins (e.g., Myc and cyclin D_1) associated with cell growth and proliferation [31, 32]. The relatively subtle actions of rapamycin on protein synthesis may explain why this drug functions primarily as a cytostatic agent, in contrast to the much more cytotoxic outcomes associated with cellular exposure to global inhibitors of protein synthesis, such as cycloheximide.

One additional response to rapamycin that bears special mention is the induction of macroautophagy (hereafter termed autophagy), a recycling mechanism that allows cells to digest macromolecules and entire organelles into metabolic or energetic precursors [33, 34]. Autophagy serves as a protective response during times of starvation or stress, but also supports many other functions that contribute to overall organismal homeostasis, including the removal of dysfunctional mitochondria and immune stimulation by intracellular pathogens [35–38]. In fed cells, mTORC1 activity is high, and autophagic activity is suppressed; during starvation, mTORC1 activity declines, leading to increased autophagy [34, 39, 40]. Rapamycin short-circuits this mTORC1-dependent regulatory mechanism, such that autophagic activity is elevated, in spite of the presence of ample supplies of nutrients. The exact mechanism through which rapamycin stimulates autophagy remains poorly understood, but recent data indicate that mTORC1 serves as an upstream regulator of a conserved component of the autophagic machinery, a protein kinase termed autophagy gene 1 (Atg1) in yeast and Unc 51-like kinase (ULK1) in mammals [41]. The stimulatory effects of rapalogs on autophagic activity have important implications for the clinical use of these drugs as anticancer and immunomodulatory agents. Furthermore, a number of neurodegenerative diseases, including Huntington's, Alzheimer's, and Parkinson's disease, are characterized by the accumulation of misfolded protein aggregates. Considerable excitement has been generated by recent reports that certain mTORC1 inhibitors increase the clearance of neurotoxic protein aggregates and ameliorate disease symptoms in preclinical models of neurodegenerative diseases [33].

IV.   Biosynthesis and Medicinal Chemistry of Rapamycin and Its Analogs

Rapamycin is a complex natural product synthesized enzymatically by the soil-dwelling bacterium *S. hygroscopicus* NRRL 5491 and is the founding member of the small class of natural products known as immunophilin ligands. The structure of rapamycin[1], as revealed by X-ray crystallography [42] and a landmark high-field NMR study [43], was strikingly unique at the

---

1. At least two numbering systems are current used for rapamycin. The literature precedent (see ref. 42) assigns C1 to an olefinic methine and proceeds around the macrocyclic ring in the direction of the pipecolic acid moiety, with the cyclohexane-ring containing side-chain and appended methyl/methoxys following. The system in wider use, and the one followed here, assigns C1 to the highest priority lactone carbon of the macrolide, and proceeds around the ring also towards the pipecolate moiety (though with the opposite sense) through to the cyclohexyl moiety, with similar assignments for the methyls/methoxys.

time that it was discovered. One of the largest macrolide rings (31 atoms) then known, the compound consists of an extended cyclohexane starter side chain, elaborated with a polyketide chain and pipecolic acid (the six-membered ring analog of proline), before terminating in a macrolactone.

Rapamycin has been the subject of extensive biosynthetic investigations since the sequencing of the biosynthetic gene cluster [44] that encodes the enzyme complexes responsible for the compound's production in *S. hygroscopicus*. Briefly, a modular (Type I) polyketide synthase (PKS) is primed with a starter unit, 4,5-dihydroxycyclohex-1-enecarboxylic acid (derived from shikimate). This enzyme complex then extends a chain of carbon atoms from this starter unit in a series of condensation reactions using either two carbon (malonyl CoenzymeA (CoA)) or three carbon (2-methylmalonyl CoA) building blocks, in a manner analogous to fatty acid synthesis (Figure 17.2). The PKS installs the correct stereochemistry and oxidation state (ketone, alcohol, olefin, or alkane) of the new so-called extender unit immediately following each condensation step, depending upon the presence and functionality (ketoreductase (KR), enoylreductase (ER), or dehydrogenase (DH)) of each module, much like an assembly line. The amino acid, L-pipecolate, is subsequently incorporated at the terminus of the linear polyketide product via the nonribosomal peptide synthetase (NRPS) gene product of *rapP*, which also catalyzes macrocyclization to yield the first enzyme free product, so-called "prerapamycin." A series of post-PKS "tailoring" steps, including cytochrome P450 catalyzed oxidation and methyl transfer, yields rapamycin. Several excellent reviews [45–47] provide many of the details behind the PKS/NRPS/tailoring steps as outlined here.

With the later discovery of the related compounds (Figure 17.3), FK-506 [48] and FK-520 [49] (the latter turning out to be identical to the previously isolated compound, ascomycin [50]), investigations into the biological activity and mechanism of action of these compounds yielded important breakthroughs. In seminal chemical biology experiments, rapamycin and FK-506 were immobilized on solid supports and used as affinity probes, or "molecular baits," to identify putative binding partners [51, 52]. These efforts led to the identification of FKBP12, the founding member of the family of FKBPs, a subfamily of the immunophilins that also includes the cyclophilins (structurally unrelated proteins that bind to the immunosuppressant natural product, cyclosporine). The immunophilins possess peptidyl-prolyl *cis–trans* isomerase (PPIase) activities [53]. Over two dozen immunophilins are expressed in human cells, and are known to associate with a diverse array of proteins including ion channels [54–57], steroid receptor complexes [58], transcription factors [59], tubulin [60], and B-cell CLL/lymphoma 2 (Bcl-2) upon conditional activation [61]. The only other known naturally occurring ligands of the FKBP subfamily are the structurally related macrolides

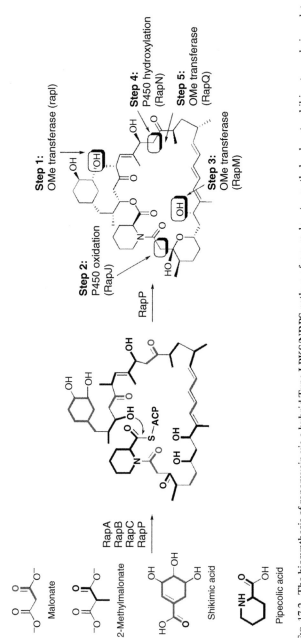

Fig. 17.2. The biosynthesis of rapamycin via a hybrid Type I PKS/NRPS pathway from malonate, methylmalonate, shikimate, and pipecolate.

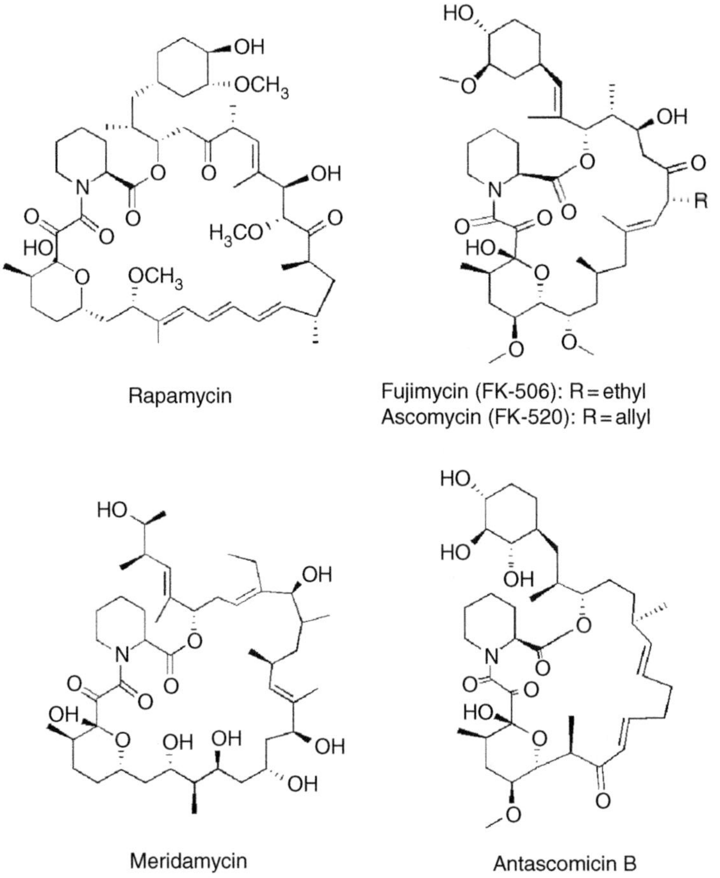

Rapamycin

Fujimycin (FK-506): R = ethyl
Ascomycin (FK-520): R = allyl

Meridamycin

Antascomicin B

FIG. 17.3. The structures of naturally occurring immunophilin ligands rapamycin, FK-506, FK-520 (ascomycin), meridamycin, and antascomicin B.

meridamycin [62] and antascomicin [63] which lack immunosuppressive activity, though 3-normeridamycin has demonstrated *in vitro* neuroprotection against 1-methyl-4-phenylpyridinium (MPP$^+$) induced toxicity in dopaminergic neurons [64].

The X-ray crystal structure [65] of rapamycin bound to FKBP12 revealed a number of key hydrogen bonds that account for the compound's high affinity ($K_d = 0.2$ nM) for this protein. The structural features common to both FK-506 and rapamycin (the pipecolic acid, tricarbonyl region from C-8 to C-10, and the lactone functionalities), make key contacts with the FKBP12-binding pocket. Specifically, important hydrogen bonds include:

the lactone carbonyl oxygen at C-1 to the backbone NH of Ile56, the amide carbonyl at C-8 to the phenolic group on the sidechain of Tyr82, and the hydroxyl proton at the hemiketal carbon, C-10, to the sidechain of Asp37.

As discussed above, binding of rapamycin to FKBP12 represents an activation step that renders the drug a far more potent inhibitor of the TOR proteins in eukaryotic cells [66]. The X-ray crystal structure [26] of the ternary complex showed that the strong binding affinity of the FRB complex for the FRB domain of mTOR (residues 2025–2114 of human mTOR) stems from several key hydrophobic contacts with the 340 Å^2 of solvent accessible surface area of rapamycin, with the notable absence of hydrogen bond-based interactions. Also, in contrast to the FKBP12·FK-506 interaction with its target protein, calcineurin [67, 68], there are no direct interactions between FKBP12 and mTOR in the rapamycin ternary complex. These hydrophobic contacts occur principally between the triene region (C-16 to C-23) of rapamycin and several aromatic residues in the FRB domain (Phe2039, Trp2101, Tyr2105, and Phe2108). A number of methyl groups also make hydrophobic contacts with residues in the FRB domain, notably the methoxy group at C-16 [69], and the methyls attached to C-23, C-29, and C-31. An extensive biophysical characterization [70] of the FRB complex with the FRB domain of mTOR has further confirmed the importance of these van der Waals contacts between mTOR and rapamycin, and corroborated the lack of any significant protein–protein interactions between FKBP12 and the FRB domain of mTOR in the absence of rapamycin. In this way, the allosteric inhibition of mTOR kinase activity occurs via binding of a specific conformation of the otherwise dynamic rapamycin macrocycle (preorganized as a result of binding to FKBP12) to the FRB domain of mTOR.

Programs that aim to identify semisynthetic or biosynthetic rapamycin analogs bearing improved pharmaceutical properties must, therefore, cope with the likelihood that modifications to the structure of rapamycin can affect binding to the FRB domain of mTOR, both directly and indirectly as a function of binding to FKBP12 [71, 72]. Moreover, the binding of the FRB complex to the FRB domain of mTOR is exquisitely sensitive to changes in the conformational flexibility of the rapamycin macrocycle, as well as the overall energetics driving the relative proportions of these conformational states. In other words, changes to one part of the molecule may alter the global conformation of the ring, leading to unpredictable effects on FKBP12- or mTOR-binding activity. Elucidation of a systematic structure–activity relationship (SAR) for the rapamycin analogs has, therefore, proven challenging. Nonetheless, a number of general trends have emerged over the course of such SAR efforts. Much of the data from the early medicinal chemistry programs resides in the patent literature, and has been reviewed elsewhere [73]. Modifications to the macrocyclic core of rapamycin invariably led to decreases in FKBP12-binding affinity, mTOR inhibition, or both parameters, with a concurrent loss of potency in cellular models of proliferation. It is tempting, therefore, to conclude that eons of evolutionary pressure yielded, in rapamycin, a parent natural product that is maximally optimized for potency against its protein target. In this context, then, every such rapamycin-based program that has led to a successful clinical candidate has explored substitution at the C-40 hydroxyl as an avenue to improve or otherwise modify the pharmacologic and pharmacokinetic parameters of the compound. In fact, all the analogs of rapamycin that are either in clinical usage or currently undergoing clinical trials (Figure 17.4) have different properties from rapamycin as a result of modification at C-40, including temsirolimus (CCI-779, Torisel®) [74], everolimus (RAD001, Affinitor®) [75], zotarolimus (ABT-578) [76–78], and ridaforolimus (deforolimus, AP23573, MK-8669) [79]. More recent reports that describe attempts to modulate the immunosuppressive activity [80] of rapamycin or improve its safety profile by reducing its systemic exposure [81] have also focused on modifications at C-40, and to a lesser extent, C-16.

Rapamycin continues to attract and challenge the skills of organic chemists, as demonstrated by the large and growing number of elegant

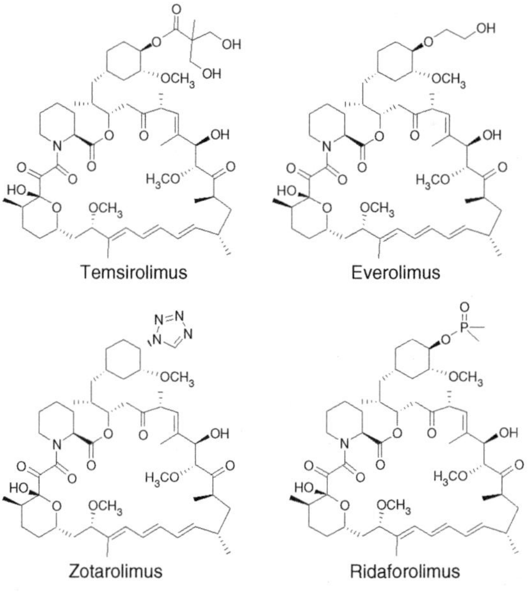

Temsirolimus

Everolimus

Zotarolimus

Ridaforolimus

FIG. 17.4. The structures of rapamycin analogs currently in clinical use or undergoing clinical trials.

total syntheses [82–88]. Rapamycin analogs have been prepared by precursor-directed biosynthesis [89, 90], biotransformation [91], and manipulation [92] of the genes responsible for rapamycin biosynthesis. More recently, an engineered mutant strain of *S. hygroscopicus*, MG2-10, in which all the genes thought to be responsible for the downstream tailoring steps of the post-PKS/NRPS product were removed by double recombination, has enabled production of a number of synthetically inaccessible rapamycin analogs [93–96].

Rapamycin and FK-506 exhibit potent neuroprotective activity *in vitro* [97] that has been attributed to FKBP interactions [98] rather than mTOR inhibition. However, it has also been demonstrated that while the immunophilin ligand FK-506 showed efficacy in a rodent model of ischemic stroke [99], rapamycin had no effect in this model, suggesting that immunophilin binding may only partially explain this neuroprotective activity. Efforts to develop therapeutically useful compounds by uncoupling neuroprotection from immunosuppression in this class of compounds have met with limited success [100]. Modification of rapamycin at the mTOR binding region yielded the novel immunophilin ligand, ILS-920, which has potent neuroprotective activity in cortical neuronal cultures, demonstrated efficacy in a rodent model for ischemic stroke, and carries significantly reduced immunosuppressive activity [101]. Affinity methods were used to identify FKBP52 and the β_1 subunit of L-type Ca^{2+} channels (CACNB1) as binding partners for the compound. Neurite outgrowth was significantly increased in FKBP52 siRNA-treated neurons, suggesting that inhibition of FKBP52 function underlies the effect of ILS-920 on this response.

In conclusion, the chemistry and biosynthesis of rapamycin and its analogs continues to inspire and challenge medicinal chemists and microbial engineers. Ongoing studies on the details and mechanism of rapamycin biosynthesis, enzymology, and synthetic biology are providing ever-deeper insights into PKS/NRPS biosynthesis and post-PKS reactions. In addition, advances in the medicinal chemistry and chemical biology of nonimmunosuppressive rapamycin analogs have revealed new opportunities for the development of rapamycin-based compounds with distinct pharmacological profiles, and potential applications in the treatment of neurodegenerative diseases and ischemic stroke.

V.   Anticancer Activities of the Rapalogs

Rapamycin's potent antiproliferative effect on T lymphocytes *in vitro* and in allograft rejection models in animals naturally led to its evaluation as an anticancer drug [102]. Early efforts revealed that considerably higher doses of rapamycin were needed to demonstrate antitumor activity in preclinical mouse models [103]. These findings discouraged further evaluation of rapamycin, due to concerns that toxicity would preclude achievement of a therapeutically efficacious dose in humans. The observation that some cancer cell lines were up to 3 orders of magnitude more sensitive to rapamycin than other lines suggested that the cytostatic effects of the drug were selective rather than global, and, in turn, that appropriate patient selection strategies might make rapamycin a safe and effective treatment for certain cancers [104]. This observation, coupled with the findings that rapamycin produced a unique pattern of responsiveness when compared to known anticancer drugs in the NCI-60 cell line panel suggested a novel mechanism of action, and rekindled interest in rapamycin as an anticancer drug [105, 106].

Rapamycin's physicochemical properties (e.g., high molecular weight and hydrophobicity) and poor oral bioavailability (10–15%) prompted the development of analogs (rapalogs) with improved pharmaceutical properties, including temsirolimus, everolimus, and ridaforolimus [107]. As discussed above, these efforts have met with limited success; however, a more pervasive problem has been the delineation of the molecular bases of drug sensitivity and resistance. In addition, this drug has proven surprisingly difficult to combine in a productive fashion with either conventional or molecular targeted anticancer drugs, due mainly to poor tolerability of drug combinations tested to date.

A. THE mTOR PATHWAY AND CARCINOGENESIS

A major breakthrough in mTOR biology as it relates to cancer was the discovery that growth factors signal to mTOR through activation of Class I PI3Ks, and that this pathway, through the intermediate activation of AKT, converged on the TSC2 (Tuberin) of the TSC heterodimer [108–111]. Phosphorylation of TSC2 by AKT relieves the negative regulatory input from the TSC2 to RHEB (Ras homolog expressed in brain), a small GTP binding protein that functions as a proximal activator of mTORC1 [112, 113]. This pathway is deregulated in the majority of solid and hematopoietic tumors by gain-of-function mutations in growth factor receptor signaling pathways, loss of tumor suppressors such as PTEN (phosphatase related to tension located on chromosome 10), and activating mutations in the PI3K catalytic subunit itself [114]. Several inheritable syndromes characterized by benign tissue overgrowth (sometimes leading to malignancy) are caused by mutations in genes encoding regulators of the PI3K → AKT → TSC2 → mTOR pathway [115]. These include: PTEN in Cowden's disease, tuberin (TSC2) or hamartin (TSC1) in tuberous sclerosis, neurofibromin (NF-1) in neurofibromatosis, and the LKB1 kinase in Peutz-Jehger syndrome. These diseases are characterized by constitutive mTORC1 activity, benign tissue overgrowths (hamartomas) or fibromas, and increased risk of malignancy [115]. Mutation or deletion of PTEN also occurs with high frequency across multiple nonhereditary tumor types [116, 117]. Similarly, LKB1 is mutated in a significant proportion of nonsmall cell lung cancers (NSCLCs) [118], endometrial cancers [119], and a variety of other cancer subtypes.

Activation of the PI3K–AKT–mTORC1 pathway enhances protein synthesis at multiple levels including translation initiation, polypeptide chain elongation, and ribosome biogenesis [120, 121]. The effects of mTORC1 signaling on translation initiation largely reflect its regulatory actions on the eukaryotic initiation factor (eIF)-4E. The eIF-4E binds to the 7-methylguanosine cap appended to the 5′ ends of most eukaryotic mRNAs, and serves as the docking site for the remaining components of the multiprotein translation initiation complex [122]. The recruitment function of eIF-4E is blocked by the binding of a small set of translational repressor proteins termed eIF-4E-binding proteins (4E-BPs) [123]. Three 4E-BPs are expressed in mammalian cells, with the 4E-BP1 isoform being the most intensively studied. Multisite phosphorylation of 4E-BP1 by mTORC1 disrupts binding to eIF-4E and derepresses the translation initiation activity of this protein [124, 125]. Interestingly, eIF-4E itself is an oncogene and is overexpressed in about 30% of human cancers including subsets of breast, NSCLC, colon, head and neck, and thyroid cancers, as well as non-Hodgkin's lymphoma (NHL) [126, 127]. The oncogenic activities of eIF-4E may be largely attributed to the fact that the expression of several critical cell growth- and proliferation-inducing genes is strongly dependent on cap-dependent translation, and, in turn, on the eIF4E-activating function of mTORC1.

B. TRANSLATIONAL REGULATION OF ONCOGENIC PROTEINS BY mTORC1

The antiproliferative effects of rapalogs are correlated with inhibition of eIF-4E-dependent translation of cell-cycle proteins (including Myc, D-type cyclins, and ornithine decarboxylase) in multiple tumor cell lines [31]. Posttranslational effects of rapamycin on the stability of proliferation-associated proteins, such as the cyclin-dependent kinase inhibitor $p27^{Kip1}$ and D-type cyclins, have also been reported [24, 128] further extending the number of mechanisms whereby mTORC1 controls cell proliferation.

In addition to the cell-cycle proteins, proangiogenic mediators, such as vascular endothelial cell growth factor (VEGF) and hypoxia-inducible transcription factors (HIFs)-1α and -2α are regulated at the translational level by mTORC1 and hence are susceptible to rapamycin exposure [129–131]. The HIFs are persistently activated by hypoxic conditions, and transiently activated by growth factors and oncogenes. Gene products regulated by the HIFs include glycolytic enzymes, tumor growth-promoting factors such as transforming growth factor-alpha (TGF-α), and proangiogenic factors such as platelet-derived growth factor (PDGF) and VEGF [132]. VEGF is also translationally regulated by mTORC1 owing to the complex secondary structure in the $5'$ untranslated region of the cognate mRNA [129]. The inhibitory effects of rapalogs on VEGF production, which drives new blood vessel formation, as well as PDGF production, which supports blood vessel maturation, strongly suggest that these mTOCR1 inhibitors are active anti-angiogenic agents [132, 133].

Components of the apoptotic machinery are also translationally regulated by mTORC1. In mouse models of lymphoma, translation of the mRNA encoding Mcl1, a prosurvival member of the Bcl2 family, was inhibited by rapamycin [134]. Similarly, levels of the antiapoptotic proteins, Bcl2 and Bcl-XL, were reduced by the rapalogs temsirolimus and everolimus [135]. In combination with dexamethasone, rapamycin significantly blocked translation of the antiapoptotic proteins, BAG3, XIAP, and c-IAP [136]. Signaling through mTORC1 may also oppose apoptosis by stimulating the transcriptional activity of NF-κB via activation of the upstream inhibitor of NF-κB kinase B isoform (IKK-β)[137]. Although mTORC1 inhibition *in vitro* rarely induces apoptosis, the demonstrated effects of mTORC1 on prosurvival proteins suggest that rapalogs may potentiate the effect of cytotoxic therapies in which resistance ensues from upregulation of prosurvival proteins [138].

C. ROLE OF mTORC2 IN CANCER

The regulatory functions of mTOR during protein synthesis are channeled largely through the mTORC1 and its major downstream targets, the 4E-BPs and the S6K1 kinase. Activation of the mTORC2 complex is poorly understood but also appears to be downstream of growth factor receptor activation and as such may play a role in tumorigenesis [139]. Rapalogs do not inhibit mTORC2 directly, but can interfere with the assembly of mTORC2 complexes in certain cell types [13]. The most compelling evidence supporting a role for mTORC2 in cancer is the observation that mTORC2-dependent phosphorylation of AKT at Ser-473 is required for full AKT activation [140]. Inasmuch as AKT is a pivotal regulator of cell-cycle proteins, cell survival pathways, and metabolism, it is not surprising that AKT activity is commonly dysregulated in human cancers [116].

Accumulating evidence suggests that the refractoriness of mTORC2 to rapalog inhibition may actually limit the anticancer activities of these selective mTORC1 inhibitors. Rapalogs inhibit S6K1 downstream of mTORC1 and abrogate a negative feedback inhibition of PI3K/AKT mediated by S6K1 [141–143]. The consequent rebound activation of AKT may supply growth and survival signals that dampen the effectiveness of rapalogs against certain tumors. Furthermore, mTORC2 may be a cancer target in its own right. A recently described subunit of both mTORC1 and mTORC2, termed DEPTOR, appears to suppress the signaling functions of both TORCs. Interestingly, DEPTOR is overexpressed in multiple mye-loma, and this overexpression upregulates mTORC2 activity, presumably by inactivation of the mTORC1-dependent feedback inhibition of PI3K [15]. In addition, studies in mouse models of prostate cancer suggest that prostate tumorigenesis driven by loss of PTEN is strongly dependent on mTORC2, whereas normal development of the prostate is mTORC2 inde-pendent [144, 145]. As mentioned previously, rapalogs interfere with mTORC2 function in a limited subset of cancer cell lines [29]. The idea that broader inhibition of mTORC1 and mTORC2 functions might boost the intrinsic antitumor activities of the rapalogs propelled the development of second-generation mTOR inhibitors targeting the mTOR kinase domain. These mTOR kinase inhibitors (MKIs) will be discussed in Section VI.F.

D. DETERMINANTS OF MTOR INHIBITOR RESPONSIVENESS IN CANCER

The critical elements that dictate tumor sensitivity and resistance to mTOR inhibitors remain surprisingly elusive, given the detailed knowledge related to mTOR regulation and function. Cancer genetic profiling also offers no direct clues regarding drug responsiveness, in that the mTOR gene (termed *FRAP1*) is neither amplified nor mutated in human cancers [10]. Moreover, phosphorylation of the mTORC1-proximal targets, 4E-BP1 and S6K1, is inhibited with essentially equal efficiency in rapalog-sensitive and -resistant tumors [104]. As such, the investigative focus on response determinants for rapalogs has shifted to parallel pathways that circumvent the effects of mTORC1 inhibitors on key cellular responses, including protein synthesis, ribosome biogenesis, and metabolic control mechanisms [146, 147]. The oncogenic Ras pathway is an example of paramount importance, based on the frequency (30–50% of all cancers) with which this pathway is deregulated in various tumors. Oncogenic Ras signaling may bypass the inhibitory effects of rapalogs on translation initia-tion by circumventing the requirement for signaling through the PI3K–AKT–mTORC1 cascade [148]. Furthermore, the Ras to extracellular sig-nal-responsive kinase (Erk) signaling cascade may partially bypass the suppressive effect of rapalogs on eIF-4E function via activation of the MAPK-regulated Mnk-1 and Mnk-2 kinases, which phosphorylate and stimu-late eIF-4E-dependent translation initiation [149]. Another MAPK-regulated kinase, the ribosomal S6 kinase (Rsk), also enhances translation initiation via phosphorylation of eIF-4B, which enhances the RNA helicase activity of the associated eIF4A protein [150]. The Ras pathway effectors, Erk and Rsk, also act upstream of mTORC1, by phosphorylating and inactivating TSC2, thereby relieving the suppressive action of the TSC1/2 complex on the mTORC1-activating Rheb GTPase [151]. In addition, signaling through the Ras to Erk cascade tips the balance between pro- and antiapoptotic factors toward the promotion of cell survival [152, 153]. These observations suggest that combinations of Ras pathway inhibitors with rapalogs might capture many tumors that do not respond to rapalogs alone, toxicity issues notwithstanding. In support of this concept, combinations of rapalogs with a MEK inhibitor have shown synergistic activities in multiple preclinical models [154–157].

Increased expression of the Myc transcription factor may also confer resistance to rapalogs. Myc stimulates the transcription of numerous genes related to cell growth and proliferation, including the translation initiation factors eIF4-E, eIF-4A, and eIF-4G, all of which are regulated through the parallel mTORC1 pathway [158]. Overexpression of Myc might therefore blunt the suppressive effects of rapalogs on translation initiation. Myc is also a centrally important promoter of both glycolysis and glutaminolysis, as well as the expression of proangiogenic genes. These Myc-regulated pathways show considerable overlap with those activated in a TORC1-dependent fashion by the HIFs [159]. Recently, Myc was shown to transcriptionally upregulate 4E-BP1 expression in prostate tumors, suggesting that Myc-driven cancers may tolerate suppression of eIF-4E activity by 4E-BP1 and rapalogs [160]. Taken together, a clinically testable model posits that tumors bearing oncogenic Ras or Myc gene signatures will prove to be more resistant to single-agent therapy with mTORC1-selective inhibitors.

E. Clinical Development of Rapalogs as Anticancer Agents

Treatment with the rapalog, temsirolimus, improved overall survival in a phase III clinical trial in renal cell carcinoma (RCC) patients with poor prognosis [161]. Shortly thereafter, everolimus was approved in second line treatment of RCC patients that had progressed on tyrosine kinase inhibitor therapy [162]. Temsirolimus also received European approval for treatment of relapsed mantle cell lymphoma (MCL) and, like everolimus, gave positive phase II clinical results in endometrial and breast cancer [163]. Positive phase II studies with impressive objective response rates have also been reported for temsirolimus in NHL. These studies have validated mTORC1 as a cancer chemotherapy target across multiple tumor types and are consistent with the broad spectrum of activity observed with rapalogs in preclinical tumor models [31].

While treatment with rapalogs has clearly shown efficacy in several tumor types, the overall effect of these drugs as single agents has been modest [164]. A consideration of the tumor subtype in which rapalogs first showed clear and reproducible therapeutic activity might be instructive in terms of patient selection in this and other malignant diseases. Most RCCs (~ 80%) are categorized as the clear cell type, and are characterized by mutation or loss of the von Hippel-Lindau (*VHL*) gene [165]. Loss of VHL function in developing RCCs leads to stabilization of HIFs, and the subsequent induction of a gene expression program geared to cope with hypoxic stress. As discussed previously, the HIF-responsive genes encode proteins that increase tumor-associated angiogenesis and promote energy production via the glycolytic pathway. Inhibition of HIF protein synthesis by mTORC1 inhibitors might contribute to the antitumor activities of these agents in RCC-clear cell subtypes [137]. Although VHL loss, leading to HIF "addiction" in RCC-clear cell tumors, offers a logical explanation for the therapeutic activity of the rapalogs in this disease, the clinical experience suggests that this model oversimplifies the actions of these drugs in RCC. Surprisingly, the efficacy of temsirolimus in the phase III RCC study was as good or better in nonclear cell RCCs bearing normal VHL protein levels [166]. Nonclear cell RCCs are mainly papillary RCCs expressing other oncogenic mutations that may result in increased HIF activities. These alterations include *MET* gene amplification, which leads to increased expression of the c-Met receptor tyrosine kinase, hyperactivation of the PI3K–AKT–mTOR pathway, and consequent upregulation of HIF protein synthesis [167]. In addition, loss-of-function mutations in enzymes involved in the tricarboxylic acid (TCA) cycle that lead to the accumulation of the TCA cycle intermediates, succinate and fumarate, promote HIF protein expression by interfering with the activities of the HIF prolyl hydroxylases, which normally mark the HIFs for degradation via the ubiquitin–proteasome pathway [168]. Thus, RCC subtypes may achieve HIF dependence through the acquisition of diverse mutational alterations that increase HIF-1/2α protein synthesis or decrease HIF-1/2α degradation. More work is clearly needed to prove that the single-agent activities of the rapalogs in RCC reflect a relatively common lineage dependence on HIF-dependent gene expression in this particular tumor type.

The issue of whether the primary antitumor effect of mTORC1 inhibition is largely cancer cell-autonomous or mediated through alterations in the tumor microenvironment (e.g., angiogenesis) has been indirectly addressed in clinical cancer trials. In RCC, treatment with either temsirolimus or the anti-VEGF antibody, bevacizumab, produced a low (~10%) objective response (tumor regression) rate. In a phase I study, the combination of the two drugs in a small number of patients induced a striking 64%

objective response rate [169]. Inasmuch as bevacizumab is largely ineffective as a single agent in any tumor type, the synergistic activity with temsirolimus suggests that the mechanism of action of temsirolimus is distinct from that of bevacizumab. Resistance to anti-VEGF therapy occurs as a function of revascularization via alternative angiogenic factors, including basic fibroblast growth factor (b-FGF) and interleukin-8 (IL-8) [170]. Rapamycin inhibits signaling through the b-FGF receptor, as well as the production of IL-8, which may partially explain the apparent synergy with bevacizumab in early clinical trials.

Additional support for a role for mTOR inhibition in the tumor microenvironment is derived from the effectiveness of rapamycin in organ transplant patients with Kaposi's sarcoma, a highly angiogenic tumor associated with AIDS and other immunodeficiency syndromes [171]. Emerging evidence suggests that rapamycin suppresses both vasculogenesis and lymphangiogenesis, the latter through multiple mechanisms involving the inhibitor of DNA binding 2 (Id2) transcriptional suppressor, the VEGF-R3 receptor, and the lymphatic vascular endothelial hyaluronan receptor-1 (LYVE-1) [172]. The inhibitory actions of mTORC1 inhibition on lymphangiogenesis may prevent lymphatic spread and subsequent outgrowth of metastases in Kaposi's sarcoma patients. These effects may explain the seemingly paradoxical clinical observations that Kaposi's patients receiving rapamycin in the organ transplant setting experience delayed progression to aggressive metastatic disease [171].

In the MCL clinical trials with temsirolimus, a strikingly high (22%) objective response rate was seen in heavily pretreated, relapsing patients [173]. MCL is a B-cell-derived neoplasm characterized by translocation of cyclin D1 to the immunoglobulin heavy chain promoter, and the established role of mTORC1 in the promotion of cyclin D1 translation provided a sound rationale for treatment of this disease with an mTOR inhibitor [174]. However, three independent studies have failed to show significant reductions in cyclin D1 protein levels in MCL cell lines, or in tumor tissues from rapalog-treated patients [175–177]. Rapamycin was effective in lowering cyclin D1 in one cell line in which cyclin D1 expression was controlled by glycogen synthase kinase-3β (GSK3β), which regulates the turnover of the cyclin D1 protein [175]. Notably, cyclin D3 expression was more consistently reduced by rapamycin, suggesting that it may be the relevant target in MCL, because normal B-cell-cycle progression is predominantly dependent on the cyclin D3 isoform. Other studies have shown that cyclin D1 binds and inactivates TSC2, which may result in hyperstimulation of mTORC1 activity when cyclin D1 is overexpressed [178]. If this scenario holds in MCL, then temsirolimus might act as a downstream inhibitor of cyclin D1-dependent oncogenic signaling, by inactivating mTORC1, an important downstream target of cyclin D1, in MCL.

A similar conundrum came to light in a phase II study in endometrial cancer where a relatively high (25%) objective response rate to temsirolimus was reported [179]. The majority of endometrial cancers are clinically denoted as Type I, and are characterized by PI3K pathway deregulation through multiple mechanisms. Multiple mechanisms lead to pathway deregulation in this disease, including PTEN loss ($\sim$80%), PI3K mutation (36%), and dysregulated IGF-1R signaling [180, 181]. PI3K hyperactivation is evident in the early hyperplastic lesions that serve as the precursors of frankly malignant tumors. Thus, tumorigenesis in endometrial tissue appears to involve early, "founder" lesions in the PI3K pathway, which suggests that the remaining genetic hits leading to malignancy are incurred on a background of prooncogenic PI3K signaling [182]. This evolutionary sequence of malignant progression may differentiate endometrial cancer from the many other tumor types (e.g., glioblastoma and melanoma) that also display frequent loss of PTEN, because this PI3K-activating event typically occurs at more advanced stages of disease in these cancer subtypes [183]. Pathway addiction and hence rapalog sensitivity might be considerably more variable in cancer cells whose genomes have not been shaped around the PI3K–AKT–mTOR signaling pathway from the early stages of oncogenesis.

F. HIGH-DOSE EFFECTS OF RAPALOGS AND DEVELOPMENT OF SECOND-GENERATION MKIS

In contrast to most conventional protein kinase inhibitors, rapalogs exert a very prolonged pharmacodynamic effect on mTORC1 signaling owing to the unique molecular mechanism of action of these drugs [107]. The trimolecular complex formed by the binding of the FKBP12·rapalog complex to mTOR is essentially irreversible, and effectively stabilizes rapamycin in the intracellular milieu [184]. Studies in cultured cell lines have shown that a 1-h pulse treatment with rapamycin causes inhibition of mTORC1 function that persists for 7 days after drug exposure [184]. Similarly, ribosomal S6 protein phosphorylation was inhibited for greater than 82 h in patients following a single 25 mg dose of temsirolimus [185]. These protracted pharmacodynamic effects of the FKBP12-bound rapamycin and its derivatives have permitted the development of intermittent dosing protocols for temsirolimus in cancer patients. In addition, the intrinsic hydrophobicity of rapalogs results in good tissue penetration and, consequently, a high volume of distribution [186]. The accumulation of rapalogs in tissues may be very significant from a pharmacological perspective, in that achievable drug concentrations (micromolar) are far in excess of the nanomolar concentrations needed to exert maximal inhibitory effects on mTORC1 via the canonical FKBP12-dependent mechanism. At nanomolar concentrations, rapamycin maximally inhibits the mTORC1-proximal targets, S6K1 and 4E-BP1, and modestly reduces protein synthesis, but generally causes only a partial inhibition of cell proliferation (40–60% reduction) across a panel of human cancer cell lines [29]. In contrast, rapamycin concentrations in the 2–20 μM range cause nearly complete proliferative arrest, accompanied by a profound inhibition of protein synthesis. This "high-dose" effect of rapalogs may provoke a more complete suppression of mTORC1 function than can be obtained with drug concentrations in the nanomolar range [29]. Alternatively, at micromolar concentrations, the rapalogs are capable of binding directly to the FRB domain of mTOR, and inhibiting mTOR kinase activity in the absence of FKBP12 [28]. An intriguing study suggests that mTOR may signal through modalities other than mTORC1 and mTORC2. In cells rendered deficient for both mTORC1 and mTORC2 by RNA interference, exposure to rapamycin still inhibited translation of mRNAs with 5′-polypyrimidine tracts, suggesting that another subpopulation of rapamycin-sensitive mTOR molecules may be present in mammalian cells [187]. Thus, a speculative possibility is that, at sufficiently high intracellular concentrations, free rapamycin captures additional mTOR signaling entities not accessible to the FKBP12·rapamycin complex.

A second generation of mTOR-selective inhibitors—compounds that bind directly to the mTOR kinase domain—are winding their way toward the oncology clinic [188–191]. Unlike rapalogs, these compounds act as classical reversible ATP-competitive kinase inhibitors. These MKIs inhibit both mTORC1 and mTORC2, and presumably any other functional forms of mTOR in the cell. Importantly, the inhibitory effect of the MKIs on mTORC2 results in at least partial inhibition of AKT, which might counteract the propensity of selective mTORC1 inhibitors to stimulate AKT activity through disruption of the mTORC1/S6K1-dependent negative feedback loop that normally limits PI3K-dependent signaling (Figure 17.5) [142, 192]. The promise that these compounds will offer a more global suppression of mTOR functions in tumor tissues is certainly exciting, particularly in light of the aforementioned evidence that mTORC2 signaling is required for the development of prostate cancer, but not for normal prostate gland development [144]. Whether the MKIs will show improved or broader anticancer efficacy relative to the current crop of rapalogs remains an open question. The small molecule MKIs do not manifest the very durable pharmacodynamic effect on mTORC1 function characteristically obtained with the rapalogs. Second, the broader systemic suppression of mTOR activities achieved with the MKIs may increase antitumor activity, but with a counter-balancing cost of increased toxicity to normal tissues. Unfortunately, these questions can only be answered by the results of forthcoming clinical trials with MKIs as single agents, and as components of combination chemotherapy regimens.

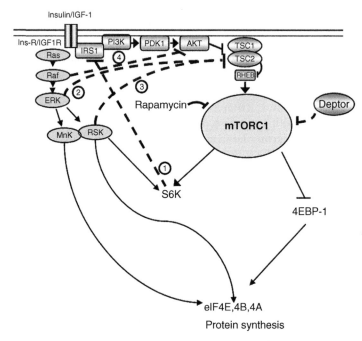

Fig. 17.5. Feedback and cross talk mechanisms that affect cellular responses to rapalogs. Four major mechanisms are depicted with boldened, dashed lines. (1) Activation of mTORC1 through growth factor receptor activation of PI3K results in activation of S6K which feeds back (dashed lines) to inhibit PI3K by phosphorylating IRS-1. Inhibition of mTORC1 by rapamycin or overexpression of DEPTOR may block feedback inhibition of PI3K and result in its hyperactivation. (2, 3) Activated Erk or Rsk activates mTORC1 via inhibitory phosphorylation of TSC2. (4) Oncogenic Ras activation cross-inhibits the AKT–mTORC1 pathway and assumes the growth-promoting functions of this pathway, thereby favoring resistance to mTORC1 inhibitors.

VI.   Effects of Rapamycin on Immunity and Longevity

This review has thus far focused largely on the effects of rapalogs on cancer cell growth and proliferation. However, these compounds exert equally interesting effects on other biological networks, such as those controlling organismal metabolism, immunity, and longevity. In this section, we will briefly discuss several noteworthy, recent findings that shed new light on the impact of rapamycin on the immune system and aging.

The immunosuppressive properties of rapamycin were recognized nearly 30 years ago [1, 3]. In the early days, many of the mechanistic studies centered on the effects of this drug on the proliferative expansion of CD4$^+$ helper T (T$_H$) lymphocytes induced by stimulation of T-cell antigen receptors (TCRs) with polyclonal activators, such as concanavalin A, or with cognate antigenic fragments displayed on the surfaces of antigen-presenting cells. Antigen-naïve T_H cells reside in G_0 phase of the cell cycle, and are induced to undergo cell-cycle entry ($G_0 \rightarrow G_1$ phase transition) by ligation of their TCRs with appropriately presented antigens, together with costimulatory signals delivered through CD28 and other coreceptors expressed on the T-cell surface. Activated T_H cells also secrete T-cell growth promoting cytokines, such as interleukin-2 (IL-2), which act in an autocrine or paracrine fashion to drive cell-cycle progression through the remainder of the mitotic cell cycle ($G_1 \rightarrow M$ phase progression). These early studies revealed that rapamycin was an extraordinarily potent inhibitor of the second, cytokine-dependent phase of T-cell proliferation, and effectively suppressed the passage of activated T cells through a late G_1-phase "restriction point" [23, 193], after which the cells were committed to enter S-phase and execute the remainder of the mitotic cell cycle. Hence, rapamycin treatment suppressed T-cell-dependent immune responses by interfering with the proliferative expansion of G_1 phase T cells. This mechanism of action on antigen-induced T_H cell responses clearly differentiated the immunosuppressive mechanism of rapamycin from that of its sister compound, FK-506, which blocked a TCR-mediated, Ca^{2+}-dependent signaling event required for the production of IL-2, but had no effect on the mitogenic response to IL-2 and other T-cell growth factors [194].

The model described above seeded many subsequent experiments regarding the effects of this drug on other cellular components of both the innate and adaptive immune systems [195, 196], and provided the backdrop for many broader studies of the roles of mTOR in growth factor-dependent cell mass accumulation and cell-cycle progression. Meanwhile, immuno-biologists began to recognize that CD4$^+$ T-cell populations contained many more functional subsets than were previously appreciated, and with this new knowledge came some surprising insights into the actions of rapamycin on T-cell biology and the adaptive immune response [196, 197]. For example, it now appears that signaling through the PI3K–AKT–mTOR pathway negatively regulates proinflammatory activities associated with the innate immune system, but does just the opposite in the adaptive immune system, supporting the expansion and functions of two major subsets of proinflammatory CD4$^+$ T cells, termed T_H1 and T_H17 cells [196]. Thus, the label "immunosuppressant," when applied to rapamycin, should not be construed as a global descriptor of this drug's action on the immune system. The dichotomous functions of mTOR during innate and adaptive immune responses suggest that rapamycin may exacerbate inflammation driven by the innate immune system, while dampening inflammatory responses stemming from the activation of CD4$^+$ effector T cells.

A particularly noteworthy insight into the impact of rapamycin on adaptive immune responses followed the discovery of regulatory T (T_{REG}) cells, which function as antigen-specific suppressors of immune responses, and protect against autoimmunity and pathologic overstimulation of effector T cells [198]. T_{REG} cells are characterized by cell surface expression of CD4 and CD25 (denoting the IL-2 receptor α polypeptide), these T cells come in two flavors: "natural" T_{REG} cells develop in the thymus, whereas "inducible" T_{REG} cells are generated as a result of immune stimulation in the periphery. Both T_{REG} subpopulations express a pivotal transcription factor, the winged-helix/forkhead family member, Fox3p. It is the inducible T_{REG} population that is dramatically altered in rapamycin-treated rodents and humans, and the direct and indirect effects of this drug on Fox3p expression explain many of the downstream consequences regarding T_{REG} cell differentiation and survival [195, 199, 200]. In postthymic, undifferentiated CD4$^+$ T cells, Fox3p expression is influenced in a time- and signal intensity-dependent fashion by the PI3K–AKT–mTOR signaling axis [200]. Costimulation of resting CD4$^+$ T cells through the TCR and CD28 coreceptor for a restricted time period (<18 h) triggers histone modifications that render the *FOX3P* gene accessible to transcription factors. Interruption of TCR-CD3 signaling during this timeframe leads to *FOX3P* transcription in a subpopulation of the activated T cells and pushes these cells into the T_{REG} differentiation pathway. The addition of rapamycin to such transiently stimulated T cell populations greatly increases the numbers of cells that express Fox3p and the inducible T_{REG} cell phenotype. Costimulation of T cells for longer times (>18 h) triggers a second wave of PI3K/AKT/mTOR activation (driven by IL-2 and related T cell growth factors) that restricts accessibility of the *FOX3P* locus and drives CD4$^+$ T differentiation away from the T_{REG} cell phenotype and toward effector T_H (e.g., T_H17)-type cells. Again, the presence of rapamycin disrupts mTORC1-dependent signaling, and skews the differentiation program toward T_{REG} cells under conditions that would normally yield predominantly immune effector T cells.

A peculiar feature of the signal transduction machinery employed by developing and fully differentiated T_{REG} cells further favors their emergence from rapamycin-treated CD4$^+$ T cell populations. These cells express higher than normal levels of the PTEN phosphatase, an effective suppressor of PI3K and, in turn, AKT–mTOR signaling [201]. Unlike other T cell subpopulations, proliferative expansion of T_{REG} cells during antigenic stimulation is not heavily reliant on the PI3K–AKT–mTOR pathway, and hence is not significantly impaired in the presence of rapamycin. The mechanism underlying this shift away from PI3K-dependent signals is not fully understood, but appears to be explained in part by the expression of the serine–threonine kinase, Pim2, which carries out functions that overlap with those of the rapamycin-sensitive mTORC1 [202]. Thus, chronic rapamycin exposure favors enrichment for CD4$^+$CD25$^+$ T$_{REG}$ cells through alteration of at least two major aspects of the CD4$^+$ T cell activation program. First, drug exposure increases the number of CD4$^+$ T cells that express Fox3p and enter the T$_{REG}$ differentiation program; second, whereas rapamycin strongly interferes with the cytokine-driven expansion of the general CD4$^+$ T cell population, the committed T$_{REG}$ cells are intrinsically wired to be less sensitive to the antiproliferative effects of this drug. Once again, the application of rapamycin as a chemical probe has yielded novel insights into T lymphocyte biology, and has generated new knowledge that may lead to improved treatment options for patients with transplanted organs, and autoimmune/chronic inflammatory diseases.

Recent results further expand the clinical possibilities for the rapalogs into some seemingly paradoxical space. At first blush, it seems that administration of a powerful immunosuppressive drug, such as rapamycin, would be an ill-advised strategy for the enhancement of vaccine efficacy. However, recent findings once again suggest that the effects of rapamycin on the immune system are broad-ranging, but remarkably idiosyncratic. Rapamycin treatment has dramatic effects on the maturation and functions of dendritic cells, which are a heterogeneous population of antigen-presenting cells that play major roles in orchestrating long-term adaptive immune responses in vaccinated rodents and human patients [203, 204]. As mentioned in Section II, rapalogs stimulate autophagy in many types of cells, including macrophages and dendritic cells. Jagannath *et al.* recently reported that treatment of mouse bone marrow-derived dendritic cells with rapamycin increased the efficiency with which these cells processed and presented *Mycobacterium bovis*-derived peptides [205]. The authors attributed this outcome to increased autophagic engulfment of the attenuated bacterial vaccine, and improved delivery of the autophagosomal cargo to the lysosome for processing into antigenic peptide fragments. These results raise the possibility that dendritic cell-based vaccines might be rendered more efficacious by *ex vivo* treatment of the patient's dendritic cells with rapamycin during the process of antigenic priming.

An even more striking application of rapamycin as a vaccine enhancer was reported by Araki *et al.* [206]. Treatment of mice with rapamycin after exposure to either live choriomeningitis virus (CMEV) or a CMEV vaccine dramatically increased the numbers of functional CD8$^+$ cytotoxic T cells, which are the major effectors of antiviral immunity. Like the T$_{REG}$ cells described above, it seems that CD8$^+$ memory T cell expansion and differentiation is actually promoted by inhibition of mTORC1 function with rapamycin *in vivo*. An elegant series of knockdown experiments established that the positive effects of rapamycin on $CD8^+$ memory T cells were directly related to drug actions on the T cells themselves.

Finally, aging research is rapidly emerging as a major area of unmet clinical need. Studies in genetically tractable model organisms, such as worms and flies, have established that caloric restriction prolongs lifespan and maintains tissue vitality [207, 208]. Numerous studies have documented that mTORC1 activity is directly related to the supply of nutrients, and that treatment of mammalian cells with rapamycin triggers a starvation-like phenotype in spite of the availability of ample supplies of nutrients [209, 210]. A landmark, multicenter study has shown that "aged," 600-day-old mice exhibit a significant prolongation of lifespan when given dietary rapamycin on a daily basis [211]. The blood concentrations of rapamycin (10–20 ng/ml) measured in these studies were comparable to those observed in transplant patients receiving rapamycin. These findings are consistent with a report that genetic deletion of the mTORC1 substrate, S6K1, also leads to lifespan extension in mice [212]. Therapeutic slowing of the aging process and age-related tissue degeneration is an exciting facet of the future in clinical medicine. We clearly have much to learn regarding the mechanisms of aging and key points of intervention, but it is highly encouraging that druggable targets such as mTORC1 and the histone deacetylase, Sirt1, have already emerged in this area. Basic scientists will need to address the mechanism(s) through which rapamycin promotes life-span extension in rodents (Figure 17.6). An obvious possibility is that chronic mTORC1 inhibition triggers a set of metabolic responses that resemble those associated with caloric restriction. Excessive caloric intake promotes a chronic inflammatory state that underlies many of the lifespan-shortening pathologies associated with the worldwide epidemic of obesity and metabolic syndrome [213]. Rapamycin may dampen certain chronic inflammatory responses, and concomitantly prevent or delay the clonal expansion of premalignant cells in aging mammals. We suspect that all three of these health-promoting actions of mTORC1 inhibition will prove to contribute to lifespan extension in humans, who live many more decades than mice.

VII.   Conclusions and Future Perspectives

This chapter has recounted some of the highlights of a 40-plus year voyage that began on Easter Island, wound through a series of chemical biology experiments, first in yeast and then in mammalian cells, then led to

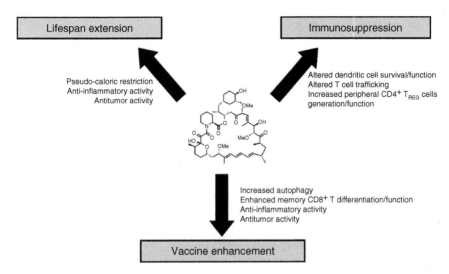

FIG. 17.6. Pleiotropic effects of rapamycin on immune responses and longevity.

the discovery of a multifunctional signaling kinase named mTOR, and finally unveiled a novel class of molecularly targeted agents (the rapalogs) with established or predicted clinical efficacies in several human diseases. This journey is far from complete, as exemplified by emerging findings that chronic mTORC1 suppression with rapamycin might prolong functional lifespan in mammals. Natural product and synthetic organic chemists have continued to build on the knowledge gleaned from years of intensive exploration of the SAR and cellular pharmacology of rapamycin. Efforts to reengineer the structure of the parent compound suggest that it is possible to introduce unanticipated pharmacological activities into the rapamycin core structure [101]. These new rapalogs will serve as useful chemical probes in their own right, and may eventually lead to effective treatments for neurodegenerative diseases and stroke. In addition, the established clinical activities of the rapalogs have spawned a massive effort to "one-up" Mother Nature, through the creation of the MKIs—highly selective, fully synthetic inhibitors of the mTOR kinase domain. The introduction of this second generation of mTOR inhibitors into clinical trials in oncology and other diseases is eagerly anticipated. Based on the history of rapamycin, we might be foolish to bet that the power of modern-day synthetic organic chemistry will overshadow the synthetic prowess of a lowly soil bacterium from Easter Island.

REFERENCES

1. Sehgal, S.N. (2003). Sirolimus: its discovery, biological properties, and mechanism of action. *Transplant Proc* 35:7S–14S.
2. Sehgal, S.N., Baker, H., and Vezina, C. (1975). Rapamycin (AY-22, 989), a new antifungal antibiotic. II. Fermentation, isolation and characterization. *J Antibiot (Tokyo)* 28:727–732.
3. Abraham, R.T., and Wiederrecht, G.J. (1996). Immunopharmacology of rapamycin. *Annu Rev Immunol* 14:483–510.
4. Sabatini, D.M., Erdjument-Bromage, H., Lui, M., Tempst, P., and Snyder, S.H. (1994). RAFT1: a mammalian protein that binds to FKBP12 in a rapamycin- dependent fashion and is homologous to yeast TORs. *Cell* 78:35–43.
5. Sabers, C.J., *et al.* (1995). Isolation of a protein target of the FKBP12-rapamycin complex in mammalian cells. *J Biol Chem* 270:815–822.
6. Brown, E.J., *et al.* (1994). A mammalian protein targeted by G1-arresting rapamycin-receptor complex. *Nature* 369:756–758.
7. Grube, E., and Buellesfeld, L. (2004). Rapamycin analogs for stent-based local drug delivery. Everolimus- and tacrolimus-eluting stents. *Herz* 29:162–166.
8. Calne, R.Y. (2003). The development of immunosuppression: the rapamycin milestone. *Transplant Proc* 35:15S–17S.
9. Abraham, R.T. (2004). PI 3-kinase related kinases: 'big' players in stress-induced signaling pathways. *DNA Repair (Amst)* 3:883–887.
10. Sabatini, D.M. (2006). mTOR and cancer: insights into a complex relationship. *Nat Rev Cancer* 6:729–734.
11. Loewith, R., *et al.* (2002). Two TOR complexes, only one of which is rapamycin sensitive, have distinct roles in cell growth control. *Mol Cell* 10:457–468.
12. Jacinto, E., and Hall, M.N. (2003). Tor signalling in bugs, brain and brawn. *Nat Rev Mol Cell Biol* 4:117–126.
13. Sarbassov, D.D., *et al.* (2006). Prolonged rapamycin treatment inhibits mTORC2 assembly and Akt/PKB. *Mol Cell* 22:159–168.
14. Thoreen, C.C., *et al.* (2009). An ATP-competitive mTOR inhibitor reveals rapamycin-insensitive functions of mTORC1. *J Biol Chem* 284:8023–8032.
15. Peterson, T.R., *et al.* (2009). DEPTOR is an mTOR inhibitor frequently overexpressed in multiple myeloma cells and required for their survival. *Cell* 137:873–886.
16. Chiang, G.G., and Abraham, R.T. (2007). Targeting the mTOR signaling network in cancer. *Trends Mol Med* 13:433–442.
17. Kim, D.H., *et al.* (2002). mTOR interacts with raptor to form a nutrient-sensitive complex that signals to the cell growth machinery. *Cell* 110:163–175.
18. Fingar, D.C., and Blenis, J. (2004). Target of rapamycin (TOR): an integrator of nutrient and growth factor signals and coordinator of cell growth and cell cycle progression. *Oncogene* 23:3151–3171.
19. Gingras, A.C., Raught, B., and Sonenberg, N. (2001). Regulation of translation initiation by FRAP/mTOR. *Genes Dev* 15:807–826.
20. Sancak, Y., *et al.* (2008). The Rag GTPases bind raptor and mediate amino acid signaling to mTORC1. *Science* 320:1496–1501.
21. Manning, B.D., and Cantley, L.C. (2003). United at last: the tuberous sclerosis complex gene products connect the phosphoinositide 3-kinase/Akt pathway to mammalian target of rapamycin (mTOR) signalling. *Biochem Soc Trans* 31:573–578.

22. Manning, B.D., and Cantley, L.C. (2003). Rheb fills a GAP between TSC and TOR. *Trends Biochem Sci* 28:573–576.
23. Morice, W.G., Brunn, G.J., Wiederrecht, G., Siekierka, J.J., and Abraham, R.T. (1993). Rapamycin-induced inhibition of p34cdc2 kinase activation is associated with G1/S-phase growth arrest in T lymphocytes. *J Biol Chem* 268:3734–3738.
24. Nourse, J., *et al.* (1994). Interleukin-2-mediated elimination of the p27Kip1 cyclin-dependent kinase inhibitor prevented by rapamycin. *Nature* 372:570–573.
25. Guertin, D.A., and Sabatini, D.M. (2009). The pharmacology of mTOR inhibition. *Sci Signal* 2:pe24.
26. Choi, J., Chen, J., Schreiber, S.L., and Clardy, J. (1996). Structure of the FKBP12-rapamycin complex interacting with the binding domain of human FRAP. *Science* 273:239–242.
27. Stan, R., McLaughlin, M.M., Cafferkey, R., Johnson, R.K., Rosenberg, M., and Livi, G.P. (1994). Interaction between FKBP12-rapamycin and TOR involves a conserved serine residue. *J Biol Chem* 269:32027–32030.
28. Leone, M., *et al.* (2006). The FRB domain of mTOR: NMR solution structure and inhibitor design(,). *Biochemistry* 45:10294–10302.
29. Shor, B., *et al.* (2008). A new pharmacologic action of CCI-779 involves FKBP12-independent inhibition of mTOR kinase activity and profound repression of global protein synthesis. *Cancer Res* 68:2934–2943.
30. Edinger, A.L., Linardic, C.M., Chiang, G.G., Thompson, C.B., and Abraham, R.T. (2003). Differential effects of rapamycin on mammalian target of rapamycin signaling functions in mammalian cells. *Cancer Res* 63:8451–8460.
31. Bjornsti, M.A., and Houghton, P.J. (2004). The tor pathway: a target for cancer therapy. *Nat Rev Cancer* 4:335–348.
32. Petroulakis, E., Mamane, Y., Le Bacquer, O., Shahbazian, D., and Sonenberg, N. (2007). mTOR signaling: implications for cancer and anticancer therapy. *Br J Cancer* 96(Suppl.): R11–R15.
33. Rubinsztein, D.C., Gestwicki, J.E., Murphy, L.O., and Klionsky, D.J. (2007). Potential therapeutic applications of autophagy. *Nat Rev Drug Discov* 6:304–312.
34. Noda, T., and Ohsumi, Y. (1998). Tor, a phosphatidylinositol kinase homologue, controls autophagy in yeast. *J Biol Chem* 273:3963–3966.
35. Lum, J.J., DeBerardinis, R.J., and Thompson, C.B. (2005). Autophagy in metazoans: cell survival in the land of plenty. *Nat Rev Mol Cell Biol* 6:439–448.
36. Hoyer-Hansen, M., and Jaattela, M. (2008). Autophagy: an emerging target for cancer therapy. *Autophagy* 4:574–580.
37. Jin, S., and White, E. (2007). Role of autophagy in cancer: management of metabolic stress. *Autophagy* 3:28–31.
38. Meijer, A.J., and Codogno, P. (2009). Autophagy: regulation and role in disease. *Crit Rev Clin Lab Sci* 46:210–240.
39. Meijer, A.J., and Codogno, P. (2004). Regulation and role of autophagy in mammalian cells. *Int J Biochem Cell Biol* 36:2445–2462.
40. Chang, Y.Y., and Neufeld, T.P. (2009). An Atg1/Atg13 complex with multiple roles in TOR-mediated autophagy regulation. *Mol Biol Cell* 20:2004–2014.
41. Jung, C.H., *et al.* (2009). ULK-Atg13-FIP200 complexes mediate mTOR signaling to the autophagy machinery. *Mol Biol Cell* 20:1992–2003.
42. Swindells, D.C.N., White, P.S., and Findlay, J.A. (1978). The x-ray crystal structure of rapamycin, c51h91no13. *Can J Chem* 56:2491–2492.
43. Findlay, J.A., and Radics, L. (1980). On the chemistry and high field nuclear magnetic resonance spectroscopy of rapamycin. *Can J Chem* 58:579–590.

44. Schwecke, T., *et al.* (1995). The biosynthetic gene cluster for the polyketide immunosuppressant rapamycin. *Proc Natl Acad Sci USA* 92:7839–7843.
45. Sattely, E.S., Fischbach, M.A., and Walsh, C.T. (2008). Total biosynthesis: in vitro reconstitution of polyketide and nonribosomal peptide pathways. *Nat Prod Rep* 25:757–793.
46. Fischbach, M.A., and Walsh, C.T. (2006). Assembly-line enzymology for polyketide and nonribosomal Peptide antibiotics: logic, machinery, and mechanisms. *Chem Rev* 106:3468–3496.
47. Staunton, J., and Wilkinson, B. (1997). Biosynthesis of erythromycin and rapamycin. *Chem Rev* 97:2611–2630.
48. Kino, T., *et al.* (1987). FK-506, a novel immunosuppressant isolated from a Streptomyces. I. Fermentation, isolation, and physico-chemical and biological characteristics. *J Antibiot (Tokyo)* 40:1249–1255.
49. Hatanaka, H., *et al.* (1988). FR-900520 and FR-900523, novel immunosuppressants isolated from a Streptomyces. II. Fermentation, isolation and physico-chemical and biological characteristics. *J Antibiot (Tokyo)* 41:1592–1601.
50. Arai, T., Kouama, Y., Suenaga, T., and Honda, H. (1962). Ascomycin, an antifungal antibiotic. *J Antibiot (Tokyo)* 15:231–232.
51. Harding, M.W., Galat, A., Uehling, D.E., and Schreiber, S.L. (1989). A receptor for the immunosuppressant FK506 is a cis-trans peptidyl-prolyl isomerase. *Nature* 341:758–760.
52. Fretz, H., Albers, M.W., Galat, A., Standaert, R.F., Lane, W.S., and Burakoff, S.J. (1991). Rapamycin and FK506 binding proteins (immunophilins). *J Am Chem Soc* 113:1409–1411.
53. Edlich, F., and Fischer, G. (2006). Pharmacological targeting of catalyzed protein folding: the example of peptide bond cis/trans isomerases. *Handb Exp Pharmacol*359–404.
54. Lam, E., *et al.* (1995). A novel FK506 binding protein can mediate the immunosuppressive effects of FK506 and is associated with the cardiac ryanodine receptor. *J Biol Chem* 270:26511–26522.
55. Sinkins, W.G., Goel, M., Estacion, M., and Schilling, W.P. (2004). Association of immunophilins with mammalian TRPC channels. *J Biol Chem* 279:34521–34529.
56. Cameron, A.M., Steiner, J.P., Sabatini, D.M., Kaplin, A.I., Walensky, L.D., and Snyder, S.H. (1995). Immunophilin FK506 binding protein associated with inositol 1, 4, 5-trisphosphate receptor modulates calcium flux. *Proc Natl Acad Sci USA* 92:1784–1788.
57. Shim, S., *et al.* (2009). Peptidyl-prolyl isomerase FKBP52 controls chemotropic guidance of neuronal growth cones via regulation of TRPC1 channel opening. *Neuron* 64:471–483.
58. Ratajczak, T., Ward, B.K., and Minchin, R.F. (2003). Immunophilin chaperones in steroid receptor signalling. *Curr Top Med Chem* 3:1348–1357.
59. Jin, Y.J., and Burakoff, S.J. (1993). The 25-kDa FK506-binding protein is localized in the nucleus and associates with casein kinase II and nucleolin. *Proc Natl Acad Sci USA* 90:7769–7773.
60. Chambraud, B., Belabes, H., Fontaine-Lenoir, V., Fellous, A., and Baulieu, E.E. (2007). The immunophilin FKBP52 specifically binds to tubulin and prevents microtubule formation. *FASEB J* 21:2787–2797.
61. Edlich, F., *et al.* (2006). The specific FKBP38 inhibitor N-(N', N'-dimethylcarboxamidomethyl)cycloheximide has potent neuroprotective and neurotrophic properties in brain ischemia. *J Biol Chem* 281:14961–14970.
62. Salituro, G.M., *et al.* (1995). Meridamycin: a novel nonimmunosuppressive FKBP12 ligand from *Streptomyces hygroscopicus*. *Tetrahedron Lett* 36:997–1000.
63. Fehr, T., *et al.* (1996). Antascomicins A, B, C, D and E. Novel FKBP12 binding compounds from a Micromonospora strain. *J Antibiot (Tokyo)* 49:230–233.

64. Summers, M.Y., Leighton, M., Liu, D., Pong, K., and Graziani, E.I. (2006). 3-normeridamycin: a potent non-immunosuppressive immunophilin ligand is neuroprotective in dopaminergic neurons. *J Antibiot (Tokyo)* 59:184–189.

65. Van Duyne, G.D., Standaert, R.F., Schreiber, S.L., and Clardy, J. (1991). Atomic structure of the rapamycin human immunophilin FKBP-12 complex. *J Am Chem Soc* 113:7433–7434.

66. Heitman, J., Movva, N.R., and Hall, M.N. (1991). Targets for cell cycle arrest by the immunosuppressant rapamycin in yeast. *Science* 253:905–909.

67. Liu, J., Farmer, J.D., Jr., Lane, W.S., Friedman, J., Weissman, I., and Schreiber, S.L. (1991). Calcineurin is a common target of cyclophilin-cyclosporin A and FKBP-FK506 complexes. *Cell* 66:807–815.

68. Griffith, J.P., *et al.* (1995). X-ray structure of calcineurin inhibited by the immunophilin-immunosuppressant FKBP12-FK506 complex. *Cell* 82:507–522.

69. Luengo, J.I., *et al.* (1995). Structure-activity studies of rapamycin analogs: evidence that the C-7 methoxy group is part of the effector domain and positioned at the FKBP12-FRAP interface. *Chem Biol* 2:471–481.

70. Banaszynski, L.A., Liu, C.W., and Wandless, T.J. (2005). Characterization of the FKBP. rapamycin.FRB ternary complex. *J Am Chem Soc* 127:4715–4721.

71. Kallen, J.A., Sedrani, R., and Cottens, S. (1996). X-ray crystal structure of 28-O-methylrapamycin complexed with FKBP12: is the cyclohexyl moiety part of the effector domain of rapamycin? *J Am Chem Soc* 118:5857–5861.

72. Sedrani, R., Jones, L.H., Jutzi-Eme, A.M., Schuler, W., and Cottens, S. (1999). Cleavage of the cyclohexyl-subunit of rapamycin results in loss of immunosuppressive activity. *Bioorg Med Chem Lett* 9:459–462.

73. Caufield, C.E. (1995). Structure-activity relationships involving modifications to the macrolides FK-506 and rapamycin. *Curr Pharm Des* 1:145–160.

74. Punt, C.J., Boni, J., Bruntsch, U., Peters, M., and Thielert, C. (2003). Phase I and pharmacokinetic study of CCI-779, a novel cytostatic cell-cycle inhibitor, in combination with 5-fluorouracil and leucovorin in patients with advanced solid tumors. *Ann Oncol* 14:931–937.

75. Kirchner, G.I., Meier-Wiedenbach, I., and Manns, M.P. (2004). Clinical pharmacokinetics of everolimus. *Clin Pharmacokinet* 43:83–95.

76. Chen, Y.W., *et al.* (2007). Zotarolimus, a novel sirolimus analogue with potent anti-proliferative activity on coronary smooth muscle cells and reduced potential for systemic immunosuppression. *J Cardiovasc Pharmacol* 49:228–235.

77. Palaparthy, R., *et al.* (2005). Pharmacokinetics and safety of ABT-578, a sirolimus (rapamycin) analogue, after single intravenous bolus injection in healthy male volunteers. *Clin Drug Investig* 25:491–498.

78. Karyekar, C.S., *et al.* (2005). A phase I multiple-dose escalation study characterizing pharmacokinetics and safety of ABT-578 in healthy subjects. *J Clin Pharmacol* 45:910–918.

79. Mita, M.M., *et al.* (2008). Phase I trial of the novel mammalian target of rapamycin inhibitor deforolimus (AP23573; MK-8669) administered intravenously daily for 5 days every 2 weeks to patients with advanced malignancies. *J Clin Oncol* 26:361–367.

80. Dickman, D.A., *et al.* (2000). Antifungal rapamycin analogues with reduced immunosuppressive activity. *Bioorg Med Chem Lett* 10:1405–1408.

81. Wagner, R., *et al.* (2005). Rapamycin analogs with reduced systemic exposure. *Bioorg Med Chem Lett* 15:5340–5343.

82. Nicolaou, K.C., Chakraborty, T.K., Piscopio, A.D., Minowa, N., and Bertinato, P. (1993). Total synthesis of rapamycin. *J Am Chem Soc* 115:4419–4420.

83. Romo, D., Meyer, S.D., Johnson, D.D., and Schreiber, S.L. (1993). Total synthesis of (-)
-rapamycin using an Evans-Tishchenko fragment coupling. *J Am Chem Soc*
115:7906–7907.
84. Hayward, C.M., Yohannes, D., and Danishefsky, S.J. (1993). Total synthesis of rapamycin
via a novel titanium-mediated aldol macrocyclization reaction. *J Am Chem Soc*
115:9345–9346.
85. Smith, I.A.B., Condon, S.M., McCauley, J.A., Leazer, J.L., Jr., Leahy, J.W., and
Maleczka, R.E., Jr. (1995). Total synthesis of rapamycin and demethoxyrapamycin.
J Am Chem Soc 117:5407–5408.
86. Nicolaou, K.C., Anthony, D.P., Peter, B., Tushar, K.C., Nobuto, M., and Kazunori, K.
(1995). Total synthesis of rapamycin. *Chemistry—A Eur J* 1:318–333.
87. Maddess, M.L., *et al.* (2007). Total synthesis of rapamycin. *Angew Chem Int Ed Engl*
46:591–597.
88. Ley, S.V., *et al.* (2009). Total synthesis of rapamycin. *Chemistry* 15:2874–2914.
89. Graziani, E.I., *et al.* (2003). Novel sulfur-containing rapamycin analogs prepared by
precursor-directed biosynthesis. *Org Lett* 5:2385–2388.
90. Ritacco, F.V., *et al.* (2005). Production of novel rapamycin analogs by precursor-directed
biosynthesis. *Appl Environ Microbiol* 71:1971–1976.
91. Nishida, H., *et al.* (1995). Generation of novel rapamycin structures by microbial manipulations. *J Antibiot (Tokyo)* 48:657–666.
92. Khaw, L.E., Bohm, G.A., Metcalfe, S., Staunton, J., and Leadlay, P.F. (1998). Mutational
biosynthesis of novel rapamycins by a strain of *Streptomyces hygroscopicus* NRRL 5491
disrupted in rapL, encoding a putative lysine cyclodeaminase. *J Bacteriol* 180:809–814.
93. Gregory, M.A., *et al.* (2004). Isolation and characterization of pre-rapamycin, the first
macrocyclic intermediate in the biosynthesis of the immunosuppressant rapamycin by
S. hygroscopicus. *Angew Chem Int Ed Engl* 43:2551–2553.
94. Gregory, M.A., *et al.* (2005). Mutasynthesis of rapamycin analogues through the manipulation of a gene governing starter unit biosynthesis. *Angew Chem Int Ed Engl*
44:4757–4760.
95. Gregory, M.A., *et al.* (2006). Rapamycin biosynthesis: elucidation of gene product function. *Org Biomol Chem* 4:3565–3568.
96. Goss, R.J., Lanceron, S.E., Wise, N.J., and Moss, S.J. (2006). Generating rapamycin
analogues by directed biosynthesis: starter acid substrate specificity of mono-substituted
cyclohexane carboxylic acids. *Org Biomol Chem* 4:4071–4073.
97. Steiner, J.P., *et al.* (1997). Neurotrophic actions of nonimmunosuppressive analogues of
immunosuppressive drugs FK506, rapamycin and cyclosporin A. *Nat Med* 3:421–428.
98. Gold, B.G., Densmore, V., Shou, W., Matzuk, M.M., and Gordon, H.S. (1999). Immunophilin FK506-binding protein 52 (not FK506-binding protein 12) mediates the neurotrophic action of FK506. *J Pharmacol Exp Ther* 289:1202–1210.
99. Sharkey, J., and Butcher, S.P. (1994). Immunophilins mediate the neuroprotective effects
of FK506 in focal cerebral ischaemia. *Nature* 371:336–339.
100. Bocquet, A., *et al.* (2001). Failure of GPI compounds to display neurotrophic activity
in vitro and in vivo. *Eur J Pharmacol* 415:173–180.
101. Ruan, B., *et al.* (2008). Binding of rapamycin analogs to calcium channels and FKBP52
contributes to their neuroprotective activities. *Proc Natl Acad Sci USA* 105:33–38.
102. Wiederrecht, G.J., Sabers, C.J., Brunn, G.J., Martin, M.M., Dumont, F.J., and
Abraham, R.T. (1995). Mechanism of action of rapamycin: new insights into the regulation of G1-phase progression in eukaryotic cells. *Prog Cell Cycle Res* 1:53–71.
103. Eng, C.P., Sehgal, S.N., and Vézina, C. (1984). Activity of rapamycin (AY-22, 989) against
transplanted tumors. *J Antibiot (Tokyo)* 37:1231–1237.

104. Dilling, M.B., Dias, P., Shapiro, D.N., Germain, G.S., Johnson, R.K., and Houghton, P.J. (1994). Rapamycin selectively inhibits the growth of childhood rhabdomyosarcoma cells through inhibition of signaling via the type I insulin-like growth factor receptor. *Cancer Res* 54:903–907.
105. Dancey, J., and Sausville, E.A. (2003). Issues and progress with protein kinase inhibitors for cancer treatment. *Nat Rev Drug Discov* 2:296–313.
106. Garber, K. (2001). Rapamycin's resurrection: a new way to target the cancer cell cycle. *J Natl Cancer Inst* 93:1517–1519.
107. Dancey, J.E. (2005). Inhibitors of the mammalian target of rapamycin. *Expert Opin Investig Drugs* 14:313–328.
108. Inoki, K., Li, Y., Zhu, T., Wu, J., and Guan, K.-L. (2002). TSC2 is phosphorylated and inhibited by AKT and suppresses mTOR signaling. *Nat Cell Biol* 4:648–657.
109. Goncharova, E.A., *et al.* (2002). Tuberin regulates p70 S6 kinase activation and ribosomal protein S6 phosphorylation. *J Biol Chem* 277:30958–30967.
110. Manning, B.D., Tee, A.R., Logsdon, M.N., Blenis, J., and Cantley, L.C. (2002). Identification of the tuberous sclerosis complex-2 tumor suppressor gene product tuberin as a target of the phosphoinositide 3-kinase/Akt pathway. *Mol Cell* 10:151–162.
111. Potter, C.J., Pedraza, L.G., and Xu, T. (2002). AKT regulates growth by directly phosphorylating Tsc2. *Nat Cell Biol* 4:658–665.
112. Li, Y., Corradetti, M.N., Inoki, K., and Guan, K.L. (2004). TSC2: filling the GAP in the mTOR signaling pathway. *Trends Biochem Sci* 29:32–38.
113. Long, X., Lin, Y., Ortiz-Vega, S., Yonezawa, K., and Avruch, J. (2005). Rheb binds and regulates the mTOR kinase. *Curr Biol* 15:702–713.
114. Yuan, T.L., and Cantley, L.C. (2008). PI3K pathway alterations in cancer: variations on a theme. *Oncogene* 27:5497–5510.
115. Inoki, K., Corradetti, M.N., and Guan, K.L. (2005). Dysregulation of the TSC-mTOR pathway in human disease. *Nat Genet* 37:19–24.
116. Keniry, M., and Parsons, R. (2008). The role of PTEN signaling perturbations in cancer and in targeted therapy. *Oncogene* 27:5477–5485.
117. Chalhoub, N., and Baker, S.J. (2009). PTEN and the PI3-kinase pathway in cancer. *Annu Rev Pathol* 4:127–150.
118. Makowski, L., and Hayes, D.N. (2008). Role of LKB1 in lung cancer development. *Br J Cancer* 99:683–688.
119. Lu, K.H., *et al.* (2008). Loss of tuberous sclerosis complex-2 function and activation of Mammalian target of rapamycin signaling in endometrial carcinoma. *Clin Cancer Res* 14:2543–2550.
120. Ma, X.M., and Blenis, J. (2009). Molecular mechanisms of mTOR-mediated translational control. *Nat Rev Mol Cell Biol* 10:307–318.
121. Proud, C.G. (2009). mTORC1 signalling and mRNA translation. *Biochem Soc Trans* 37:227–231.
122. Gingras, A.C., Raught, B., and Sonenberg, N. (1999). eIF4 initiation factors: effectors of mRNA recruitment to ribosomes and regulators of translation. *Annu Rev Biochem* 68:913–963.
123. Sonenberg, N. (2008). eIF4E, the mRNA cap-binding protein: from basic discovery to translational research. *Biochem Cell Biol* 86:178–183.
124. Beretta, L., Gingras, A.C., Svitkin, Y.V., Hall, M.N., and Sonenberg, N. (1996). Rapamycin blocks the phosphorylation of 4E-BP1 and inhibits cap-dependent initiation of translation. *EMBO J* 15:658–664.
125. Brunn, G.J., *et al.* (1997). Phosphorylation of the translational repressor PHAS-I by the mammalian target of rapamycin. *Science* 277:99–101.

126. Lazaris-Karatzas, A., Montine, K.S., and Sonenberg, N. (1990). Malignant transformation by a eukaryotic initiation factor subunit that binds to mRNA 5' cap. *Nature* 345:544–547.
127. De Benedetti, A., and Graff, J.R. (2004). eIF-4E expression and its role in malignancies and metastases. *Oncogene* 23:3189–3199.
128. Hashemolhosseini, S., Nagamine, Y., Morley, S.J., Desrivières, S., Mercep, L., and Ferrari, S. (1998). Rapamycin inhibition of the G1 to S transition is mediated by effects on cyclin D1 mRNA and protein stability. *J Biol Chem* 273:14424–14429.
129. Thomas, G.V., *et al.* (2006). Hypoxia-inducible factor determines sensitivity to inhibitors of mTOR in kidney cancer. *Nat Med* 12:122–127.
130. Zhong, H., *et al.* (2000). Modulation of hypoxia-inducible factor 1alpha expression by the epidermal growth factor/phosphatidylinositol 3-kinase/PTEN/AKT/FRAP pathway in human prostate cancer cells: implications for tumor angiogenesis and therapeutics. *Cancer Res* 60:1541–1545.
131. Semenza, G.L. (2009). Regulation of cancer cell metabolism by hypoxia-inducible factor 1. *Semin Cancer Biol* 19:12–16.
132. Guba, M., *et al.* (2002). Rapamycin inhibits primary and metastatic tumor growth by antiangiogenesis: involvement of vascular endothelial growth factor. *Nat Med* 8:128–135.
133. Kim, W.Y., and Kaelin, W.G. (2004). Role of VHL gene mutation in human cancer. *J Clin Oncol* 22:4991–5004.
134. Mills, J.R., *et al.* (2008). mTORC1 promotes survival through translational control of Mcl-1. *Proc Natl Acad Sci USA* 105:10853–10858.
135. Wangpaichitr, M., *et al.* (2008). Inhibition of mTOR restores cisplatin sensitivity through down-regulation of growth and anti-apoptotic proteins. *Eur J Pharmacol* 591:124–127.
136. Yan, H., *et al.* (2006). Mechanism by which mammalian target of rapamycin inhibitors sensitize multiple myeloma cells to dexamethasone-induced apoptosis. *Cancer Res* 66:2305–2313.
137. Dan, H.C., Cooper, M.J., Cogswell, P.C., Duncan, J.A., Ting, J.P., and Baldwin, A.S. (2008). Akt-dependent regulation of NF-{kappa}B is controlled by mTOR and Raptor in association with IKK. *Genes Dev* 22:1490–1500.
138. Wei, G., *et al.* (2006). Gene expression-based chemical genomics identifies rapamycin as a modulator of MCL1 and glucocorticoid resistance. *Cancer Cell* 10:349–351.
139. Huang, J., and Manning, B.D. (2009). A complex interplay between Akt, TSC2 and the two mTOR complexes. *Biochem Soc Trans* 37:217–222.
140. Sarbassov, D.D., Guertin, D.A., Ali, S.M., and Sabatini, D.M. (2005). Phosphorylation and regulation of Akt/PKB by the rictor-mTOR complex. *Science* 307:1098–1101.
141. Tremblay, F., and Marette, A. (2001). Amino acid and insulin signaling via the mTOR/p70 S6 kinase pathway. A negative feedback mechanism leading to insulin resistance in skeletal muscle cells. *J Biol Chem* 276:38052–38060.
142. O'Reilly, K.E., *et al.* (2006). mTOR inhibition induces upstream receptor tyrosine kinase signaling and activates Akt. *Cancer Res* 66:1500–1508.
143. Um, S.H., *et al.* (2004). Absence of S6K1 protects against age- and diet-induced obesity while enhancing insulin sensitivity. *Nature* 431:200–205.
144. Guertin, D.A., *et al.* (2009). mTOR complex 2 is required for the development of prostate cancer induced by Pten loss in mice. *Cancer Cell* 15:148–159.
145. Nardella, C., *et al.* (2009). Differential Requirement of mTOR in Postmitotic Tissues and Tumorigenesis. *Sci Signal* 2:ra2.
146. Shaw, R.J., and Cantley, L.C. (2006). Ras, PI(3)K and mTOR signalling controls tumour cell growth. *Nature* 441:424–430.

147. Deberardinis, R.J., Lum, J.J., Hatzivassiliou, G., and Thompson, C.B. (2008). The biology of cancer: metabolic reprogramming fuels cell growth and proliferation. *Cell Metab* 7:11–20.

148. Anjum, R., and Blenis, J. (2008). The RSK family of kinases: emerging roles in cellular signalling. *Nat Rev Mol Cell Biol* 9:747–758.

149. Silva, R.L., and Wendel, H.G. (2008). MNK, EIF4E and targeting translation for therapy. *Cell Cycle* 7:553–555.

150. van Gorp, A.G.M., *et al.* (2008). AGC kinases regulate phosphorylation and activation of eukaryotic translation initiation factor 4B. *Oncogene* 28:95–106.

151. Ballif, B.A., Roux, P.P., Gerber, S.A., MacKeigan, J.P., Blenis, J., and Gygi, S.P. (2005). Quantitative phosphorylation profiling of the ERK/p90 ribosomal S6 kinase-signaling cassette and its targets, the tuberous sclerosis tumor suppressors. *Proc Natl Acad Sci USA* 102:667–672.

152. von Gise, A., *et al.* (2001). Apoptosis suppression by Raf-1 and MEK1 requires MEK-and phosphatidylinositol 3-kinase-dependent signals. *Mol Cell Biol* 21:2324–2336.

153. She, Q.-B., Solit, D.B., Ye, Q., O'Reilly, K.E., Lobo, J., and Rosen, N. (2005). The BAD protein integrates survival signaling by EGFR/MAPK and PI3K/Akt kinase pathways in PTEN-deficient tumor cells. *Cancer Cell* 8:287–297.

154. Lasithiotakis, K.G., *et al.* (2008). Combined inhibition of MAPK and mTOR signaling inhibits growth, induces cell death, and abrogates invasive growth of melanoma cells. *J Invest Dermatol* 128:2013–2023.

155. Bertrand, F.E., Spengemen, J.D., Shelton, J.G., and McCubrey, J.A. (2005). Inhibition of PI3K, mTOR and MEK signaling pathways promotes rapid apoptosis in B-lineage ALL in the presence of stromal cell support. *Leukemia* 19:98–102.

156. Yu, K., Toral-Barza, L., Shi, C., Zhang, W.G., and Zask, A. (2008). Response and determinants of cancer cell susceptibility to PI3K inhibitors: combined targeting of PI3K and Mek1 as an effective anticancer strategy. *Cancer Biol Ther* 7:307–315.

157. Kinkade, C.W., *et al.* (2008). Targeting AKT/mTOR and ERK MAPK signaling inhibits hormone-refractory prostate cancer in a preclinical mouse model. *J Clin Invest* 118:3003–3006.

158. Robert, F., and Pelletier, J. (2009). Translation initiation: a critical signalling node in cancer. *Expert Opin Ther Targets* 13:1279–1293.

159. Dang, C.V. (2007). The interplay between MYC and HIF in the Warburg effect. *Ernst Schering Found Symp Proc* 4:35–53.

160. Balakumaran, B.S., *et al.* (2009). MYC activity mitigates response to rapamycin in prostate cancer through eukaryotic initiation factor 4E-binding protein 1-mediated inhibition of autophagy. *Cancer Res* 69:7803–7810.

161. Hudes, G., *et al.* (2007). Temsirolimus, interferon alfa, or both for advanced renal-cell carcinoma. *N Engl J Med* 356:2271–2281.

162. Motzer, R.J., *et al.* (2008). Efficacy of everolimus in advanced renal cell carcinoma: a double-blind, randomised, placebo-controlled phase III trial. *Lancet* 372:449–456.

163. Meric-Bernstam, F., and Gonzalez-Angulo, A.M. (2009). Targeting the mTOR signaling network for cancer therapy. *J Clin Oncol* 27:2278–2287.

164. Hait, W.N. (2009). Targeted cancer therapeutics. *Cancer Res* 69:1263–1267.

165. Pantuck, A.J., Thomas, G., Belldegrun, A.S., and Figlin, R.A. (2006). Mammalian target of rapamycin inhibitors in renal cell carcinoma: current status and future applications. *Semin Oncol* 33:607–613.

166. Dutcher, J., *et al.* (2009). Effect of temsirolimus versus interferon-α on outcome of patients with advanced renal cell carcinoma of different tumor histologies. *Med Oncol* 26:202–209.

167. Migliore, C., and Giordano, S. (2008). Molecular cancer therapy: can our expectation be MET? *Eur J Cancer* 44:641–651.
168. Bratslavsky, G., Sudarshan, S., Neckers, L., and Linehan, W.M. (2007). Pseudohypoxic pathways in renal cell carcinoma. *Clin Cancer Res* 13:4667–4671.
169. Merchan, J.R., *et al.* (2007). Phase I/II trial of CCI-779 and bevacizumab in stage IV renal cell carcinoma: phase I safety and activity results. *J Clin Oncol (Meeting Abstracts)* 25:5034.
170. Rini, B.I., and Atkins, M.B. (2009). Resistance to targeted therapy in renal-cell carcinoma. *Lancet Oncol* 10:992–1000.
171. Hansen, A., Boshoff, C., and Lagos, D. (2007). Kaposi sarcoma as a model of oncogenesis and cancer treatment. *Expert Rev Anticancer Ther* 7:211–220.
172. Stallone, G., *et al.* (2009). ID2-VEGF-related pathways in the pathogenesis of Kaposi's sarcoma: a link disrupted by rapamycin. *Am J Transplant* 9:558–566.
173. Hess, G., *et al.* (2008). Phase III study of patients with relapsed, refractory mantle cell lymphoma treated with temsirolimus compared with investigator's choice therapy. *J Clin Oncol (Meeting Abstracts)* 26:8513.
174. Costa, L.J. (2007). Aspects of mTOR biology and the use of mTOR inhibitors in non-Hodgkin's lymphoma. *Cancer Treat Rev* 33:78–84.
175. Dal Col, J., *et al.* (2008). Distinct functional significance of Akt and mTOR constitutive activation in mantle cell lymphoma. *Blood* 111:5142–5151.
176. Hipp, S., Ringshausen, I., Oelsner, M., Bogner, C., Peschel, C., and Decker, T. (2005). Inhibition of the mammalian target of rapamycin and the induction of cell cycle arrest in mantle cell lymphoma cells. *Haematologica* 90:1433–1434.
177. Yazbeck, V.Y., *et al.* (2008). Temsirolimus downregulates p21 without altering cyclin D1 expression and induces autophagy and synergizes with vorinostat in mantle cell lymphoma. *Exp Hematol* 36:443–450.
178. Zacharek, S.J., Xiong, Y., and Shumway, S.D. (2005). Negative regulation of TSC1-TSC2 by mammalian D-type cyclins. *Cancer Res* 65:11354–11360.
179. Oza, A.M., *et al.* (2006). Molecular correlates associated with a phase II study of temsirolimus (CCI-779) in patients with metastatic or recurrent endometrial cancer—NCIC IND 160. *J Clin Oncol* 24:3003.
180. Gadducci, A., Tana, R., Cosio, S., Fanucchi, A., and Genazzani, A.R. (2008). Molecular target therapies in endometrial cancer: from the basic research to the clinic. *Gynecol Endocrinol* 24:239–249.
181. Delmonte, A., and Sessa, C. (2008). Molecule-targeted agents in endometrial cancer. *Curr Opin Oncol* 20:554–559.
182. Weinstein, I.B., and Joe, A.K. (2006). Mechanisms of disease: oncogene addiction-a rationale for molecular targeting in cancer therapy. *Nat Clin Pract Oncol* 3:448–457.
183. Abraham, R.T., and Gibbons, J.J. (2007). The mammalian target of rapamycin signaling pathway: twists and turns in the road to cancer therapy. *Clin Cancer Res* 13:3109–3114.
184. Hosoi, H., *et al.* (1999). Rapamycin causes poorly reversible inhibition of mTOR and induces p53-independent apoptosis in human rhabdomyosarcoma cells. *Cancer Res* 59:886–894.
185. Boni, J., Burns, J., Hug, B., and Sonnichsen, D. (2007). mTOR inhibition following a single intravenous infusion of temsirolimus in healthy individuals. In AACR-NCI-EORTC International Conference on Molecular Targets and Cancer Therapeutics, San Francisco, CA.
186. Atkins, M.B., *et al.* (2004). Randomized phase II study of multiple dose levels of CCI-779, a novel mammalian target of rapamycin kinase inhibitor, in patients with advanced refractory renal cell carcinoma. *J Clin Oncol* 22:909–918.

187. Patursky-Polischuk, I., *et al.* (2009). The TSC-mTOR pathway mediates translational activation of TOP mRNAs by insulin largely in a raptor- or rictor-independent manner. *Mol Cell Biol* 29:640–649.
188. Feldman, M.E., *et al.* (2009). Active-site inhibitors of mTOR target rapamycin-resistant outputs of mTORC1 and mTORC2. *PLoS Biol* 7:e38.
189. Thoreen, C.C., *et al.* (2009). An ATP-competitive mammalian target of rapamycin inhibitor reveals rapamycin-resistant functions of mTORC1. *J Biol Chem* 284:8023–8032.
190. García-Martínez, J.M., *et al.* (2009). Ku-0063794 is a specific inhibitor of the mammalian target of rapamycin (mTOR). *Biochem J* 421:29–42.
191. Yu, K., *et al.* (2009). Biochemical, cellular, and in vivo activity of novel ATP-competitive and selective inhibitors of the mammalian target of rapamycin. *Cancer Res* 69:6232–6240.
192. Tremblay, F., *et al.* (2005). Overactivation of S6 kinase 1 as a cause of human insulin resistance during increased amino acid availability. *Diabetes* 54:2674–2684.
193. Pardee, A.B. (1989). G_1 events and the regulation of cell proliferation. *Science* 246:603–640.
194. Schreiber, S.L., and Crabtree, G.R. (1992). The mechanism of action of cyclosporin A and FK506. [Review]. *Immunol Today* 13:136–142.
195. Thomson, A.W., Turnquist, H.R., and Raimondi, G. (2009). Immunoregulatory functions of mTOR inhibition. *Nat Rev Immunol* 9:324–337.
196. Weichhart, T., and Saemann, M.D. (2009). The multiple facets of mTOR in immunity. *Trends Immunol* 30:218–226.
197. Zhou, L., Chong, M.M., and Littman, D.R. (2009). Plasticity of CD4+ T cell lineage differentiation. *Immunity* 30:646–655.
198. Feuerer, M., Hill, J.A., Mathis, D., and Benoist, C. (2009). Foxp3+ regulatory T cells: differentiation, specification, subphenotypes. *Nat Immunol* 10:689–695.
199. Haxhinasto, S., Mathis, D., and Benoist, C. (2008). The AKT-mTOR axis regulates de novo differentiation of CD4+Foxp3+ cells. *J Exp Med* 205:565–574.
200. Sauer, S., *et al.* (2008). T cell receptor signaling controls Foxp3 expression via PI3K, Akt, and mTOR. *Proc Natl Acad Sci USA* 105:7797–7802.
201. Bensinger, S.J., *et al.* (2004). Distinct IL-2 receptor signaling pattern in CD4+CD25+ regulatory T cells. *J Immunol* 172:5287–5296.
202. Basu, S., Golovina, T., Mikheeva, T., June, C.H., and Riley, J.L. (2008). Cutting edge: Foxp3-mediated induction of pim 2 allows human T regulatory cells to preferentially expand in rapamycin. *J Immunol* 180:5794–5798.
203. Granucci, F., Zanoni, I., and Ricciardi-Castagnoli, P. (2008). Central role of dendritic cells in the regulation and deregulation of immune responses. *Cell Mol Life Sci* 65:1683–1697.
204. Pulendran, B. (2004). Modulating vaccine responses with dendritic cells and Toll-like receptors. *Immunol Rev* 199:227–250.
205. Jagannath, C., Lindsey, D.R., Dhandayuthapani, S., Xu, Y., Hunter, R.L., Jr., and Eissa, N.T. (2009). Autophagy enhances the efficacy of BCG vaccine by increasing peptide presentation in mouse dendritic cells. *Nat Med* 15:267–276.
206. Araki, K., *et al.* (2009). mTOR regulates memory CD8 T-cell differentiation. *Nature* 460:108–112.
207. Campisi, J., and Yaswen, P. (2009). Aging and cancer cell biology, 2009. *Aging Cell* 8:221–225.
208. Fontana, L. (2009). The scientific basis of caloric restriction leading to longer life. *Current Opinion in Gastroenterology* 25:144–150, 110.1097/MOG.1090b1013e32831ef32831ba.
209. Polak, P., and Hall, M.N. (2009). mTOR and the control of whole body metabolism. *Curr Opin Cell Biol* 21:209–218.

210. Wullschleger, S., Loewith, R., and Hall, M.N. (2006). TOR signaling in growth and metabolism. *Cell* 124:471–484.
211. Harrison, D.E., *et al.* (2009). Rapamycin fed late in life extends lifespan in genetically heterogeneous mice. *Nature* 460:392–395.
212. Selman, C., *et al.* (2009). Ribosomal protein S6 kinase 1 signaling regulates mammalian life span. *Science* 326:140–144.
213. Hotamisligil, G.S., and Erbay, E. (2008). Nutrient sensing and inflammation in metabolic diseases. *Nat Rev Immunol* 8:923–934.

Author Index

Numbers in regular font are reference numbers and indicate that an author's work is referred to although the name is not cited in the text. Numbers in italics refer to the page numbers on which the complete reference appears.

A

Abedin, M., 214, *226*
Abraham, R. T., 10, 13–14, *18*, 103, 113, *122*, *126*, 274, 279, *282*, *284*, 330–332, 340, 348, 350–351, *356–357*, *360*, *364*
Acevedo, D., 41, *53*
Acosta-Jaquez, H. A., 13, *19*, 62, 72, 113, *126*
Adam, S. A., 136, *144*
Adami, A., 10, *18*
Aebersold, R., 201, *222*
Agredano-Moreno, L., 287, *300*
Aguilar, P. S., 190, *197*
Ahsan, U. S., 187, *196*
Ai, W., 206, 209, *224*, 265, *269*
Aiba, K., 259, 262, 264, *268*, 272, 274, *281*
Aicher, L. D., 306–307, 309, *325*
Akiyoshi, Y., 241, *250*
Alaoui, H., 216, *226*
Alarcon, C. M., 8, *18*
Albers, M. W., 335, *358*
Albig, A. R., 162, *174*
Albright, C. F., 235, *247*, 255, *267*
Aldea, M., 167, *175*
Alessi, D. R., 26, 34, *36*, 75, *85*, 96, *100*, 102, 105, 107, 109–112, 119–121, *122–125*, *127*, 133, 137, *142*, *144*, 294, *301*
Ali, S. M., 26, 28–29, 34, *36*, 49, 56, 58, *70*, 76, *86*, 90, *98*, 107, 109, *123–124*, 179, *193*, 258–259, *268*, 272, 276, 278, *281*, *283*, 343, *362*
Allen, D. A., 180, *193*
Alrubaie, S., 96, *100*
Al-Saleem, T., 256, *267*

Alvarez, B., 231–232, 235, 238, 244, *245*, 253–254, 259, *266*, 274–277, *283*, 293, *301*
Amann, G., 305, *323*
Ammerer, G., 133, *142*, 201, 207–208, *222*, *224*
Amri, A., 136, *144*
Anand, V., 202–203, *223*, 276, *283*
Ananthanarayanan, B., 109, *124*
Andersen, D., 138, *144*, 262, *269*
Anderson, C. M., 279, *284*
Anderson, G. H., 244, *250*, 289, 293–295, *300–301*
Anderson, S., 149, 152, *162*, *164*, 171–172, 179–180, 190, 201–202, *221–222*
Andes, D. R., 213, 219, *226–227*
Andjelkovic, M., 105, *123*
Andrabi, K., 58, *70*, 107, *123*
Andrade, M. A., 6, *17–18*, 272, *281*
Angeles de la Torre-Ruiz, M., 180, *193*
Anjum, R., 63, 72, 114–115, *126*, 305, *324*, 344, *363*
Anraku, Y., 185, *196*, 204, *223*
Anrather, D., 133, *142*, 201, *222*
Anthony, D. P., 340, *360*
Anthony, J. C., 305, *324*
Anthony, R., 294, *301*
Anthony, T. G., 59, *70*
Aoki, M., 15, *20*
Aono, T., 231, 239, *245*
Aponte, A. M., 140, *145*
Apsel, B., 64, 72, 133, *142*
Arai, T., 335, *358*
Araki, K., 353, *365*

367

Index

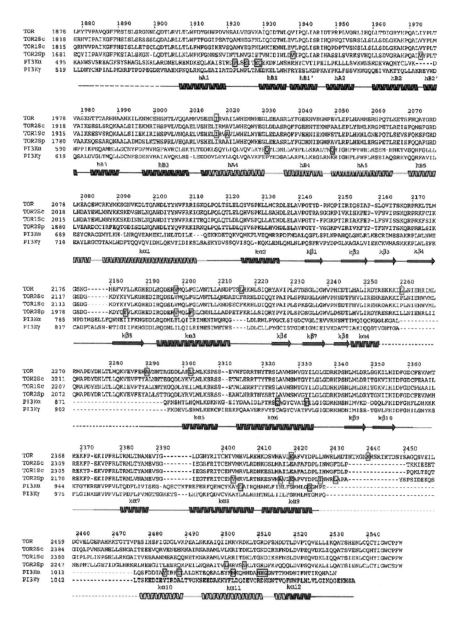

Vittoria Zinzalla *et al.*, Fig. 1.3, page 7. Structural alignment of TORs and PI3Ks. The TORs are human TOR (TOR), *S. cerevisiae* TOR1 and TOR2 (TOR1Sc and TOR2Sc), and *S. pombe* TOR2 (TOR2Sp). The PI3Ks, both class I PI3Ks, are pig PI3Kγ and human PI3Kα. The structural elements, shown below the alignment, are drawn according to the crystal structure of PI3Kγ [40]. The color code of the domains is as follows: FAT (shown is residues 1906–2014 which is only part of the FAT), blue; FRB (residues 2015–2114), green; kinase (residues 2115–2426), red; FIT (residues 2427–2516); FATC (shown is residues 2517–2526 which is only part of the FATC), cyan. Sites of activating mutations in TOR and oncogenic mutations in PI3Kα [41–44] are shown in red and blue, respectively. Oncogenic mutations are particularly frequent in codons E542 and E545 (between helices hA1 and hB1) and H1047 (helix kα11), and these sites thus constitute two oncogenic hotspots. This figure is modified from Figure 1 of Ref. [45].

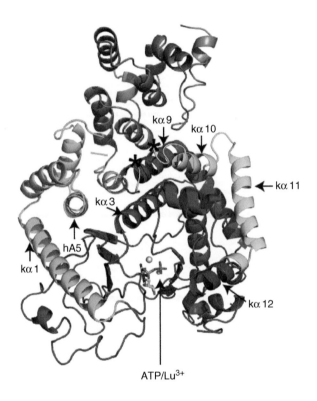

VITTORIA ZINZALLA *ET AL.*, FIG. 1.4, PAGE 9. Model of the catalytic region of human TOR. The modeled catalytic region extends from the FAT domain to near the end of the FATC domain [45]. The predicted helices are shown as ribbons colored according to domains. The color code for domains is as in Figure 1.3. ATP is shown as a backbone stick structure. Two Lu^{3+} ions are shown as two small gray spheres. The ATP and Lu^{3+} ions are positioned according to the crystal structure of PI3Kγ [40]. Asterisks indicate the insertion site of a so-far structurally undefined portion of the FIT domain (the undefined portion of the FIT domain is residues 2427–2477). The model is drawn using MacPymol (Delano Scientific).

A

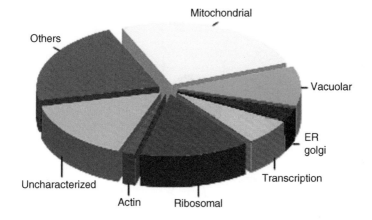

B

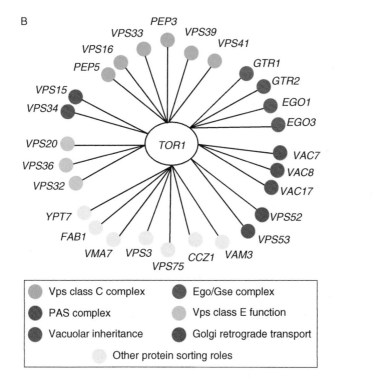

ROBERT J. BASTIDAS AND MARIA E. CARDENAS, FIG. 11.1, PAGE 203. Schematic representation of the genetic synthetic Tor1 interaction network. (A) Functional categories represented by the genes that when mutated in combination with *tor1* mutation confer synthetic lethal or decrease fitness defects. (B) The genetic synthetic interaction Tor1 network involved in vacuolar, vesicular trafficking, and protein sorting functions includes distinctive protein complexes. Mutation of 23 out of 27 of these genes results in rapamycin hypersensitivity [25–27]. Modified from Figure 2 [79].

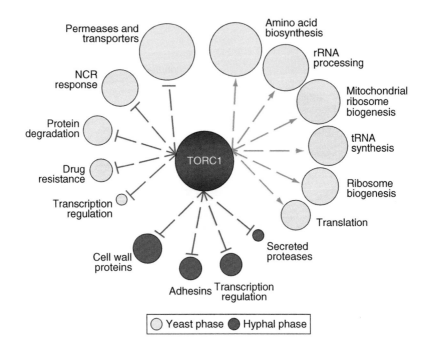

ROBERT J. BASTIDAS AND MARIA E. CARDENAS, FIG. 11.2, PAGE 210. Transcriptional programs governed by TORC1 in *C. albicans*. TORC1 stimulates transcriptional programs devoted to protein synthesis, similar to the Ribi regulon governed by TORC1 in *S. cerevisiae*. TORC1 also inhibits the NCR transcriptional response and expression of genes coding for permeases and transporters. In contrast to *S. cerevisiae*, TORC1 in *C. albicans* modulates the expression of hyphae-specific genes coding for cell wall constituents, including several adhesins and their transcriptional regulators, as well as key virulence factors such as secreted proteases. Adapted from Ref. [59], Supplementary Tables S1, S2, and S4.

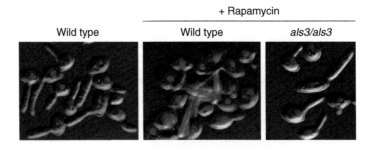

ROBERT J. BASTIDAS AND MARIA E. CARDENAS, FIG. 11.3, PAGE 213. Rapamycin stimulates surface expression of the Als3 adhesin in *C. albicans* hyphal cells. Exposure of wild-type hyphal cells to rapamycin results in increased expression of the Als3 adhesin at the surface of *C. albicans* hyphae, as assayed by indirect immunofluorescence (red) with an Als3 antiserum. Als3 resembles mammalian cadherins and mediates adherence of *C. albicans* hyphal cells to N-cadherins in endothelial cells and E-cadherins in oral epithelial cells [115]. Als3 is also required for cell–cell adhesion during biofilm development [92]. Adapted from Ref. [59], Figure 2C.

Printed and bound by CPI Group (UK) Ltd, Croydon, CR0 4YY

08/05/2025

01864953-0003